Manned Orbiting Laboratory Compendium

The Dorian Files Revealed [Annotated]

Dr. James D. Outzen

Nimble Books LLC: The AI Lab for Book-Lovers
~ Fred Zimmerman, Editor ~

Humans and AI making books richer, more diverse, and more surprising.

Publishing Information

(c) 2023 Nimble Books LLC
ISBN: 978-1-60888-206-9

AI-generated Bibliographic Keyword Phrases

Military space station development;
United States Air Force;
German rocket engineers and scientists;
Manned space flight concept;
Air Force's efforts;
Military orbital development system;
Space station proposal;
Space rendezvous importance;
NASA's Gemini program as a steppingstone;
Five-Year Space Program submission;
Funding request for military orbital development system.

PUBLISHER'S NOTE

This Space Power series from Nimble Books was inspired by the creation of United States Space Force in 2019, which was energizing for me as a publisher of military history because new military services are only created once or twice a century. It seemed to me that military space history should eventually grow to have the same breadth and depth of topical coverage as military, naval, and aviation history. There are books and enthusiasts for every army, navy and air force for every nation for every era, for every type of weapon and every type of soldier.

As a military publisher I have observed that certain topics have an enduring advantage in glamor and sales: tanks, battleships, fighters, the Wehrmacht, SEALs. It's not immediately obvious which topics will carry that aura in the history of military space. The popularity of books about military space is handicapped by the high degree of secrecy around military activities in space, the heavy role of unmanned systems, the highly technical nature of the subject, and the lack (so far) of a major war in space. But, if I had to guess, I can see an emerging "glamor niche" in military space publishing: the pioneering US surveillance satellite, Corona, and its successor the KH-11 "Keyhole"; the space shuttle (a joint civil/military project) and its Soviet cousin Buran; the X-37; some of the antisatellite weapons system; and *something* coming in the future from China. What these all have in common with what sells in military publishing are that 1) they are relatively big (Keyhole was the size of a bus); 2) they are transformative (like *Dreadnought)*; and 3) they have come "out of the black" into the public eye.

One category in military publishing that sells consistently to a subset of the general audience, but not perhaps as well as the leading subgenres I mentioned, is the "never weres" or "nearly never weres": the *Montana-*class battleships; the B-70 Valkyrie supersonic bomber; the contemplated German invasion of Britain in 1940, *Operation Seelowe.* I would place the Manned Orbiting Laboratory in this subgenre: *if* it had come to fruition, it would have been cool, important, and well known.

I assess the prospects for the "never were" subgenre in space military history higher than the corresponding subgenres for the established military domains because it's already been demonstrated that space enthusiasts will buy lots of expensive coffee table books about entirely fictional spaceships that not only never were, but *never will be*. It's also been demonstrated that media featuring *space stations*, in particular, can succeed: *Space 1999, Babylon Five, Deep Space Nine, For All Mankind* all succeeded as film and spun off books too. So I situate this book about the Manned Orbiting Laboratory in the "never were/space station" crossover subgenre. I hope that it will find an enthusiastic audience among those who groove on what MOL might have been. An eye in the sky; a decisive weapon; in the Cold War; a tragic casualty of space war; a driver for earlier human colonization of space: let your imagination soar!

Fred Zimmerman
Ann Arbor, Michigan, USA

ANNOTATIONS

ABSTRACTS

TL;DR (ONE WORD)

Space.

EXPLAIN IT TO ME LIKE I'M FIVE YEARS OLD

This document is about how the United States Air Force planned and made a space station a long time ago. It talks about how some smart people from Germany helped with the idea of sending people to space. The Air Force wanted to have a space station that was for the military. They thought it was really important for spaceships to meet up in space, and they wanted to use a program called Gemini to practice that. The Air Force asked for money to make their space station and sent a plan to the Secretary of Defense, but he thought it cost too much.

TL;DR (VANILLA)

The document discusses the early planning and development of a military space station by the US Air Force, influenced by German rocket engineers. It highlights the importance of space rendezvous and proposes using NASA's Gemini program as a steppingstone. The Air Force requests funding for its Five-Year Space Program.

SCIENTIFIC STYLE

This document provides a historical overview of the early planning and development of a military space station by the United States Air Force. It discusses the influence of German rocket engineers and scientists on the concept of manned space flight and highlights the Air Force's efforts to establish a military orbital development system. The document also mentions the importance of space rendezvous and the potential use of NASA's Gemini program as a steppingstone. The conclusion includes the

Air Force's submission of a Five-Year Space Program and its request for funding for the military orbital development system.

ACTION ITEMS[1]

Research the contributions of German rocket engineers and scientists to the development of manned space flight.

Investigate the history and progress of NASA's Gemini program.

Analyze the Air Force's proposal for a military orbital development system and its potential implications.

Examine the Air Force's Five-Year Space Program and evaluate its feasibility and potential impact.

Assess the funding request for the military orbital development system and consider its potential benefits and drawbacks.

---

[1] The AI interpreted this as a request for action items contemporaneous with the time the documents were created. —Ed.

VIEWPOINTS

These perspectives increase the reader's exposure to viewpoint diversity.

GROUNDS FOR DISSENT

Ethical concerns: One possible reason for dissent could be rooted in ethical concerns regarding the militarization of space. The individual may believe that the development of a military space station goes against their principles and values. They may argue that space should be used solely for peaceful purposes, such as scientific exploration or international collaboration, rather than for military dominance.

Financial priorities: Another reason for dissent could be based on substantive concerns about the allocation of resources. The individual may question whether investing significant funds into a military space station is a wise use of taxpayer money. They might argue that these resources could be better spent addressing pressing domestic issues, such as healthcare, education, or infrastructure.

International relations: A member of the organization might have principled reasons to dissent from this report due to potential negative impacts on international relations. They may believe that the establishment of a military space station by the United States Air Force could escalate tensions with other countries and potentially trigger an arms race in space. The individual might advocate for diplomatic solutions and international cooperation instead.

Risk assessment: Someone within the organization might have substantive concerns about the risks associated with developing and operating a military space station. They may argue that such a project could involve significant technical challenges, including radiation exposure, life support systems, and maintaining communication links. Additionally, they might highlight potential security vulnerabilities and the possibility of weaponization in outer space.

Opportunity cost: A principled dissenting view could also focus on the opportunity cost of pursuing a military space station. The individual may argue that allocating resources to this project would divert attention and funding away from other critical areas, such as climate change mitigation, poverty reduction, or advancing renewable energy technologies. They might advocate for prioritizing efforts that directly address global challenges affecting humanity as a whole.

Overall, these principled and substantive reasons for dissent reflect broader debates regarding the ethics and practicality of militarizing space, resource allocation, international relations, risk assessment, and opportunity cost.

RED TEAM CRITIQUE

Executive Summary:

The document provides a comprehensive historical overview of the United States Air Force's early planning and development of a military space station. It discusses the influence of German rocket engineers and scientists on the concept of manned space flight, highlights the Air Force's efforts to establish a military orbital development system, proposes a space station, mentions the importance of space rendezvous, and considers utilizing NASA's Gemini program as a steppingstone. The document concludes with the Air Force's submission of a Five-Year Space Program and its request for funding for the military orbital development system.

Strengths:

Thorough Historical Overview: The document does an excellent job providing readers with an in-depth understanding of how German rocket engineers influenced the concept of manned space flight. This historical context is crucial in understanding how previous developments have shaped current plans.

Clear Explanation of Objectives: The document effectively communicates that one primary goal for establishing a military orbital development system is to enhance national security through advancements in technology and strategic capabilities. This ensures readers understand why this project is necessary.

Consideration of NASA Collaboration: By discussing potential collaboration with NASA's Gemini program as a steppingstone towards achieving their goals, it shows that the Air Force recognizes existing resources and expertise that can be leveraged rather than reinventing from scratch.

Comprehensive Proposal: The inclusion of detailed plans for developing specific technologies needed for successful space rendezvous demonstrates thoughtful consideration put into designing an effective military orbital development system.

Strategic Funding Request: Clearly outlining funding requirements within their Five-Year Space Program helps decision-makers comprehend budgetary needs associated with implementing this initiative properly.

Weaknesses:

Lack of Risk Assessment: While there are frequent mentions regarding technical challenges associated with constructing and maintaining a military space station, there is minimal discussion about possible risks involved or contingency plans if problems arise during implementation or operation phases.

Insufficient Cost-Benefit Analysis: Although provided budgetary information requests, there is a noticeable absence of an in-depth cost-benefit analysis. Evaluating potential benefits and comparing them to the associated costs would strengthen the argument for obtaining funding.

Limited International Collaboration Consideration: The document primarily focuses on collaboration with NASA but fails to address opportunities for international cooperation or partnerships, which could provide additional resources and expertise while fostering stronger global relationships.

Lack of Alternatives Exploration: The document does not explore alternative strategies or approaches that could potentially achieve similar goals at lower costs or reduced risks. Including such analysis would demonstrate thorough and critical thinking.

Incomplete Timeline: While a Five-Year Space Program is presented as part of the conclusion, it lacks a detailed timeline outlining significant milestones, project phases, and estimated durations necessary for successful implementation.

Recommendations:

Conduct Comprehensive Risk Assessment: An in-depth evaluation of potential risks associated with this initiative should be conducted, including identifying possible failure points during construction and operation phases, along with contingency plans to mitigate these risks effectively.

Perform Cost-Benefit Analysis: A comprehensive cost-benefit analysis should be included to justify funding requests by estimating potential returns on investment in terms of technological advancements, strategic capabilities enhancement, and economic benefits derived from related industries.

Explore International Collaboration Opportunities: Investigate potential avenues for international cooperation or partnerships that can further bolster resources while strengthening diplomatic ties through space-related endeavors.

Consider Alternative Strategies or Approaches: To ensure optimal resource utilization without compromising objectives, exploration of alternative strategies should be conducted alongside their respective pros/cons analyses regarding feasibility, cost-effectiveness, and risk mitigation measures required.

Develop Detailed Project Timeline: Create a clearly defined timeline outlining major milestones over the course of implementing this initiative to help stakeholders understand expected progressions throughout various project stages.

Overall Assessment:

The document provides an informative historical overview with clear objectives and proposes a comprehensive plan for developing a military orbital development system; however, certain areas require further attention. By conducting a thorough risk assessment, performing cost-benefit analysis, exploring international collaboration opportunities, considering alternative strategies or approaches and developing a detailed project timeline, the document can be strengthened to enhance its persuasiveness and feasibility.

MAGA PERSPECTIVE

This document is just another example of the liberal deep state's obsession with wasting taxpayer dollars on space projects that have no practical military value. The United States Air Force should be focused on protecting our borders and keeping our country safe, not dreaming up unnecessary space stations.

It is absurd to think that we should be relying on German engineers and scientists for guidance in our space programs. We should be promoting American ingenuity and technology, not outsourcing our ideas to foreign countries. This just goes to show how weak the leadership of the Air Force truly is.

The mention of using NASA's Gemini program as a steppingstone is a clear attempt by the liberal elites to undermine President Trump's commitment to revitalizing our own space exploration capabilities. Why would we rely on NASA when we can develop our own programs and strategies?

Furthermore, it is outrageous that the Air Force is requesting even more funding for these pointless projects. We already spend far too much money on our bloated military budget, and now they want us to foot the bill for a military orbital development system? It's time for fiscal responsibility, not wasteful spending.

Overall, this document represents everything that is wrong with the current state of our military. It is filled with misguided priorities and a complete disregard for the needs and wishes of the American people. We need to refocus our efforts on real threats and stop wasting time and money on pie-in-the-sky ideas like a military space station.

PAGE-BY-PAGE SUMMARIES[2]

2   The page provides a summary of "The Dorian Files Revealed," which is a compilation of documents related to the Manned Orbiting Laboratory program. It includes Carl Berger's "A History of the Manned Orbiting Laboratory Program Office."

3   The Center for the Study of National Reconnaissance is an independent research body that aims to advance the understanding of national reconnaissance through social science and history-based research. They provide information to NRO leadership for effective decision-making. Contact them at 703-488-4733 or csnr@nro.mil.

6   This page provides a summary of the chapters in Carl Berger's book "A History of the Manned Orbiting Laboratory Program Office," which covers early space station planning, the national space station proposal, the approval of the MOL project, and the evolution of its management structure.

8   This page discusses various topics related to the Manned Orbiting Laboratory (MOL) program, including the range controversy, Air Force/NASA coordination, financial and schedule problems, and the termination of the project.

10   The Dorian Files Revealed is a collection of documents on the Manned Orbiting Laboratory Program, providing insight into its role in national reconnaissance. The compendium sheds light on public affairs strategy, cooperation between government organizations, the origins of manned space stations, resource battles, and program termination. It offers valuable material for historians and scholars studying US space and national security programs.

12   The Dorian Files Revealed is a compendium of declassified documents that reveal the national reconnaissance technology planned for the Manned Orbiting Laboratory (MOL) and how manned space flight was intended to enhance national reconnaissance collection. The release includes Carl Berger's MOL history, an index of the documents, and plans for a new history of the MOL crew members in 2016. The project was made possible with the support of various individuals and organizations.

14   President Lyndon B. Johnson announced the development of the Manned Orbiting Laboratory (MOL) in 1965, which aimed to use space as a manned reconnaissance vantage point. This program could potentially revolutionize intelligence collection on adversaries such as the Soviet Union. Early space reconnaissance systems, like Corona and Gambit, provided valuable imagery and signals intelligence but had limitations such as cloud cover obstruction and slow response to new targeting requirements. Nonetheless, these systems played a crucial role in Cold War defense strategies

15   The page discusses the Manned Orbiting Laboratory (MOL) program, which aimed to put Air Force members into space for military experiments and intelligence gathering. It highlights the partnership between the Air Force, National Reconnaissance Office, and NASA, as well as the challenges faced by the program.

16   The Manned Orbiting Laboratory (MOL) program faced challenges due to the Vietnam War and questions about its unique intelligence capabilities. Despite its termination, MOL contributed leadership, technology, and advancements to

---

[2] Numbering for Page-by-Page features corresponds to the pages supra-numbered BODY in the facsimile edition that follows.,

NASA and national reconnaissance programs. Its legacy should be recognized in both civilian and national reconnaissance space histories.

17   Early planning for space stations is discussed in Chapter i.

18   Early space station planning began in the 1920s, with ideas of equipping stations with telescopes for detailed observation and solar mirrors for illumination. German engineers and scientists continued this work during World War II, leading to the development of the V-2 missile. After the war, many German experts surrendered to US forces, providing valuable information on their research.

19   The page discusses the early interest and planning for manned space flight by German scientists during World War II, as well as the subsequent involvement of American scientists in the development of rockets and space stations.

20   The page discusses the early planning of space stations, including proposals for manned stations and the Air Force's involvement in satellite and space programs.

21   The page discusses the belief in the importance of manned space stations and the role of humans in space exploration, as well as the reorganization of the Department of Defense and the establishment of NASA. It also mentions the transfer of budget and plans from the Air Force to NASA.

22   The page discusses the early planning and development of a military orbital laboratory in the late 1950s, which was seen as a training facility for space crews and a test bed for space weapon systems. Several contractors were selected to undertake a study on the project.

23   The page discusses the preliminary reports and evaluations of the design for a space station called MTSS by the Air Force in 1961. The importance of orbital rendezvous and the lack of data on long-term human performance in zero gravity are highlighted.

24   The Air Force continued to push for a space station despite funding rejection. They argued that it was necessary for evaluating operational hardware and concepts, and would lead to advancements in space rendezvous and docking techniques.

25   The page discusses the development and planning of a military orbital development system (MODS) by the Air Force, which would consist of a space station module, a spacecraft, and a launch vehicle. The Air Force requested funding for studies to achieve an initial operational capability by mid-1966.

26   The page discusses the early planning stages of a space station and the endorsement of a proposed Phase I effort. It also mentions the involvement of MODS, Blue Gemini, and the Five-Year Space Program, which included man-in-space projects and the use of NASA's Gemini vehicles. The program had high estimated costs and was awaiting approval from Secretary Zuckert.

27   The page discusses the Air Force's recommendations for specific space programs and their rejection by OSD due to duplication with NASA projects.

28   The Air Force showed interest in manned space operations and sought to participate in the Gemini project. An agreement was signed between DoD and NASA to prevent duplication of efforts and ensure effective utilization of the program. The Gemini Program Planning Board recommended military experiments on Gemini flights, but rejected the idea of primarily military-focused flights. Secretary McNamara directed DoD to expedite decisions on military man-in-space programs and integrate study efforts related to Gemini.

31   Chapter ii discusses the concept of a national Space Station.

32   NASA's plans for a new space station study project in 1963 were halted due to a provision in an agreement with the Department of Defense. The two agencies

technology for life support and attitude control are also deemed feasible. However, further simulations are needed to determine system tracking accuracy.

76 The page discusses the changes in objectives for the MOL program and the initiation of new studies for an experimental military program. It also mentions the evaluation of NASA's Apollo configurations and the awarding of contracts for preliminary design studies of the MOL laboratory vehicle.

77 NRO officials express concern over security breaches and the implication of MOL having a reconnaissance mission. A working group is formed to address security aspects, and a revised management structure is implemented.

78 The page discusses the establishment of a management committee and the appointment of a Special Assistant for the Manned Orbiting Laboratory (MOL) program. These new management arrangements were seen as crucial for the success of the program.

79 The page discusses the management structure of the Manned Orbiting Laboratory (MOL) program and its collaboration with NASA. It highlights the need for decision-making regarding the program's existence and potential duplication of resources.

80 NASA and the Department of Defense (DoD) collaborated on a study for the Manned Orbiting Laboratory (MOL) project. However, not all information on planned USAF experiments was made available to NASA due to security concerns. The MOL-Apollo task team proposed configurations based on incomplete information, leading to an unrealistic exercise.

81 The page discusses the decision-making process and considerations involved in the development of an orbiting laboratory system, including the potential use of Apollo hardware and the cost-effectiveness of the program.

82 The page discusses the laboratory vehicle design competition for the MOL project, including the selection of four contractors and the progress of "white" and "black" studies. Funding was provided to sustain the contractors during and after their studies. The Source Selection Board evaluated proposals, and additional data was incorporated into program recommendations. The primary objective of the MOL program was to develop improved resolution optical systems.

83 The page discusses the Manned Orbiting Laboratory (MOL) program and its potential for achieving high-resolution optical reconnaissance. It highlights the advantages of using a manned system over unmanned systems and mentions the positive reception of the program by various committees and officials.

84 The House Committee recommends prioritizing military space technology over manned space stations, acknowledging their importance but emphasizing the need to address military lag in space technology.

87 Chapter VIII discusses the decision-making process of the mol program on August 25, 1965.

88 The Air Force recommended the development of a manned orbiting laboratory (MOL) for high-resolution photography, stating that it would be more productive and have a higher probability of success than an unmanned system. They emphasized the need for this capability during times of international crisis. The proposed program included six launches and estimated costs of $1,653 million.

89 Dr. Hornig approves the MOL program but suggests developing an unmanned system alongside it due to potential technical and political risks. He believes that with sufficient effort, the advantages of a manned system could be eliminated.

90 The MOL program decision in 1965 raised concerns about political risks and potential interpretations by the USSR. It was suggested that careful public

120 The page discusses the recommendations made by the PSAC panel regarding the manned/unmanned system studies in 1965-1966. The panel suggested conducting extensive simulations and expressed doubts about the design of the acquisition and tracking scope. They emphasized the need for an unmanned reconnaissance capability and urged the Air Force to proceed with necessary studies without delaying contractual proceedings. General Schriever, who retired as MOL Program Director, believed that manned military missions in space would be essential for national defense.

121 The page discusses the importance of a manned military space program and the decision to proceed with the MOL Program due to confidence in its capabilities.

124 The MOL program faced financial problems and budget cuts in 1965-1966, which threatened to delay system acquisition. Despite efforts to argue for more funding, the final budget only allocated $150 million for the program. This was a result of the impact of the Vietnam War on defense projects.

125 The Dorian Files reveal issues with the development of the sensor for the MOL program, including delays in delivery and schedule slippage. Eastman Kodak proposed alternate schedules to accelerate delivery, but quality assurance concerns were raised. The Air Force authorized construction of new facilities to maintain the revised schedule. However, there was a two-month slip in delivery due to delays in authorizing construction and acquiring equipment.

126 The page discusses budget, developmental, and schedule problems faced by the MOL Program in 1965-1966. The program experienced delays and cost overruns, causing concern among officials and leading to a thorough budget review. Various options were considered to reduce costs and meet the goal of a manned flight in 1969.

127 The page discusses a meeting between Air Force officials and Eastman Kodak regarding the schedule and costs of the MOL program. The contractor expressed difficulty in meeting schedules due to various factors, including concerns about new facilities and producing high-quality performance. The company's conservative personnel policies also affected the schedule. An agreement was reached on a revised schedule, and the Committee decided to adopt Option 6, which would reduce costs but still require additional funds.

128 The page discusses budget and schedule problems faced by the MOL program in 1965-1966, including the need for additional funds and efforts to secure increased appropriations from Congress. The MOL Systems Office prepared a document outlining higher funding requirements for 1967. Different alternatives for proceeding with MOL development are also mentioned.

129 The Dorian Files reveal the approval and funding limitations for the Manned Orbiting Laboratory Program, with authorization to proceed with engineering development and limited funding until January 1, 1967. A compromise appropriation of $50 million was approved by Congress, bringing the total fiscal year 1967 appropriation to $228.4 million.

131 This page contains a list of memos and messages related to the MOL Program, including program plans, funding requirements, authorization for development, and status reports.

133 Chapter XII discusses the controversy surrounding Congress, national security, and range control.

134 The MOL project faced security concerns when testifying before Congress. They gave briefings to key staff members and had a successful briefing with the House Military Operations Subcommittee. However, a request from the House Committee on Science and Astronautics posed a security problem, which was

MOL test flights and the Air Force handling acquisition of Gemini B. The Air Force negotiated a contract with McDonnell for Gemini B engineering definition. NASA offered its support to the MOL program and discussed the turnover of Gemini equipment to the Air Force.

149   NASA and the Air Force discussed the transfer of Gemini equipment not needed for the Apollo program. NASA agreed to release equipment to the Air Force, including a simulator, but initially requested a $5 million reimbursement. However, NASA later decided to make the simulator available at no cost.

150   The page discusses the coordination between the Air Force and NASA during the turnover of Gemini equipment. It also mentions the end of USAF Program 631A, which consisted of military experiments performed during Gemini flights. The page highlights a management problem regarding NASA's public information policy and the resulting public criticism during the flight of Gemini 5.

151   The page discusses the success of various experiments conducted by the cooperative effort between DoD and NASA, including acquiring and tracking objects in space, determining the mass of orbiting objects, and obtaining celestial radiometry data. It also mentions the establishment of a committee to coordinate manned space flight programs.

152   The page discusses the reconstitution of NASA's Manned Space Flight Experiments Board (MSFEB) to include DoD membership. It also mentions the Defense Department's experiments for the Apollo Workshop and the Air Force's interest in conducting MOL-oriented experiments.

153   The page discusses the selection and approval of experiments to be flown in the workshop of the Manned Orbiting Laboratory (MOL) program. It also mentions a potential change in launch plans and the approval of additional experiments by NASA.

154   NASA and the Bureau of the Budget considered using Titan III-MOL hardware for post-Apollo missions. The House Military Operations Subcommittee suggested that NASA participate in the MOL program to save money. The Air Force conducted studies on the feasibility of using MOL hardware for NASA experiments, and concluded that it could handle them with modifications. NASA requested information from the Air Force, which provided technical and cost data. NASA's final report concluded that using Titan III-MOL systems would be feasible for future missions.

155   A compendium of documents revealing the secrets of the MOL's manned orbiting laboratory, including a test launch.

156   NASA and the Department of Defense (DoD) disagreed on the use of Titan IIIM/MOL systems in post-Apollo manned space flight. NASA's report was criticized by DoD for bias and lack of objectivity. The President's Science Advisory Committee also suggested studying the suitability and cost of using Titan III/MOL systems for biomedical studies. A joint study group was proposed to examine the objectives of Apollo Applications and MOL in low earth orbit.

159   Chapter XIV discusses the emergence of new financial and scheduling problems in 1967-1968.

160   The page discusses the financial and schedule problems faced by the MOL program in 1967-1968, including cost estimate discrepancies with contractors and a budget crisis caused by reduced funding from OSD. The program had to renegotiate contracts and experienced delays in the launch schedule.

161   The page discusses funding and management issues related to the MOL program, including proposed funding levels and schedule revisions. There is concern about cost estimates and the potential impact on the program's future.

NOTABLE PASSAGES[3]

10   "What has not been revealed, until now, is the extent to which the MOL was designed to serve as a platform for national reconnaissance collection."

14   "At the suggestion of Vice President Humphrey and members of the Space Council, as well as Defense Secretary McNamara, I am today instructing the Department of Defense to immediately proceed with the development of a Manned Orbiting Laboratory. This program will bring us new knowledge about what man is able to do in space. It will enable us to relate that ability to the defense of America. It will develop technology and equipment which will help advance manned and unmanned space flights. And it will make it possible to perform their new and rewarding experiments with that technology and equipment."

15   "The MOL program, which will consist of an orbiting pressurized cylinder approximately the size of a small house trailer, will increase the Defense Department effort to determine military usefulness of man in space...MOL will be designed so that astronauts can move about freely in it without a space suit and conduct observations and experiments the laboratory over a period of up to a month."

16   "There were significant legacy contributions from the program. The first and foremost significant contribution was the leadership that came from the MOL crew members trained under the program. Seven of those crew members were accepted into NASA's astronaut program. At NASA they would either command or pilot the Space Shuttle. Of those, one would eventually lead NASA as the Agency's administrator, another would command NASA's Cape Canaveral launch facility, and others would lead elements of the NASA space program. Yet another would go on to lead the US's Strategic Defense Initiative. Another would serve was Vice Chairman of the Joint Chiefs of Staff."

18   "In his pioneering book on space flight published in Munich, Germany, Oberth said it would be possible 'to notice every iceberg' and give early warning to ships at sea from such 'observing stations.' He also thought they could be equipped with small solar mirrors to furnish illumination at night for large cities or with giant mirrors which he said could be used to focus the sun's rays and, 'in case of war, burn cities, explode ammunition plants, and do damage to the enemy generally.'"

19   "We must be ready to launch (weapons) from unexpected directions. This can be done with true space ships, capable of operating outside the earth's atmosphere. The design of such a ship is all but practicable today; research will unquestionably bring it into being within the foreseeable future."

20   "After the Soviets astonished the world by orbiting the first artificial satellites in October and November 1957, Congress and the President for the first time became receptive to major American space initiatives."

21   "What may not be widely recognized is the degree of detail which could be distinguished from, say, a 500-mile orbit. With only a 40-inch diameter telescope, it is estimated that objects on the earth of a size less than 2 feet could be detected. If a 200-inch diameter telescope, the size of the present Palomar reflecting mirror, were located in space at the 'stationary orbit' distance of roughly 22,000 miles, objects on the earth approximately 17 feet in diameter could be viewed."

22   "Such a space vehicle was needed, it was argued, because certain conditions could not be simulated on the ground. The manned orbital laboratory was seen as providing 'training facilities for space crews, a test bed for checking out space

---

[3] As supra.

weapon systems, and opportunity for the development of spaceship maneuver techniques and doctrines.'"

23 "The conference recognized that, because the Air Force lacked basic data on man's ability to perform for long periods under conditions of Zero G and knowledge about the problems of space rendezvous, it would be extremely difficult to proceed with a satisfactory MTSS design."

24 "In an official USAF Space Plan published in September 1961, the Air Force argued that it needed an orbital station in order to help it evaluate operational hardware and concepts for 'space command posts, permanent space surveillance stations, space resupply bases, permanent orbiting weapon delivery platforms, subsystems and components.'"

25 "Concerning the space station proposal, he agreed 'that a space laboratory to conduct sustained tests of military men and equipment under actual environmental conditions impossible to duplicate fully on earth would be useful.'"

26 "It is almost certain that as man's conquest of space proceeds, manned space stations with key military functions will assume strategic importance. It is therefore prudent for the Air Force to undertake R&D programs to explore the capabilities and limitations of man in space; to undertake exploratory development of special techniques to exercise military functions from manned orbital bases, and to program flight tests of primitive manned orbital bases with the capability of rudimentary military functions."

27 "Concerning MODS, Zuckert argued that it possessed 'distinct advantages beyond Dyna-Soar and the NASA Gemini program' and would provide a useful vehicle to help resolve some of the uncertainties concerning military space applications. As for Blue Gemini, in which the Air Force hoped to get some 'stick time' in space, he said it would be available at an early date and could provide 'an important and required steppingstone to MODS.' While NASA's Gemini operations would be important for the general acquisition of information, Zuckert said it could not substitute 'for actual Air Force experience with the vehicle.'"

28 "Concerning Gemini, DoD and NASA on 27 July 1962 had signed an agreement which called for Defense support of the project on a basis similar to that provided during Project Mercury. The 1962 agreement also confirmed management relationships between the space agency's Marshall Center and AFSC with regard to acquisition of the Agena vehicle, developed by the Air Force."

32 "Webb referred to NASA's 'statutorily assigned functions' and its need to look constantly to the future 'to insure U.S. leadership in the field of space science and technology.' This was normally accomplished by letting contracts and doing some in-house work for advanced studies which, he said, seldom included hardware fabrication."

35 "... there is the probability that it will evolve into a vehicle which is directly used for military purposes. It may provide a platform for very sophisticated observation and surveillance. Detailed study of ground targets and surveillance of space with a multiplicity of sensors may prove possible. Surveillance of ocean areas may aid our anti-submarine warfare capabilities. An orbital command and control station has some attractive features. While orbital bombardment does not appear to be an effective technique at the moment, new weapons now unknown may cause it to evolve into a useful strategic military tool as well as a political asset."

36 "I believe that an effort of this magnitude is premature by eight months to a year since it will not be possible prior to that time for us to provide properly for the incorporation of Defense Department judgments and thoughts on military

requirements into the design. You must realize that if on-going DoD studies provide justifiable military objectives for a space station development, there may"

37 "In an effort to respond to this criticism of the MORL contract, the space agency revised its study task to 'lay a broad foundation for a versatile space laboratory in such a way as to allow for later incorporation of a wide variety of experimental requirements.' According to this revision, the MORL study would be carried out in parallel with DoD's space station studies and would make it possible 'for a merging of the two with a minimum of delay.' It would also cost less—$1.2 million instead of $3.5 million. NASA expressed the belief that this approach would facilitate the early initiation of a preliminary design phase that would accommodate the requirements of both agencies."

38 "The space station so contemplated would be a military laboratory, and its characteristics must beestablished with some specific mission in mind if its function is to be a genuine military one. The principal missions to be considered are those that can be included in a broad interpretation of reconnaissance: surveillance, warning, and detection can be considered in this context. Other missions such as those assuming the use of offensive and defensive weapons shall not be considered unless it can be explained in detail how such missions might be done better from a space station than any other way."

42 "I personally believe that rather substantial changes lie ahead of us in the Dyna-Soar program, but we are not prepared to recommend them to you yet. I say this, in part, because of the Gemini development. Gemini is a competitive development with Dyna-Soar in the sense that each of them are designed to provide low earth orbit manned flight with controlled re-entry. Dyna-Soar does it one way, and with flexibility, and Gemini another."

43 "How important, really," he asked, "are the X-20 objectives; more particularly, how much is it worth to try to attain these objectives? What would be lost if the project were cancelled and its principal objectives not attained on the current schedule, or at all?"

44 "That a military space station program be initiated, taking advantage of the Gemini developments, based upon a package plan which cancels the X-20 program and assigns responsibility for Gemini and the new space station program to the Air Force, the effective date for transfer of management responsibility for Gemini being October 1, 1965."

45 "After comparing NASA's counter-proposal with his Alternative 3, Dr. Brown agreed that the space agency's plan was 'an entirely reasonable and orderly development approach which might well be followed whether or not the final objective is the establishment of a space station.' However, he thought that while much valuable military testing could be accomplished using NASA's approach, it was not fully equivalent to a space station because it lacked 'the operations of rendezvous, docking, resupply and crew rotation.'"

47 "At a press conference on 10 December, Secretary McNamara formally announced that the Defense Department intended to build and launch a two-man orbital laboratory into space in late 1967 or early 1968 "to determine military usefulness of man in space." At the same time he announced cancellation of Dyna-Soar, stating that the substitution of MOL for it would save $100 million in the budget scheduled to be sent to Congress in January."

48 "DoD would initiate, under USAF management, a MOL program directed toward determining the military utility of man in orbit."

52 "(The) astronaut will carry out scientific observations of both space and earth. He will adjust equipment to ensure its maximum performance. He will maintain the

repair equipment. He will be measured to see if he is capable of coping with the unusual—either in his observation or in his equipment operation. Indeed, it is planned that he will be challenged so severely that room in the laboratory must be planned to provide minimum elements of personnel comfort such as rest, exercise, and freedom from the confinement of a space suit."

53   "We must move out immediately and aggressively on the MOL Program for which we have waited and prepared for so long. I cannot overemphasize the national importance of this military manned space undertaking and am confident that we can rise to meet the difficult challenge it presents."

54   "General Momyer said the Board has a number of questions about the proposed plan. One involved the launch schedule, which the Board members felt should be moved up in view of NASA's plans to launch its three-man Apollo spacecraft in early 1967. The Board also was concerned about 'putting all the Air Force man-in-space eggs in the reconnaissance basket' and recommended reexamining the mission area."

55   "Such was the situation when Secretary McNamara announced the MOL project and discussions ensued about its surveillance-reconnaissance mission. Concerned about the security aspects of the new program, the military Director of the NRO Staff**, Brig Gen John L. Martin, Jr., on 14 January 1964 reminded McMillan that the entire U.S. satellite reconnaissance effort was being conducted in the 'black' and had been a forbidden subject within the Air Force since late 1960."

56   "Such a system is a 60-inch diameter cassegrainian type telescope with diffraction limit optics over the useful field of view. This caliber of optics using high resolution film such as S0132 (Eastman Kodak 4404) and the man to adjust the image motion compensation to better than- 0.1 percent, will yield ground resolutions of {better than one foot} from 100 nautical miles altitudes with 20 degree sun angle light conditions, neglecting degradations caused by atmospheric seeing. Under low light levels associated with 5-degree sun angles such as would be useful over the Soviet Union during the winter months, ground resolutions of {better than one foot} could be realized."

57   "In March 1964 there was still another group which recommended that manned space reconnaissance be pursued. A panel of Project Forecast, established the previous spring by Secretary Zuckert and General LeMay with General Schriever as its Director, declared that the areas of most promise for manned reconnaissance were 'those of high resolution photography, infrared imagery, and the all-weather capabilities of the synthetic array side-looking radar.' The panel estimated that high resolution camera systems could be built within a few years that would 'yield ground resolutions of less than {one foot}. It believed the systems could be enhanced by using man to point at the proper targets and adjust for image motion compensation."

58   "To obtain authoritative data, in an economical way, on the possible contributions of man to the performance of military missions in space, and to obtain data on man's performance sufficient to form a basis for design and evaluation of manned systems."

59   "After eliminations and consolidations, 59 experiments were identified. These were further scrutinized, evaluated, and finally reduced to 12 primary and 18 secondary MOL experiments which were incorporated into the preliminary technical development plan submitted to AFSC on 1 April."

60   "The advantages of having a man in space vehicle were in his ability to recognize patterns, interpret them in real time, and report the results, and his ability to point a sensor (telescope-camera) and provide image motion compensation. He

said the proposed MOL experiments should provide answers to the question whether better results could be obtained by using a man as compared to an unmanned system of the same weight."

61   "The objective," he said, "is to determine what man can do in acquiring and tracking and compensating for image motions. Design of orbital gear is incidental to a third phase of the task. The first two phases are simulation and aircraft tests, which are prerequisite to, rather than concurrent with, the third phase." Consequently, he asked that the plan be rewritten to clearly establish the main objective."

64   "USA officials believed that development of a national space station would require an effort comparable to 'the Manhattan project, our ICBM program, and the Lunar program' and they felt it essential that the Air Force be chosen executive manager."

65   "In January 1964, at the request of headquarters USAF, AFSC prepared a MOL management paper which General Schriever submitted to Dr. McMillan on the 20th. The stated objective was to provide 'continuing positive direction and control of the program by the Secretary of the Air Force while assuring the necessary flexibility at the operating management level.' To achieve that objective, AFSC recommended placing General Schriever at the head of a 'MOL Special Program Office' to be located in the Washington area, preferably in the Pentagon. It would be responsible for overall review and program control, report directly to the Secretary in directing the project and implementing his decisions."

66   "For the reasons outlined...above, this office must be specifically and directly responsive to the requirements of the Secretary of the Air Force, as well as to the Chief of Staff."

68   "if the time allotted for project definition could be sharply curtailed, MOL experimental test flights could begin 18 months after contractor go-ahead, a MOL with limited subsystems suitable for manned flight could be made available within 24 months, and one with complete subsystems in about 32 months."

69   "These delays were exceedingly frustrating to USAF officials. On 13 November General Schriever complained to Dr. McMillan that AFSC had been left 'without current direction or intention on which to base the allocation of command manpower and resources to meet MOL milestones. If it was OSD's intention to revert to the original development schedule, ''program ramifications'' must be recognized, he said. He referred specifically to the launch dates in the late 1960s, which would place the Air Force 'in a poor competitive posture with NASA's current and extended Apollo programs.' He reiterated his strong desire 'to undertake a more progressive MOL program.'"

70   "In a new and rapidly developing field such as astronautics wherein new opportunities as well as perhaps constraints and limitations are revealed almost from day to day, it seems to me that we should not attempt rigidly to interpret or classify programs in terms of possible undertakings in the future. In the area of manned spaceflight, both in potential scientific and military applications, I view Gemini, Apollo, and the DoD MOL all as important contributors to the ultimate justification and definition of a national space station."

71   "The development of technology- contributing to improved military observational capability for manned or unmanned operation. They saw this as possibly including intermediate steps toward operational systems."

72   "All data to date has confirmed our original assumptions that man could correct the line of sight to an accuracy of less than 0.1 degree and that he can reduce image motion or smear rate to a value of less than 0.1 percent of V/H (velocity over

height). The data indicates that the target acquisition and centering task ranges from approximately 3 to 8 seconds with a mean resultant displacement error in the line of sight of 0.06 degree when performing at five times magnification and 0.02 degree when performing at 45 times. We have consistently demonstrated the ability to reduce the residual tracking rate error to values of .025 percent of V/H. This level of rate performance is accomplished within the first 2 seconds

73   "The results of studies of MOL experiments P-1, -2, and -3 and contractor simulation tests demonstrated 'that man can accomplish IMC to better than .2% consistently and was limited only by the quality and magnification of the optics and the inherent stability of the vehicle.' Extensive B-47 flights conducted with a modified bombsight and using two cameras also had verified that man had the ability 'to acquire unknown targets as small as trucks and trains and make an accurate count of the total present.'"

76   "In accordance with the agreements reached at the budget meetings on 7-8 December 1964, DDR&E submitted new instructions to the Air Force which formally changed MOL program objectives. On 4 January 1965 he directed Dr. McMillan to initiate additional studies for an experimental military program which would contribute 'to improved military observational capability for manned or unmanned operation' and to development and demonstration of manned assembly and servicing of structures in orbit with potential military applications such as a telescope or radio antenna."

77   "The overall impression created in the minds of unwitting people involved has been that MOL has finally been assigned a reconnaissance mission."

78   "The objective, of course," Vance wrote McMillan on 7 January, "is the creation of a system which will allow the exercise of firm control which will unquestionably be needed to prevent the program from becoming prohibitively complex and costly, and at the same time to deal effectively with the many governmental elements that are involved in such a large program, particularly during the early stages."

79   "As I see it, from the point of view of the Department of Defense, the central question relative to the Manned Orbiting Laboratory is one of existence: the question whether or not to proceed with a major program of manned military space flight. This is a question to which the Secretary of Defense must develop an answer. Furthermore, before any such program is undertaken... he must reach agreement with the President's Science Advisor and the Director of the Bureau of the Budget that the program of military, engineering and scientific experiments and steps toward operational capability is worth the cost and does not duplicate approved programs in any other agency."

80   "Even before most of this data was in hand, NASA organized a MOL-Apollo task tea to prepare the space agency report. It also contacted its various centers for assistance and let three contracts (to Grumman, Boeing, and North American Aviation) to help identify and define proposed scientific experiments which might be conducted aboard the MOL. A total of 84 NASA experiments were identified and a report describing them was sent to OSD on 17 March 1965."

81   "Several days later he and NASA Administrator Webb issued joint statement pledging close cooperation and coordination of each other's space projects. The primary purpose of the statement was to provide a basis for the impending budget message 'and Congressional testimony and public remarks of all officials concerned.' The two agency chiefs said that they intended to avoid 'duplicative programs' and that any manned space flights undertaken in the years ahead by DoD or NASA would 'utilize spacecraft, launch vehicles, and facilities already available or now under development to the maximum degree possible.'"

83   "a large optical telescope could be built for manned orbital operations, that man could plan a useful role in the alignment and checkout of large structures in orbit, and that the program could be justified in terms of the high resolution obtainable {1 foot or less} through employment of man in orbit."

84   The "compelling need of the moment is to overcome a military lag in space technology."

88   "The Air Force emphasized that the optics and optical technology to be developed for MOL would be directly applicable to unmanned systems. It planned to pursue development of elements such as image trackers, which were crucial to the performance of large unmanned systems. It said, however, that the development of MOL would produce a resolution of {better than one foot} much sooner" and with a higher probability of initial success" than a development based on an unmanned configuration."

89   "...for equal total weights and total volumes, the manned system does have an advantage over the unmanned system and can be expected to provide a higher average resolution at an earlier time than the unmanned system. I therefore would support approval of the MOL program."

90   "We should give consideration at the highest level to the contingencies which may occur so that one day we are not caught by surprise by the intensity of the reaction abroad as we were when the U-2 was shot down over the USSR."

91   "I consider it most important that to the extent it can be controlled, everything said publicly about the MOL project emphasize its experimental and research nature, and that statements and implications that MOL constitutes a new military operational capability in space, or an intermediate step toward such a capability, be rigorously avoided."

92   "Having coordinated with all key individuals and agencies, Dr. Brown and Colonel Battle put the finishing touches to McNamara's memo to the President. The Defense Secretary reviewed the final draft on 24 August, made several minor language changes, and that same day carried it over to the White House where he recommended to the President that they proceed with MOL project definition beginning in fiscal year 1966."

93   "After discussion with Vice President Humphrey and members of the Space Council as well as Defense Secretary McNamara, I am today instructing the Department of Defense to immediately proceed with the development of a manned orbiting laboratory. This program will bring us new knowledge about what man is able to do in space. It will enable us to relate that ability to the defense of America. It will develop technology and equipment which will help advance manned and unmanned space flight and it will make it possible to perform very new and rewarding experiments with that technology and equipment..."

94   "It is possible that MOL will demonstrate the feasibility of a few American and Soviet space men in their respective spacecraft operating a continuous space watch. If it does, and if both nations exercise restraint, it could have a stabilizing effect, as have our mutual unmanned reconnaissance satellites. If man can be an efficient observer in orbit for extended periods, the time may come when the U.S. should invite the United Nations to maintain a continuous space control, with a multinational crew to warn of any impending or surprise attack."

95   "Since I have mentioned the Manned Orbiting Laboratory, it is worth pausing right now to challenge forthrightly those who have asserted or intimated that it has something to do with a weapons race. We expect misrepresentations of that sort to came from unfriendly countries and sometimes from ignorant domestic critics. However, I was disappointed to find that a few otherwise well informed

publications and individuals have asserted that MOL is a weapons carrier and a project contrary to our peaceful progress in space."

98   "During the spring and early summer of 1965, General Evans' staff undertook to draft a paper on the proposed USAF management structure. It proposed the creation of a 'strong, autonomous, integrated program implementation office' on the West Coast, headed by a general officer to be known as the Deputy Director, MOL. He would report to the MOL Program Director (General Schriever), who would be responsible to the Secretary of the Air Force, the Under Secretary, and the Director, NRC, for 'total program direction.' The Deputy Director, MOL, would be given 'full procurement authority necessary to conduct both 'black' and 'white' procurement of the MOL program from funds provided him from higher authority.'"

99   "I think it is important that any time anything goes on in the sensor area important enough to talk to your boss [DNRO], I should also be informed. We must not keep secrets from one another."

100   "Since security dictated that there continue to be a visible MOL program, with certain aspects of its mission kept under wraps, it appeared necessary that DDR&E remain in an authoritative position to justify, review and approve various funding requests."

101   "The fact that he had little control over the payload portion of the program remained troubling to General Schriever. In October, at his direction, Evans and his staff drafted a paper on the MOL organization to 'clarify' the management principles outlined in the 24 August plan. Their view was that, unlike previous 'black' projects, where the reconnaissance sensor itself was 'the major element around which overall system integration is postured,' MOL was different because of the introduction of man into the system and because of 'the currently expressed national policy of overt and unclassified admission of the existence of MOL.' Consequently, the suggestion to conduct MOL as a covert program was denied, although 'conduct of covert activities within the program itself' was

102   "Under terms of this agreement the NRO Comptroller and the MOL Program Office would work together to prepare current and future year cost estimates of MOL black requirements. These would be reviewed and approved by both the Director, NRO, and Director, MOL, before issuance. The responsibilities of the Director of Special Projects would include providing 'black' cost estimates, coordinating with the Deputy Director, MOL, and forwarding them to the NRO Comptroller and the MOL Program Office. Authority to obligate the 'black' funds would be issued by the NRO Comptroller directly to the Director of Special Projects, who would be held accountable for them."

103   "Emphasizing the need for 'vertical organization to totally support the MOL program,' Schriever said that all elements of the various corporate divisions 'must be responsive to the MOL Director commensurate with the unique Air Force management structure.'"

104   "...it may be particularly advantageous to draw on the capabilities of two outstanding contractors to accomplish the task originally envisioned for a single laboratory vehicle contractor. I have carefully reviewed the report and findings of the selection board and have completed an additional examination of contractor past experience and performance, and of security factors pertinent to the MOL Program. The Douglas Company offered the best overall technical and management approach. Its past experience and performance as a system integrator on weapons such as Thor, Genie and Nike Hercules/Zeus is good and considerably broader than that of the General Electric Company. General Electric,

on the other hand, showed superiority in important aspects that bear on mission capability."

110   "A solution to these problems would permit the unmanned system, operating with essentially the same camera, to achieve the same ground resolution on prescribed targets as the manned system. It would also contribute significantly to the manned operation by relieving the observer of much of the routine tracking and identification task, and making the pointing and selection of area of interest less critical."

111   "After reviewing the above memorandum, Dr. Hornig asked DDR&E to meet with him on 23 August to discuss the issues PSAC had raised. During this meeting Drs. Hornig and Brown agreed that the Department of Defense would undertake to develop MOL with a {better than 1 foot} capability, either manned or unmanned. They also agreed that a flight demonstration of the unmanned system would be conducted nine months after the first manned flight."

112   "The panel believed that a {better than 1 foot} resolution could be obtained by a properly designed unmanned as well as manned system. It thought that MOL officials should pursue an operational program which could use both elements of the system."

114   This suggestion was formally embodied in a Schriever directive to General Evans on 17 January 1966, instructing him to initiate a study which would bring into sharper focus man's role in MOL. In particular, Evans was to consider NASA's experience with manned space flight and the Air Force's extensive accomplishments "in the effective utilization of man in the performance of unique and highly complex functions under conditions of extreme stress"—as typified by the F-12, X-15, XB-70 and other flight test programs. The AFSC Commander thought a fresh look at this problem might suggest actions "that we should take to exploit more completely man's contributions in the conduct of MOL missions, and in particular the high resolution optical reconnaissance mission

115   "But contrary to the group's expectation, it found as it completed its work 'that the argument for man is as strong now or even stronger than it was when the program was first approved.'"

116   "We believe the essence of today's argument is that, from a current program viewpoint, inclusion of man will virtually guarantee an earlier {better than 1 foot} resolution capability—and earlier useful 'take'—even for the unmanned MOL configuration than would be possible in a wholly unmanned system. Further, we believe that a system capable of {better than 1 foot} resolution will be more cost-effective in a manned configuration if, 'in fact, {better than one foot} resolution is possible at all with an unmanned system."

117   "The results of the study show that, with the astronauts performing a weather avoidance role, the manned system will successfully photograph significantly more intelligence targets than will the unmanned system on a comparable mission. Various cases were examined and the improved factor of the manned system over the unmanned ranged from 15 to 45 percent."

118   "In his memorandum, Brown repeated the earlier conclusion that either the automatic version of MOL or a completely unmanned configuration potentially could give the same resolution as a manned system. On the other hand, he noted that many of the automatic devices had never before been used in an orbital reconnaissance system, and while it was believed they ultimately could be made to perform reliably, there was uncertainty how long it might take. For this reason, the Air Force was convinced that the risk against early achievement of {better than one foot} resolution was 'considerably greater with an unmanned vehicle';

that is, to the extent that man's participation in the development proved effective, 'the {better than one foot} resolution unmanned capability should be achieved earlier in the automatic mode of M

119 "The results of operator-reaction tests conducted on a laboratory-simulator showed, he said, that 'crew participation in target selection could yield almost three times as many photographs of high-intelligence-value targets as could be taken by an unmanned system on the same mission.'"

120 "The panel wanted the Air Force to proceed with the various studies needed to answer the several questions raised during the meeting. However, he concluded (most importantly from the Air Force viewpoint) that the panel also was adamant 'that we should not hold up any contractual proceedings while these questions were being settled.'"

121 "It is my firm conviction that conduct of a vigorous manned military space program is essential in preparing to respond to hostile activity in the space environment. As operational space functions become more complex and more sophisticated with time, the need for the development of truly effective manned systems emerges with increasing urgency. There is no true alternative for a manned system..."

124 "Even as Air Force officials were reacting to PSAC's insistence that they incorporate an unmanned configuration into the MOL program, a severe financial problem arose that threatened and finally delayed early system acquisition."

125 "On 9 December, the Eastman Kodak Company, the DORIAN sensor contractor, dispatched a letter to Gen. Martin advising that the firm would be unable to fulfill its original commitment to deliver the first optical sensor in January 1969 for a planned April 1969 first manned launch‡. Company officials stated they would require a 10-month extension, with delivery of the first flight optics taking place about 15 October 1969 and the first manned launch slipping to mid-January 1970."

126 "MOL Program officials were thus faced with the fact that less than eight months after the President had announced the first manned launch would take place in late calendar year 1968, it had slipped into calendar year 1970. This situation was particularly embarrassing to those OSD and Air Force officials who had recently testified before Congress. On 23 February 1966, for example, Secretary Brown told the House Appropriations Subcommittee: 'Our best estimate at this time is that the first manned flight will not occur prior to mid-1969, which is a slip of about nine months from what we said last year.' On 8 March Dr. Foster, also advised the Senate Armed Services Committee that the first manned launch would take place 'about mid-calendar

127 "First, he has undoubtedly factored into his planning the bitter experience he is presently having in attempting to meet G3 schedules**. Secondly, he is undoubtedly concerned about the availability on schedule of the large new facility and the unknowns facing him in the area of simulated zero gravity testing of 72" light weight mirrors. Thirdly, he must produce specification performance {better than one foot} resolution) on the first flight."

128 "That is, he proposed that the Air Force: Proceed initially with...the recommended program schedule [the first manned flight in December 1969] with the proviso to reschedule the MOL Program no later than-January 1967 based on the realities of negotiated contract prices and FY 67 fund availability. Also this approach would allow the subsequent reprogramming action to take into account the level of FY 69 funds provided in the [impending] FY 68 budget. The merit of this approach is that it affords the least disruption to the program"

129   "Based on the approval already granted, Evans suggested there was clear 'intent and willingness' on the part of OSD 'for the Air Force not only to proceed with the Engineering Development Phase but also to protect development lead-time where necessary.'"

134   "When the MOL project received Presidential approval in August 1965, Air Force officials recognized they would soon be called to testify before Congress. To smooth their path on Capitol Hill, they worked closely with representatives of the Office of Legislative Liaison, OSAF, in particular with Col William B. Arnold, who was extraordinarily helpful to the program."

135   "In a lengthy front page story, the Orlando Sentinel on 4 February castigated Air Force planners for 'cutting loose completely' from the Eastern Test Range and saddling the U.S. taxpayer with unnecessary costs. It derided the Air Force for claiming it was necessary to launch MOL into polar orbit from Vandenberg, and cited unnamed 'veterans of the space program' as declaring such a requirement was 'nonsense.' Fulminating against 'certain Air Force space empire builders,' the Sentinel urged Florida's Congressmen and Senators, state and local government officials and the citizens of the state 'to stop this threatened waste of national resources.'"

136   "What is the ultimate purpose of MOL and why is it that everything the Air Force is doing cannot be done by NASA?" Schriever replied that the mission was military in nature, was not of interest to NASA, and did not fall within the space agency's area of responsibility. At this point Chairman Miller returned to the hearing room and remarked: "It is not necessary to ask this type of question if you have confidence in the U.S. military."

137   "The people, officials and news media of central Florida are complaining vigorously about this proposal which they tell me will cost our country unnecessarily many millions of dollars of added expense and will deteriorate the fine joint effort of NASA and the Air Force which has been conducted so effectively at Cape Kennedy. They also feel that such a move would cause unnecessary hardship to many families not settled in the Cape Kennedy area. They feel, and strongly assert, that there is no sound reason whatever for making this proposed move."

138   "While it is true that some polar launches have been conducted from the Eastern Test Range using the 'dog leg' maneuver, this does result in a reduction of physical capacity. In the case of the Manned Orbiting Laboratory, the 10-15 percent loss in payload required by performing this maneuver is sufficient to jeopardize seriously the success of the program. Furthermore, there is a risk in the case of failures of impacting classified military payloads in areas where classified information might be compromised."

139   "To satisfy these requirements, there is no question but that we must place the MOL payload in near polar orbits. Orbital inclinations from 800 to 100°are considered mandatory."

140   "Thus, during a floor debate in the House of Representatives on 3 May on NASA's authorization bill, Colonel Arnold observed that 'not one of the nine Members of the Florida Delegation present rose to protest the planned use of the Western Test Range for MOL.'"

144   "Duplicative programs will be avoided and manned space flight undertaken in the years immediately ahead by either DoD or NASA will utilize spacecraft, launch vehicles, and facilities already available or now under active development to the maximum degree possible."

145 "On 3 March 1965, Dr. Brown wrote to Seamans about Gemini B. Referring to their agreement of the previous year, he advised that—in order 'to preserve the option of proceeding at a later date with a configuration based upon Gemini B and Titan IIIC'—DoD planned to negotiate a second contract with McDonnell for design definition of Gemini B 'to the point of engineering release.' In response, Seamans reminded DDR&E that their 1964 agreement required the space agency's approval of any such follow-on contract. A second contract, he said, was 'a matter of direct concern to us because of the possible effect it might have on the fulfillment of NASA's Gemini Contract by McDonnell.'"

148 "During the next several days, Generals Evans and Jones worked out the details of the responsibilities of the two agencies. In their preliminary draft agreement, the Air Force assigned to NASA the responsibility for engineering, contract management, and procurement associated with refurbishment and modification of the GT-2 spacecraft and Static Article #4. The Air Force also agreed to provide several highly qualified personnel to participate in the above work. As for Gemini B, the Air Force alone would be responsible for its acquisition and would contract directly with McDonnell."

149 "During the meeting at Houston on March 28, the MSC representatives proposed that the Air Force reimburse NASA $5 million in return for the mission simulator at Cape Kennedy. This came as quite a surprise in view of Dr. Seaman's statement to the Senate Aeronautical and Space Sciences Committee on February 24 that crew trainers and simulators would be made available to the MOL program 'as soon as they can be scheduled for this purpose.' We have not been advised of the terms of this latter qualification. In the same context, Dr. Seamans expressed a view of equipment availability on a nonreimbursable basis except where modification costs are incurred on a NASA contract as in the case of the HSQ spacecraft."

150 "When information was released on the DoD photographic experiments, 'a hue and cry about NASA's peaceful image vs the military spy-in-the-sky implications' arose."

151 "Experiments D-1, D-2, and D-6 clearly demonstrated the capability of man to acquire, track, and photograph objects in space and on the ground. Experiment D-3 showed it was feasible to determine the mass of an orbiting object (in this case, an Agena target vehicle) by thrusting on it with a known thrust and then measuring the resulting change in velocity."

152 "Once docked, the airlock unit will provide ingress-egress capability, life support, electrical power, and the necessary environmental control required for pressurizing and maintaining the S-IVB stage hydrogen tank so that astronauts may work inside in a shirt-sleeve environment during a 14-day or greater mission."

153 "Even so, he stated that the mission would be flown 'no later than July 1968' and he urged the Air Force to continue its experiment development 'at the maximum pace possible.'"

154 "In a report published on 21 March 1966, the House Military Operations Subcommittee complained about 'unwarranted duplication' between AAP and MOL and suggested that the greatest potential savings 'would come from NASA participation in the MOL program.' It noted that both NASA and the Air Force had talked about the possibility of accommodating NASA experiments on a non-interference basis on MOL 'but to date little has been done to achieve this goal.' Instead, the subcommittee said, NASA was proceeding with 'a similar near-earth manned space project which will also explore the effects on man of long duration

space flights and the capability of man to perform useful functions in space.' The House unit urged the NASA and the Air Force to get together in a joint

156 "There is no doubt that technically, if given sufficient resources and time, Apollo systems could be used in MOL. Similarly, under the same assumption, MOL systems could also be used in AAP. However, the assessment of the desirability of use of one specific system hardware in another program must consider all cost effectiveness factors, principally those associated with performance, schedule, and cost."

160 "As noted, from the earliest days of the program Air Force officials had felt a sense of urgency about getting the MOL into orbit at an early date. They were motivated in part by their experiences with 'the B-70 and Dyna-Soar programs, which had been dragged out interminably—mainly due to lack of administration support— until finally cancelled. In both cases, the Air Force rationale had been disputed and neither was supported by the White House."

161 "The present management structure is incapable of producing a well-integrated, well-managed large program such as MOL."

162 "There must be some schedule slip at which it is cheaper to stop some efforts, but we are informed that this is impossible because it would preclude 'orderly' development of everything."

164 "The MOL report concluded, therefore, that the DORIAN system would be {many more} times as productive as photos in the 12 inch or higher class."

165 "I believe the present MOL Program approach is worth the cost in terms of assurance of meeting the resolution goal and returning a worthwhile product at the earliest reasonable date, plus the verification and exploration of additional manned reconnaissance contributions such as target verification, target selection, weather avoidance, etc."

166 "Satellite photography with {the best possible} to 12 inch resolution would help identify a larger number of small items or features beyond existing capabilities. It would increase U.S. confidence in identifying items 'we can now [only] discern' and would reduce the error of measurement of such items. Higher resolutions also would improve U.S. understanding of some operating procedures and construction methods at Soviet military installations and technical processes and the capacities of certain industrial facilities."

167 "During his presentation, Carter referred at one point to the development of GAMBIT 3 and declared that its design goal was {better than one foot} ground resolution. The system's current performance, he said, was at the 13 to 15 inch level. At this point, Chairman Mahon remarked that the products looked so good that 'we ought to be able to slip the MOL.'"

168 "By year's end, Air Force officials had successfully navigated MOL through the financial shoals of 1968 and still had a program which they considered viable. At this time, they were watching with interest the activities of the new President-elect, Richard M. Nixon, as he undertook to organize his administration. They were hopeful they would receive the DORIAN support of the new Chief Executive, whose campaign literature had pledged a strengthened military space program."

172 "MOL's very high resolution (VHR) photography would 'improve the accuracy and timeliness of performance estimates of enemy weapon systems over that provided by HR photography produced by a mature KH-8 system.' MOL, it said, would produce photos containing sufficient detail to determine the performance characteristics, capabilities and limitations of important enemy weapons. It also could provide intelligence of {a highly important intelligence target} and

contribute 'to the monitoring of any arms limitation agreement.' In periods of international crisis, it would prove especially helpful."

173 "Both during Stewart's presentation and the ensuing discussion, Foster, Brown, Flax, and Carroll expressed favorable opinions on the value to DoD of the information 'derivable from very high resolution photography' and strongly supported the existing MOL program for that purpose. They concurred that very high resolution photography "is of significant value to DoD in [making] multi-billion dollar R&D and force structure decisions." Stewart said that MOL was the best way to have a VHR photographic capability at an early date and that the Air Force had proceeded very deliberately to insure very high confidence in an operational system."

174 "A program to minimize FY 70 funding might entail a 50 percent reduction in [the] work force, new material purchases, etc. Approximately $275-400 million would be required in FY 70 to maintain personnel competency and fast readiness. A delay of more than one year in development prior to the first manned flight would result and total program costs would increase more than $360 million."

175 "That is, it became clear to them that additional resolution was required to define the capabilities and characteristics of the missiles."

176 "MOL photography alone will enable the production of performance estimates of foreign weapon systems that are {several} times more accurate and 2-3 years sooner than from current all-source intelligence. Certain important performance parameters and characteristics of foreign weapons, systems, facilities and equipment can be derived with reasonable accuracy, timeliness and confidence from VHR imagery alone... MOL photography will be of considerable value in any strategic arms limitation agreement (along the lines of those now under discussion with the USSR) to provide very high confidence that the Soviets either are adhering to or violating the terms of the Treaty, and further to provide additional technical intelligence on subtle weapon improvements."

177 "Your expressed desire, as reported by Mr. Mayo, that we fund MOL at less than the $525 million now requested of the Congress for FY 1970 has resulted in our making a careful reappraisal of the program. I conclude that we either should fund MOL at a level commensurate with reasonable progress for the large amounts involved, or terminate the overt manned MOL program and continue only the covert very high resolution (VHR) camera system toward future use in an unmanned satellite..."

178 "It is my view that the MOL... has been underfunded the past few years. It is very difficult to run a program on a reduced budget and still have it meaningful, and it is even more difficult when the budgets are continually reduced to change the program to suit the budgetary needs. I believe that if the funding is reduced much below the present level, it would be very difficult to maintain progress and to keep up morale and achieve any meaningful results."

179 "The Dorian Files revealeD: a CompenDium oF The nro's manneD orbiTing laboraTory DoCumenTs which followed and led to President Johnson's 1965 go-ahead decision. He reported about $1.3 billion had been spent on the MOL program to date, that another $1.9 billion would complete it, and that about 65,000 people were involved in the program (including the Associate Contractors and subcontractor personnel)."

180 "I wish to make two final points for the record. It should be clearly understood that termination is not in any sense an unfavorable reflection on MOL contractors. They have all worked very hard and have achieved excellent results. Likewise, MOL termination should not be construed as a reflection on the Air Force. The

MOL goals were practical and achievable. Maximum advantage was being taken of hardware and experience from NASA and other Department of Defense projects, and the program was well managed and good progress was being made. Under other circumstances, the continuance would have been fully justified."

181 "Someday the Department of Defense is going to find that it needs a manned military equipped space station positioned so that it can watch our adversaries 24 hours each day. We will spend billions for unmanned space-based detection and monitoring systems and Earth and space-based warning systems only to find that in the long run it will be more economical and reliable to place manned systems in fixed synchronous orbits over viewing our adversaries..."

184 "The MOL program has been discontinued. I don't understand why and how the government can do something like that—cancel something which has taken years to start, that has taken so much money to continue and time from men who could have been more secure in another area of work. The past four years have been a waste to every man involved in the MOL program. How can the government say— all right, no more, find something else to do? I don't notice anyone cancelling the government."

186 "In mid-July, however, the MOL Systems Office advised Colonel Ford that the contractors' initial claims totaled $137 million. On his instructions, the Systems Office rejected their demands for full fees and, by December 1969, the sum required had been reduced to $128 million."

188 "In the case of the unclassified equipment, the group recommended—and Dr. Seamans authorized—the transfer to NASA of the MOL Laboratory Module Simulator (developed by McDonnell Douglas) and its specially modified IBM 360/65 computer. This equipment would be used in the space agency's AAP Workshop. Its original cost was approximately $30 million."

189 "After MOL's demise, there was a post-mortem within and outside the Program Office on what steps might have been taken to save the project. One view— strongly held by some individuals—was that the Air Force managers had made a serious error trying to proceed with a full- equipped, "all-up" MOL system. That is, they argued the program might have survived if General Evans' suggestion of March 1966 had been pursued— decoupling the optics from the first manned flight in order to fly the "man-rated system" alone at an early date. If MOL had been flying, they believed, it might have had a better chance of surviving."

190 "I can understand the decision to postpone, but I did not know we had totally cancelled all military manned exploratory use of space. Because of what man is now doing in space, the control, knowledge, and utilization of space may well determine the course of future wars."

THE DORIAN FILES REVEALED:
A COMPENDIUM OF THE NRO'S
MANNED ORBITING LABORATORY DOCUMENTS

Edited by James D. Outzen, Ph.D.

Including Carl Berger's

"A History of the Manned Orbiting Laboratory Program Office"

MOL Program Office Department of the Air Force Washington, D.C.

CENTER FOR THE STUDY OF
NATIONAL RECONNAISSANCE

AUGUST 2015

The Center for the Study of National Reconnaissance (CSNR) is an independent National Reconnaissance Office (NRO) research body reporting to the Director, Business Plans and Operations. The CSNR's primary mission is to advance and shape the Intelligence Community's understanding of the discipline, practice, and history of national reconnaissance through research and analysis. Our methodology is social science and history based. Our objective is to make available information that can provide NRO leadership with the analytic framework and historical context to make effective policy and programmatic decisions. The CSNR accomplishes its mission by chronicling the past, analyzing the present, searching for lessons for the future, and identifying models of excellence that are timeless.

Contact Information: To contact the CSNR, please phone us at 703-488-4733 or e-mail us at csnr@nro.mil

To Obtain Copies: Government personnel can obtain additional printed copies directly from CSNR. Other requestors can purchase printed copies by contacting:

Government Printing Office
732 North Capitol Street, NW
Washington, DC 20401-0001
http://www.gpo.gov

Published by
National Reconnaissance Office
Center for the Study of National Reconnaissance
14675 Lee Road
Chantilly, Virginia 20151-1715

Printed in the United States of America
ISBN: 978-1-937219-18-5

MOL ASSEM
INTEGRATION BUILDIN

MATERIAL

PROPERTY OF THE
ITED STATES GOVERNMENT

F FOUND, DO NOT OPEN

BASELINE MOL MANNED MODE

USAF MOL/KH-10

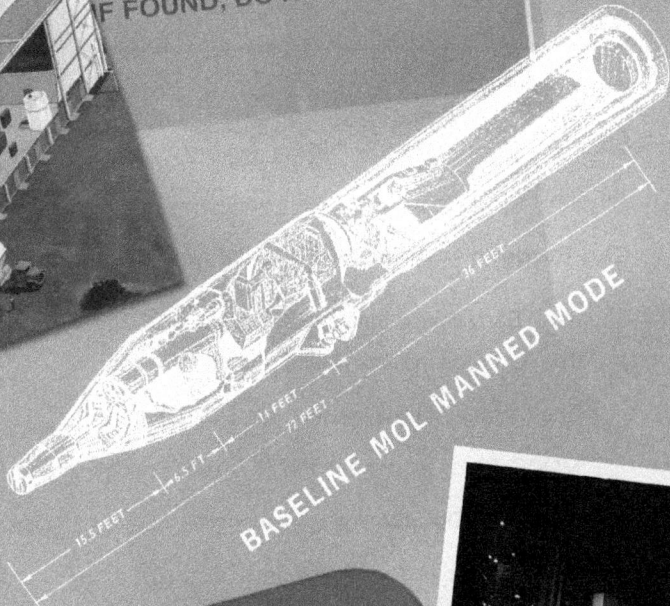

THE DORIAN FILES REVEALED:
A COMPENDIUM OF THE NRO'S MANNED ORBITING LABORATORY DOCUMENTS

CONTENTS

CONTENTS

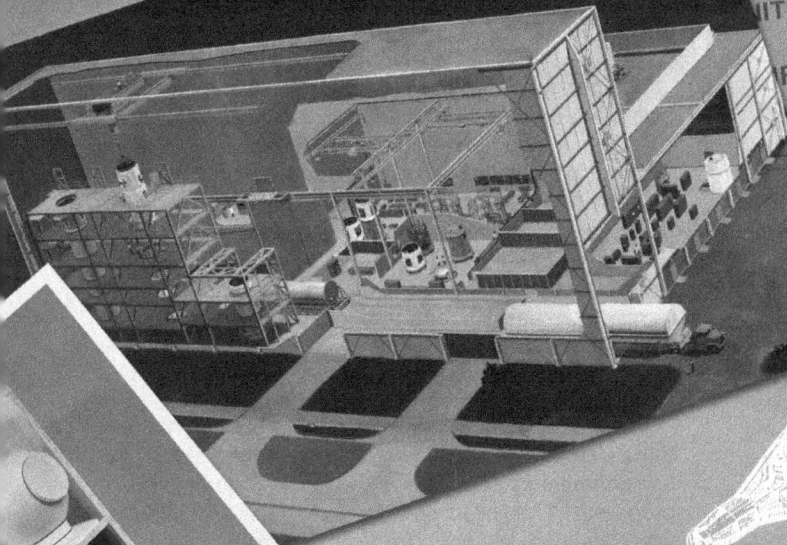

MOL ASSEM[...]
INTEGRATION BUILD[...]

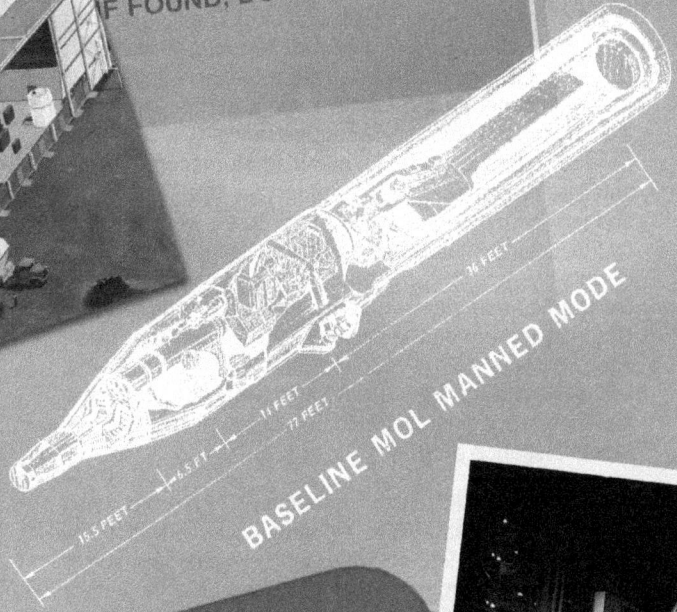

[...]
MATERIAL

PROPERTY OF THE
[U]NITED STATES GOVERNMENT

[I]F FOUND, DO NOT OPEN

BASELINE MOL MANNED MODE

15.5 FEET   6.5 FT   11 FEET   77 FEET   36 FEET

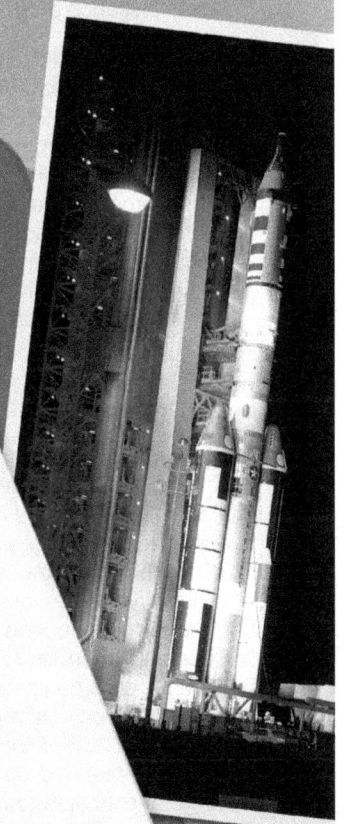

USAF MOL/KH-10

THE DORIAN FILES REVEALED:
A COMPENDIUM OF THE NRO'S MANNED ORBITING LABORATORY DOCUMENTS

FOREWORD

I am pleased to see the publication of *The Dorian Files Revealed: The Secret Manned Orbiting Laboratory Documents Compendium*. This collection joins two others released in the last five years by the Center for the Study of National Reconnaissance (CSNR)—one on the Gambit and Hexagon Photoreconnaissance satellite programs and the other on the Quill radar imagery experimental program. All three were inspired by the Central Intelligence Agency's Center for the Study of Intelligence Corona photoreconnaissance satellite program compendium—released in conjunction with the Corona program declassification. We believe that these compendiums, with a historical essay on the programs, are well suited to help the American public understand the importance and contributions of the nation's national reconnaissance programs.

The Manned Orbiting Laboratory Program was publically disclosed from its early inception—first by the Air Force in 1963 and later by President Johnson in 1965 when the program was described as a means for advancing the military's use of space. Many elements of the program have been well known, including the identities of the men selected to serve as MOL crew members, the configuration of the launch vehicle used to place the MOL in orbit, and general details of some of the experiments that were planned for the vehicle. What has not been revealed, until now, is the extent to which the MOL was designed to serve as a platform for national reconnaissance collection.

Readers of this compendium will find a remarkable collection of documents. The collection has a number of themes. For instance readers will find documents on the public affairs strategy for explaining a military program in space. This was, and remains, a sensitive subject especially as adversaries seek advantages offered through space reconnaissance and technical programs. Readers interested in cooperation between US government organizations will note the efforts necessary to accommodate different objectives between the US Air Force, NASA, and the NRO. Readers interested in the origins of manned space stations will discover a wide range of concepts to assure continued presence of US military crews in space. Readers will find concepts born in the MOL program take remarkable shape in programs matured under NASA manned space programs. Readers will also gain insight into the resource battles that occur as administrations weigh the advantages and tradeoffs of programs competing for the same pool of scarce resources. Finally, the document collection provides insight into how a large program is terminated and closed out.

In the many years since MOL's termination a dedicated group of space enthusiasts have discussed what could have been had the program continued. Perhaps a different perspective is to question the contributions of the program in terms of expertise that was carried to other space and national defense programs by those who participated in MOL and the development and transfer of technology from the MOL program. On these terms, MOL has a strong and important legacy here at the National Reconnaissance Agency and elsewhere in federal space and national defense enterprises.

This compendium will provide a large body of material that historians can use to better understand the development of US space and national reconnaissance programs. The collection will also be useful for scholars who describe lessons learned from past space and national security programs for application to present and future challenges. We look forward to continuing to share documentation that explains the invaluable contributions of the nation's national reconnaissance programs and their unique legacies.

Robert A. McDonald, Ph.D.
Director,
Center for the Study of National Reconnaissance
National Reconnaissance Office
Chantilly, VA

MOL ASSEM...
INTEGRATION BUILD...

...MATERIAL

PROPERTY OF THE
...NITED STATES GOVERNMENT

...F FOUND, DO NOT OPEN

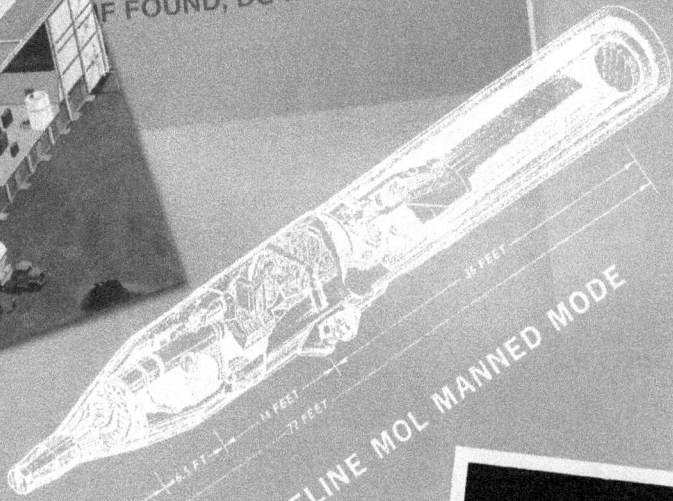

BASELINE MOL MANNED MODE

USAF MOL/KH-10...

THE DORIAN FILES REVEALED:
A COMPENDIUM OF THE NRO'S MANNED ORBITING LABORATORY DOCUMENTS

PREFACE

The *Dorian Files Revealed* is the third compendium of declassified documents the Center for the Study of National Reconnaissance (CSNR) has released in the past five years. We previously published a compendium of documents associated with the Gambit and Hexagon photoreconnaissance programs declassified in 2011 and the Quill radar imagery experimental program declassified the following year. This release of documents contains some 20,000 pages of material. These declassified materials will reveal the national reconnaissance technology planned for the Manned Orbiting Laboratory (MOL) as well as how manned space flight was intended to enhance national reconnaissance collection.

We have chosen to publish Carl Berger's MOL history, which records the administrative efforts to develop and sustain the MOL program. Virtually all the details in the history are being released to the public. I have chosen to substitute language in the redacted areas to smooth the flow of the history for reading purposes. The substitute language is bracketed to allow the reader to know where redactions occurred. A PDF of the unedited version is also available on the documents DVD included with this collection.

We have included only an index to the documents in this compendium and have chosen to leave documents themselves in a PDF format. The index is arranged chronologically with document titles that summarize document content. We believe that this is the most convenient way for compendium readers to locate documents that are of interest to them. We are also including a copy of the index in the DVD so that readers can easily search that resource as well.

We anticipate releasing a new history of the MOL crew members in 2016. A CSNR oral historian is preparing the history based on her interview of MOL crew members and other documentary research. We believe that this history, in conjunction with the Berger history, will provide more insight into the MOL program, especially with respect to the human involvement in the program.

As is always the case, this project would not be possible without the support of many individuals. They include the outstanding NRO declassification staff led by Patty Cameresi, who is Chief of the NRO's Information Review and Release Group. They enthusiastically embraced our request to review this outstanding trove of documents for release and worked meticulously to prepare the documents for release. The release efforts were also sustained by a number of security officers and officials at the NRO. We also appreciate the efforts of the National Museum of the United States Air Force, who agreed to host an event for this document release. We also appreciate

the generosity of the MOL crew members who joined in a panel discussion in conjunction with the release event. Many CSNR hands are responsible for assisting in this compendium including our outstanding graphics artist, Chuck Glover, our dedicated oral historian, Courtney Homer, and our accomplished associate historian, Mike Suk. Finally, this project was supported and sustained by Dr. Bob McDonald, the Director of the CSNR and the nation's foremost national reconnaissance scholar.

James D. Outzen
Chief, Historical Documentation and Research
Center for the Study of National Reconnaissance
Chantilly, VA

MOL ASSEM
INTEGRATION BUILD

MATERIAL

PROPERTY OF THE
NITED STATES GOVERNMENT

F FOUND, DO NOT OPEN

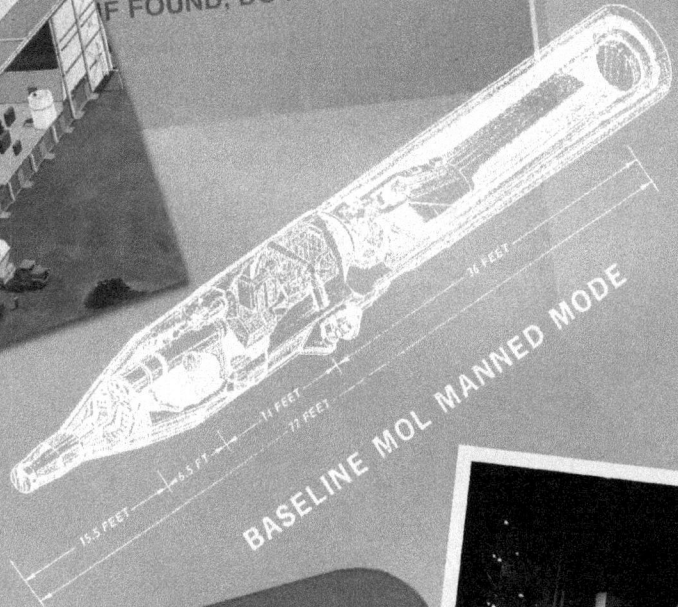

BASELINE MOL MANNED MODE

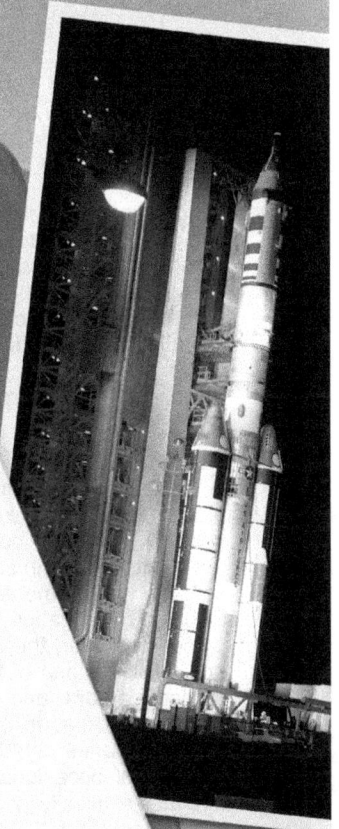

USAF MOL/KH10

THE DORIAN FILES REVEALED:
A COMPENDIUM OF THE NRO'S MANNED ORBITING LABORATORY DOCUMENTS

INTRODUCTION

BODY13

President Lyndon Baines Johnson was not afraid to embrace government programs that might bring about significant change if successful. On 25 August 1965, he announced the following to the American Public:

> At the suggestion of Vice President Humphrey and members of the Space Council, as well as Defense Secretary McNamara, I am today instructing the Department of Defense to immediately proceed with the development of a Manned Orbiting Laboratory.
>
> This program will bring us new knowledge about what man is able to do in space. It will enable us to relate that ability to the defense of America. It will develop technology and equipment which will help advance manned and unmanned space flights. And it will make it possible to perform their new and rewarding experiments with that technology and equipment.

The Manned Orbiting Laboratory, or MOL as it was known, promised to use space for the first time as a manned reconnaissance vantage point. If successful, the program could dramatically change the way the United States collected intelligence on its adversaries, including the nation's main foe, the Soviet Union.

ORIGINS OF NATIONAL RECONNAISSANCE

In order to gain both tactical and strategic intelligence on foes, nations have turned to the skies to gain a better vantage point for collecting intelligence. The United States developed in earnest active technical intelligence collection programs after World War II. The early efforts involved modification of military aircraft to fly near, and sometimes over, the denied areas of the Soviet Union and allied nations of the Soviets. The modified aircraft carried camera and signals collection equipment to capture activities in these closed areas. Unfortunately, US adversaries could down these aircraft, and did so on several occasions. Undeterred, the US developed aircraft specifically for airborne reconnaissance—first the U-2 and later the CIA's A-12 and the Air Force's variant, the SR-71. Both became obsolete for reconnaissance over the Soviet Union as Soviet air defenses improved as was manifested by the May 1960 downing of an U-2 over the Soviet Union, piloted by Francis Gary Powers.

Since 1946, the United States defense community had considered outer space as a vantage point for gaining intelligence. In that year, a think tank that would become the Rand Corporation issued a report on the feasibility of using space for defense purposes. Rand would continue to advocate for space based defense systems through

the 1950's until the U.S. Air Force funded a satellite reconnaissance development program in the mid-1950's known as Samos. The Samos program included both imagery and signals collection satellite designs, but the program faced daunting technical challenges. In the interim, President Eisenhower approved a smaller scale imagery satellite program and assigned responsibility for development to the Central Intelligence Agency. The program, known as Corona, navigated 13 failed attempts to operate before succeeding in August 1960 with a return of the first man made object from space, and in late August, the return of imagery from space. Two months prior, the US also successfully launched the Galactic Radiation and Background satellite, collecting signals intelligence from space for the first time. These programs demonstrated that technical intelligence could be collected from space and opened new horizons for intelligence collection.

LIMITATIONS OF EARLY SPACE RECONNAISSANCE COLLECTION

The Corona imagery satellites proved to be a reliable means for gathering imagery of large areas of the Soviet Union and other areas where the United States had limited access. The imagery was essential for verifying the strategic posture of US adversaries including the Soviet Union's development of strategic nuclear weapons delivery systems including Intercontinental Ballistic Missiles (ICBMs) and long range bombers. Gambit proved equally reliable for gaining high resolution imagery—with better than one foot resolution—that allowed the United States to identify key characteristics of weaponry and other targets. On the Sigint side of the house, Grab and its successor program, Poppy, helped the US identify Soviet radar coverage and other information necessary to understand defenses of US adversaries. Together, these and other national reconnaissance systems, helped the United States gain far more insight into pace and aggregate development of combat capabilities of US adversaries than had previously been available. These insights were key for determining the US's own development pace for and investment in its Cold War national defense systems.

Despite the successes of early space reconnaissance systems, they faced key limitations. For the imagery systems, they often returned imagery that was obscured by cloud cover. In common, the systems could not quickly respond to changes in targeting, especially when new requirements arose. Finally, as might be expected with revolutionary technology in space, the systems were sometimes beset by technical malfunctions. While ground crews were able to make a number of amazing fixes, the inaccessibility of the space vehicles on orbit meant that many other malfunctions could not be remedied. Despite these limitations though, the systems still provided an extraordinary amount of information critical for waging the Cold War.

THE DORIAN PROGRAM

In the early 1960's the Air Force began efforts to put Air Force members into space by developing the Manned Orbiting Laboratory. The Air Force described the MOL program as follows in its initial December 1963 press release announcing the project:

> The MOL program, which will consist of an orbiting pressurized cylinder approximately the size of a small house trailer, will increase the Defense Department effort to determine military usefulness of man in space...MOL will be designed so that astronauts can move about freely in it without a space suit and conduct observations and experiments the laboratory over a period of up to a month.

In the same release, the Air Force announced the cancellation of the X-20 Dynasoar vehicle that was intended to fly from the earth to space and return. The MOL program was described as a less expensive option that would allow the Air Force to "conduct military experiments involving manned use of equipment and instrumentation in orbit and, if desired by NASA, for scientific and civilian purposes." From the beginning of the program, however, US officials questioned the need for the MOL in addition to the US's civilian space program.

Unbeknown to the public, the MOL program included a highly secret set of experiments and capabilities to gain intelligence from space. Information about MOL's secret planned capabilities was strictly protected under a security compartment known as Dorian. The capabilities developed under the Dorian project would result in the United States using the MOL as a manned reconnaissance station in space, collecting both imagery and signals intelligence. If achieved, the MOL would allow the US to overcome the limitations of the already successful Corona and Gambit satellite reconnaissance programs.

The Dorian camera system was developed by Eastman Kodak, the same company that developed the high-resolution camera system used on the Gambit photoreconnaissance satellite. The Dorian Camera system would have some unique capabilities. First, it had a longer focal length and other improvements, permitting better resolution than the first generation of Gambit satellites. Second, the camera system would be used after MOL crew members used a spotting scope system to determine whether or not targets were clear for imagery. Third, imagery targeting priorities could more readily be changed to meet unexpected imagery opportunities. And fourth, the MOL crew members would be trained to repair the Dorian system in the event that

there were malfunctions preventing successful imaging. Together, these capabilities mitigated the shortcomings of the Corona and Gambit photoreconnaissance satellites.

NECESSITY OF INTRA-GOVERNMENT COOPERATION

In order for the MOL program to reach implementation, it required a unique partnership between the Air Force, the National Reconnaissance Office, and NASA. Since the MOL was a manned space fight program, the program required a safe and effective means for taking and returning crew members to and from space. The Air Force turned to NASA to obtain such a space flight capability by securing space capsules developed for NASA's Gemini program. The Gemini capsule was designed to ride atop the larger MOL vehicle, carrying the crew members. Once in space, the MOL crew members would open a hatch on the bottom of the Gemini capsule and travel through a passageway to the laboratory section of the MOL vehicle. They would stay in the laboratory section until returning to the Gemini capsule, with the imagery film, for the reentry through the earth's atmosphere. The Air Force also depended heavily on training procedures and facilities developed for NASA's manned space program.

The NRO contributed the reconnaissance systems that became the primary purpose for developing the MOL. By the time that the MOL program initiated development, the NRO had already developed a number of camera and signal collection sensors for gaining intelligence from space. The Air Force turned to the NRO to obtain the imagery and sensor systems necessary to use MOL as a reconnaissance platform. The MOL program contracted with Eastman Kodak to develop the camera system that was similar in form to the highly successful KH-7 and KH-8 systems also developed for the NRO's Gambit photoreconnaissance satellite. These relationships were critical for keeping the MOL program on its proposed schedule and controlling costs of an already complex and expensive program.

MOL PROGRAM CHALLENGES

The MOL program faced a number of program challenges. Although humans had successfully flown in space since 1961 with Yuri Gagarin's flight, the missions were measured in hours-- not days, weeks, and months proposed under the MOL program. In order to have 60 day missions, a whole host of technical challenges confronted the MOL program staff. The challenges included creating a safe and reliable environment to host MOL crew in space. Additionally, the MOL program required advances in use of technical collection to assure that crew members could both target and obtain intelligence within the limited time over target from space. The MOL program conducted a wide range of technical studies dealing with these challenges as well as determining the crew's physical and mental viability in space for what were considered at the time very lengthy missions. The MOL program was nearly a decade ahead of the first space station mission of the Soviet's in 1971.

The MOL program also faced challenges of a terrestrial nature—the foremost being the Vietnam War. The challenge was that as the MOL program evolved so did the Vietnam War. The Johnson administration was trying to carry out its "Great Society" domestic programs as well as sustain an escalating war abroad. The prosecution of the Vietnam War directly resulted in delays in the MOL schedule in order to reduce costs in the early years of the program and spread them into the future. President Johnson's continued hope during his administration was that the Vietnam War could be concluded and American troops withdrawn to reduce costs associated with the war. Johnson failed to entice the North Vietnam government and its allies in South Vietnam to the negotiation table in order to achieve this end. As a consequence, national security programs not associated directly with the Vietnam program suffered in resource appropriation. MOL was one such program.

Another significant challenge faced by the MOL program was the lingering question of whether or not it really brought unique intelligence collection capabilities. The NRO had already demonstrated that space could be used successfully as a reconnaissance platform through the Corona, Gambit, Grab, and Poppy programs. At the time MOL was proposed, the NRO already had plans for a more powerful high-resolution Gambit program and the CIA was in the early stages of developing a satellite to supersede the Corona program, and they hoped, the Gambit program too. That program evolved into the NRO's Hexagon program. The Hexagon program was designed to carry an immense film load, allowing it to stay on orbit for six months or more. It would also carry an improved targeting system. It promised versatility that called into question MOL's necessity. Eventually, Hexagon and the improved Gambit-3 system would suffice in the Nixon administration's view, leading to the MOL's termination in June, 1969.

THE MOL PROGRAM LEGACY

Because of the MOL program size, complexity, and time in existence, it consumed many millions of dollars in funding before termination. This begs the question of what if anything did the United States gain from the program? There were significant legacy contributions from the program. The first and foremost significant contribution was the leadership that came from the MOL crew members trained under the program. Seven of those crew members were accepted into NASA's astronaut program. At NASA they would either command or pilot the Space Shuttle. Of those, one would eventually lead NASA as the Agency's administrator, another would command NASA's Cape Canaveral launch facility,

and others would lead elements of the NASA space program. Yet another would go on to lead the US's Strategic Defense Initiative. Another would serve was Vice Chairman of the Joint Chiefs of Staff. Many would also play important roles in corporations supporting national defense and space programs. Other engineers, scientists, and staff would play key roles in other national reconnaissance programs, drawing on their experiences and insights gained from the MOL program.

The MOL program would also make important contributions to national reconnaissance and space exploration programs. The Dorian camera system was to be preserved and studied for possible incorporation into Hexagon program. One of the options for reducing costs of the MOL program was a series of unmanned missions. Those missions would carry multiple film-return capsules in a configuration that closely resembled the configuration eventually developed for the Hexagon program. The MOL program also included a segmented mirror technology that was eventually used in a domestic space observatory. Segmented mirrors offered additional advances in space exploration with MOL advancing this important technology.

Finally, MOL helped advance the technology and science necessary for longer space missions. For example, the MOL program required its crew members to travel through a narrow tube or tunnel from the Gemini capsule to the laboratory section once the vehicle was on orbit. This in turn required a flexible space suit—more so than what NASA had developed at the time. The advancements in space suits under the MOL program were transferred to NASA. MOL also included proposals for more than one space module being launched and then linked on orbit. This concept would be critical for the development of today's multi module space craft on orbit such as the International Space Station. The research and technology developed under the MOL program for sustaining crew members on orbit was also transferred to NASA, undoubtedly aiding NASA's advancements in manned space flight.

There is often a misplaced assumption that a cancelled program has no important legacy. This should not be said of the Manned Orbiting Laboratory program. For the reasons listed above, and others contained in the some 20,000 pages of documents associated with this compendium, the important contributions of MOL are clear. The MOL program should be recognized for its rich legacy in both civilian and national reconnaissance space histories.

James D. Outzen
Chief, Historical Documentation and Research
Center for the Study of National Reconnaissance
Chantilly, VA

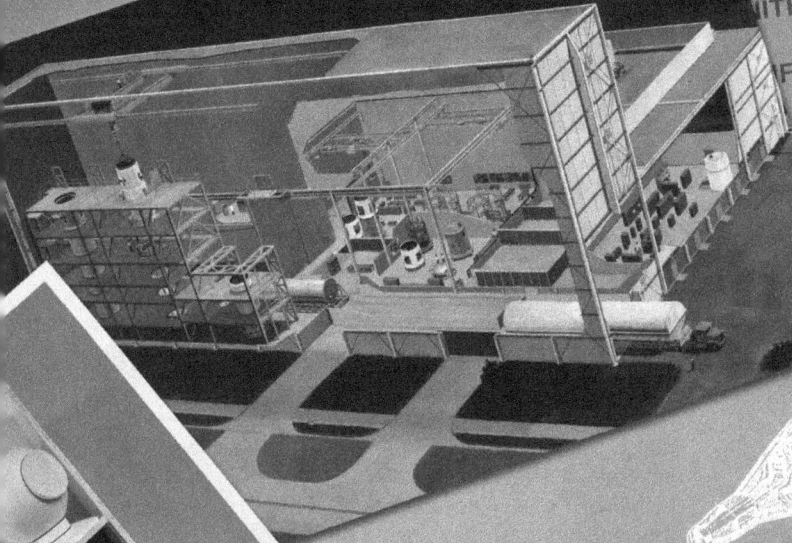

MOL ASSEM[...]
INTEGRATION BUILD[...]

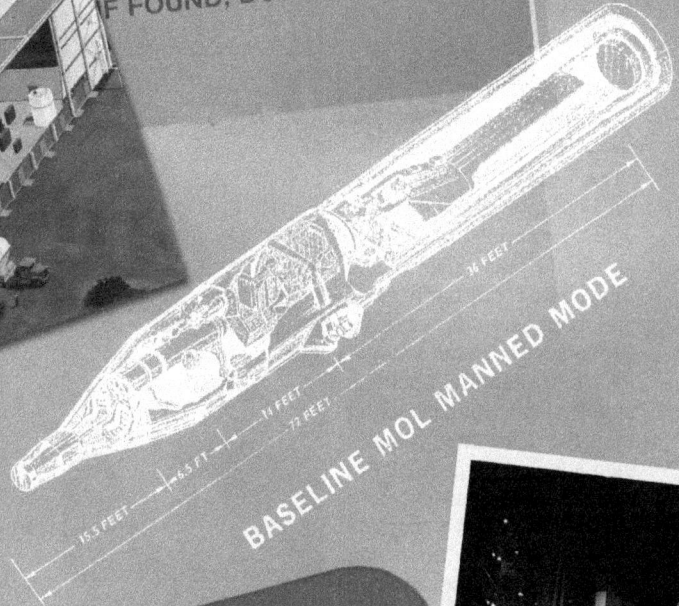

BASELINE MOL MANNED MODE

USAF MOL/KH-10[...]

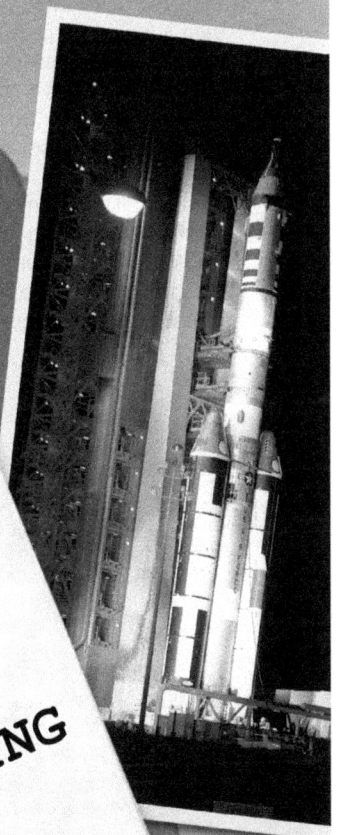

THE DORIAN FILES REVEALED:

A COMPENDIUM OF THE NRO'S MANNED ORBITING LABORATORY DOCUMENTS

CHAPTER I:
EARLY SPACE STATION PLANNING

Early Space Station Planning

The idea of equipping an orbital space station with powerful telescopes so that man might see "fine detail on earth" was first suggested in 1923 by Professor Hermann Oberth. In his pioneering book on space flight published in Munich, Germany, Oberth said it would be possible "to notice every iceberg" and give early warning to ships at sea from such "observing stations." He also thought they could be equipped with small solar mirrors to furnish illumination at night

Figure 1. Hermann Oberth
Source: CSNR Reference Collection

for large cities or with giant mirrors which he said could be used to focus the sun's rays and, "in case of war, burn cities, explode ammunition plants, and do damage to the enemy generally." [1]

Figure 2. V-2 Rocket
Source: CSNR Reference Collection

Oberth's theoretical writings on rockets, space ships and stations, and interplanetary travel were familiar to the German engineers and scientists who, beginning in the 1930s, initiated development of the V-2 missile—the first man-made object to fly through space. During World War II, even as they worked feverishly to

perfect their war rockets at Peenemunde*, these experts still found time to draft plans for future space travel. When word of their extra-curricular activities reached the German secret police in March 1944, several of Peenemunde's technical staff—including its engineering director, Wernher von Braun—were arrested and charged with concentrating on space travel to the detriment of vital missile programs. Von Braun paced a cell in a Stettin prison for two weeks

Figure 3. Wernher Von Braun
with Ferry Rocket
Source: CSNR Reference Collection

before Gen Walter Dornberger, chief of the German Army's rocket development program at Peenemunde, obtained his release by swearing that he was essential to the success of the V-2 program. [2]

Following the military collapse of Hitler's regime in the spring of 1945, many leading German rocket engineers and scientists—including Von Braun, Dornberger, and Professor Oberth—voluntarily surrendered to or were swept up by advancing U. S. Army forces. The Americans seized many of the Peenemunde documents, including drawings of Oberth's space mirror concept.[†] The Allies, who were interested in gathering all the information they could about

Figure 4. Henry H. (Hap) Arnold
Source: CSNR Reference Collection

the deadly V-2's, organized a number of interrogation teams at the detention camps. The American and British officers, as it turned out, were greatly handicapped by their lack of knowledge of German technical advances.

---

* Several thousand V-2's were launched against London and Antwerp in the final months of World War II.

† Life magazine published the Peenemunde drawings on 23 May 1945 under the heading, "German Space Mirror: Nazi Men of Science Seriously Planned to Use Man-Made Satellites as a Weapon of Conquest."

"They didn't know what to ask," Dornberger said later. "It was like they were talking Chinese to us!" The Allied officers also were skeptical about the German captives talk about manned space flight. At their request, Von Braun and Dornberger in May 1945 wrote several papers on possible future technological advances in which they expressed their strong conviction that "a complete mastery of the art of rockets" would lead eventually to orbiting space stations and ultimately to flights to the moon and planets.[3]

The information obtained from the Germans was sufficiently intriguing to the Army Air Force (AAF) for it to incorporate many of their projections into its planning documents. One interesting consequence of this receptivity was that the AAF Commander, General H. H. Arnold—in his final war report on 12 November 1945—became the first official in any branch or department of the American Government to speak of space ships and orbital weapons. In a chapter of his report devoted to future technical developments, General Arnold declared: "We must be ready to launch (weapons) from unexpected directions. This can be done with true space ships, capable of operating outside the earth's atmosphere. The design of such a ship is all but practicable today; research will unquestionably bring it into being within the foreseeable future."[4]

Even as the Arnold report was being drafted, Von Braun and a small party of V-2 experts were settling down in the United States at Fort Bliss, Texas, under contract to the Army to continue work on ballistic missiles. By early 1946 more than 110 members of the Peenemunde team had joined Von Braun and, during the next several years, they helped launch several dozen V-2's at the White Sands Proving Ground. Their presence in the country remained shrouded in secrecy until December 1946, when the Army issued a press release on their activities. The news that "nazi scientists" were working in the United States touched off a wave of criticism of the Government. Among the eminent scientific figures who protested directly to President Harry S. Truman were Dr's. Albert Einstein and Vannevar Bush. A news blackout was re-imposed on the Germans' activities and the furor in time faded away.

By 1950 the attitude of the public had changed sufficiently to enable Von Braun to surface on 3 March at a University of Illinois space medicine symposium in Chicago, where he presented a paper on the construction and launching of multistage rockets and orbiting space stations. Von Braun described how a space station might be constructed in orbit with materials sent up by rocket. He said it could be used as a bomb carrier and as an observation post "for

both military and civilian purposes." Using high-powered telescopes, he said, it would be possible to see people moving about on the face of the earth.[5]

In 1955 Von Braun, now a naturalized American citizen, invited Professor Oberth to join him in the United States‡. Both before and after his stay in the United States, Oberth continued to refine his ideas on space vehicles and travel. In a new book written in 1956, he discussed—among other things—the use of a space telescope to observe the earth. If the station were placed into polar orbit at an altitude of 375 miles, Oberth said:

> ...the crew will have every point on the earth's surface within view at least twice a day...A telescope with the magnification of a million times at a distance of 37,500 (3,250 miles) on the so-called stationary orbit space station...would make the earth appear to be only 37 M (120 ft.) from the observer. This is an almost terrifying power of observation which would make any kind of "Iron Curtain" completely senseless.[6]

THE AIR FORCE INITIATES SPACE STATION PLANNING

The writings of Oberth, Von Braun, and many others about manned space flight stimulated a small group of USAF planners at the Wright Air Development Center (WADC) to begin preliminary studies of possible military applications of satellites and space stations. On 2 January 1957, the Deputy Commander for Research at WADC prepared general guidelines for these studies. He said that the primary goal should be an Air Force space program leading to development of "manned space vehicles and stations" with the emphasis on military reconnaissance.[7] In July 1957 WADC published a technical note on "the functional areas of employment for space vehicles." One of the vehicles discussed in this note was a manned space station with an orbital weight of approximately 17,000 pounds, which would enable the use of "even sizeable astronomical telescopes and observation devices.[8]

All this preliminary USAF planning, however, had little practical meaning at this time since the only approved American space project in 1957 was the Navy-managed Vanguard program, which aimed at putting a scientific

---

‡   Oberth lived at Feucht near Nuremberg, Germany. By 1955 he had received worldwide recognition for his theoretical writings on space science.

Figure 5. Vanguard Launch
Source: CSNR Reference Collection

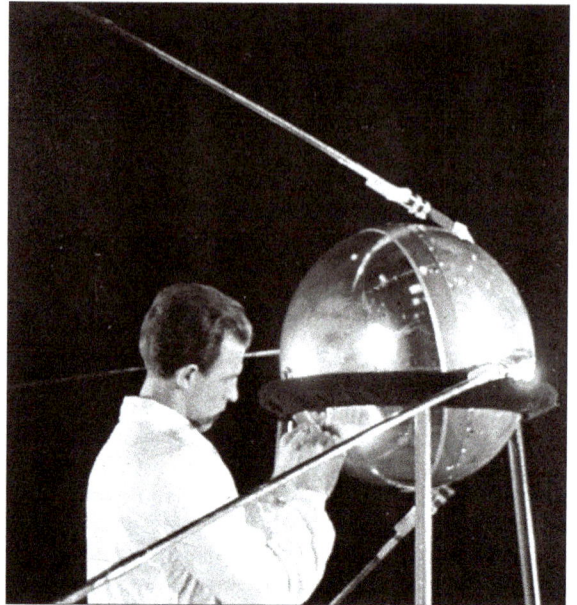

Figure 6. Sputnik
Source: CSNR Reference Collection

satellite into orbit in connection with the International Geophysical Year (IGY). However, after the Soviets astonished the world by orbiting the first artificial satellites in October and November 1957, Congress and the President for the first time became receptive to major American space initiatives. The Air Force immediately initiated studies of ways and means to counter the great political and psychological impact of the Russian achievement.§

---

§ After the failure to launch the first Vanguard satellite in late 1957, the President authorized the Army's Redstone team, led by Von Braun, to prepare to launch a U.S. satellite, which it successfully accomplished on 31 January 1968.

By the end of 1957 the Air Force also had received a dozen unsolicited contractor proposals, several of them dealing with manned space stations. One contractor suggested launching a "manned earth- satellite terminal" as the orbiting station. Another outlined a plan for constructing a four-man USAF station at an altitude of 400 miles, using Atlas ICBM's as building blocks.[9]

On 24 January 1958, in response to a request from the Office of the Secretary of Defense (OSD), the Air Force submitted its proposals and recommendations for an expedited U.S. satellite and space program. Among the projects listed was an Air Force "Manned Strategic Station," which would be assigned missions of weapons delivery and reconnaissance. Several weeks later the Air Research and Development Command (ARDC) incorporated a "USAF Space Research and Space Station" task as part of a proposed study of advanced systems and space vehicles. The task called for an exploratory analysis and design of "a general purpose space technology laboratory orbiting in the cislunar environment" to satisfy military and civilian research and test requirements.[10] Although there was no specific response from OSD, its Advanced Research Projects Agency (ARPA) several months later initiated a study of a space station, which it called "Suzanno."

On 23 April 1958, Brig Gen H. A. Boushey, USAF, Deputy Director for Research and Development, testified before a congressional committee on the status of the

U.S. space program. Among other things, Boushey emphasized ''the tremendous improvement in telescopic and photographic resolution'' which would be possible from a manned orbiting space station. He said:

> What may not be widely recognized is the degree of detail which could be distinguished from, say, a 500-mile orbit. With only a 40-inch diameter telescope, it is estimated that objects on the earth of a size less than 2 feet could be detected. If a 200-inch diameter telescope, the size of the present Palomar reflecting mirror, were located in space at the "stationary orbit" distance of roughly 22,000 miles, objects on the earth approximately 17 feet in diameter could be viewed.

Figure 7. Curtis E. LeMay
Source: CSNR Reference Collection

General Boushey also expressed his belief that man would be an "essential element" in such an orbital station. ''Even the problem of deciding where to look," he said, "is a formidable one, and if left to a mechanical device the chances of profitable search and detailed scrutiny would be far less than if under the direct supervision of an intelligent operator who could immediately exercise the faculties of suspicion, comparison, and reason."[11]

USAF opinion was unanimous in 1958 that man would have a key role to play in space. Hoping to initiate a project to get a man into space "soonest," the Vice Chief of Staff, Gen Curtis E. LeMay, in February 1958, directed ARDC to prepare and submit a development plan. Unfortunately, during this early post-sputnik period, the American failure to launch a satellite ahead of the Russians was wrongly blamed on inter-service rivalry[¶]. One result of the general outcry against the services was the 1958 reorganization of the Department of Defense (DoD), by which Congress and the President greatly strengthened the hand of the Secretary of Defense. Another was the President's decision, acquiesced in by the Congress, to establish a civilian agency—the National Aeronautics and Space Administration(NASA)—to carry out the primary mission of the peaceful exploration of outer space.

As a consequence, after President Eisenhower on 29 July 1958 signed the bill creating NASA, the Air Force was directed to transfer $53.8 billion budgeted for its space projects to the space agency. By this time the Air Force had published seven manned military space system development plans, several of its contractors had prepared studies on ways to get a man into space, and one had built a mockup of a manned space capsule. The USAF plans, as well as the contractor studies, were turned over to NASA.[12] The Air Force was left with limited space development assignments directly applicable to known defense requirements (i.e., satellite reconnaissance)[**], but it also was authorized to pursue in-house studies of advanced spacecraft which might have military significance.

Among the proposed military vehicles which were identified in early 1959 as possible subjects for investigation was a "satellite command post." An Air Force Scientific Advisory Board (SAB) panel visualized such a command post as being permanently manned, supplied, and re-manned by logistic vehicles, and "possessed of comprehensive communication facilities,

¶ The real blame must be attributed to the original political decision that Project Vanguard would not use military missiles to launch a satellite, but should develop its own "peaceful" booster
** An unmanned DoD satellite reconnaissance project was initiated in early 1958 under Air Force cognizance.

(and) reconnaissance and surveillance devices capable of exploiting its unique qualities, but carrying no weapons except for its defense.[13]

Figure 8. Dwight D. Eisenhower
Source: CSNR Reference Collection

THE MILITARY ORBITAL DEVELOPMENT SYSTEM

In March 1959, Gen Thomas S. White, the USAF Chief of Staff, instructed his Director of Development Planning, to prepare a long-range plan for an Air Force space program. The purpose was to provide guidance to the Air Staff in this general area. The Director and his staff, with the assistance of Analytic Services, Incorporated, completed the work eight months later. The results were presented in a series of briefings to the Air Council, the Under Secretary of the Air Force, USAF Commanders, and the Director of Defense Research and Engineering (DDR&E.) One project identified in the Directorate's planning document was a "manned orbital laboratory."[††] Such a space vehicle was needed, it was argued, because certain conditions could not be simulated on the ground. The manned orbital laboratory was seen

†† This 1959 phrase is the first known use of the term.

as providing "training facilities for space crews, a test bed for checking out space weapon systems, and opportunity for the development of spaceship maneuver techniques and doctrines.[14]

While work on this planning document neared completion, ARDC on 1 September 1959 issued a system study directive to the Aeronautical System Division (ASD) at Wright-Patterson AFB, Ohio, requesting a formal investigation of a military test space station (MTSS). The stated objective was to obtain preliminary designs for an orbital station where tests could be conducted in the actual space environment. As a first step, the Division asked the various ARDC sub-commands to identify tests they thought should be performed in the space station. Eventually more than 125 ideas were submitted to ASD, ranging from experiments to check electronic equipment operations in space to tests of man's ability to perform in a weightless state.[15]

Figure 9. Early Space Laboratory Concept
Source: CSNR Reference Collection

After the submissions were analyzed and collated, a statement of work and requests for proposals (RFP's) were prepared and submitted to industry on 19 February 1960. Twelve contractors made proposals. After a USAF board evaluated them, five firms were selected on 15 August to undertake the MTSS study, at a cost of $574,999. These funds were the first expended in studies which years later contributed to the MOL Program. The contractors were General Electric, Lockheed Aircraft, Martin-Denver, McDonnell Aircraft, and General Dynamics (the last performing an unfunded study).

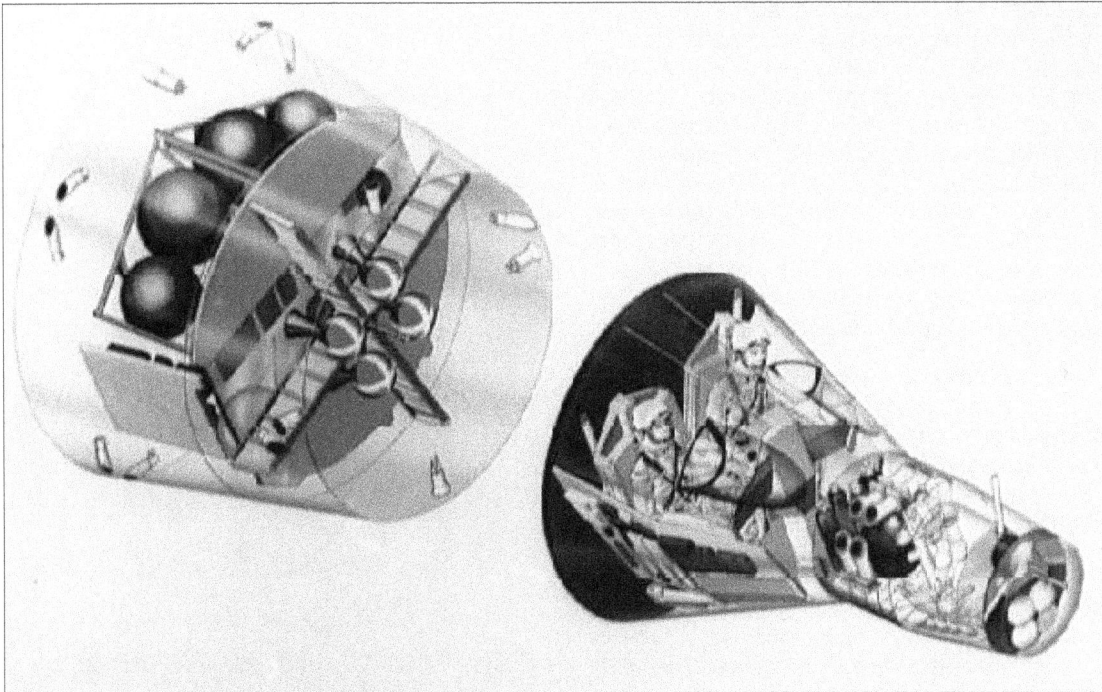

Figure 10. Mercury Mark II
Source: CSNR Reference Collection

In January 1961 the contractors submitted preliminary reports to the Air Force, describing their progress in defining designs for an MTSS, and in February they made oral presentations to a USAF-sponsored conference. Later the Aeronautical Systems Division, with the help of other Air Force agencies, evaluated the interim reports and, on the basis of their comments, a design was developed for a relatively simple space station. ASD proposed a development which would lead to the launching of a three-man ballistic capsule plus a module or station where the crew would live and function for a period of up to 30 days. The ASD concept called for the station to be abandoned when the time came for the crew to return to earth in its capsule. ASD's preliminary evaluation was submitted on 30 April 1961 to the newly-formed Air Force Systems Command (AFSC), successor to ARDC.[16]

By early July the six contractors had completed their studies and submitted final reports. Their conclusions were sufficiently encouraging for Headquarters USAF in mid-July to establish the MTSS as an active project under its newly-organized Directorate of Advanced Technology. A month later, on 16 August 1961, the Air Force submitted a Program Package VI element to OSD requesting an allocation of $5 million in fiscal year 1963 to begin space station studies. When OSD's budget guidelines were released in September, however, the proposed USAF project was left unfunded. A reclama was subsequently rejected.[17]

Meanwhile, representatives of the Air Staff, six major USAF commands, several AFSC divisions, and the RAND and Aerospace Corporations, attended a final MTSS evaluation conference on 12-15 September. They reviewed the contractors' reports and agreed that, while the individual designs differed in detail, all emphasized the importance of orbital rendezvous, not only for supply purposes but also to initially activate the station. The conference recognized that, because the Air Force lacked basic data on man's ability to perform for long periods under conditions of Zero G and knowledge about the problems of space rendezvous,[‡‡] it would be extremely difficult to proceed with a satisfactory MTSS design. They saw some hope of acquiring the necessary information from NASA's newest man-in-space project (originally called Mercury Mark II, later re-designated Gemini), one of whose major objectives was to achieve and demonstrate orbital rendezvous.[18]

[‡‡] Only two men, Soviet cosmonauts Yuri Gagarin and Gherman Titov, had flown in orbit by September 1961. Titov's flight lasted 25.3 hours. When the Russians finally released some data on these flights, they indicated Titov became disoriented. And, of course, the first orbital rendezvous between two space vehicles was still some years off.

Despite OSD's rejection of its request for 1963 study funds, the Air Force continued to push for a space station. In an official USAF Space Plan published in September 1961, the Air Force argued that it needed an orbital station in order to help it evaluate operational hardware and concepts for "space command posts, permanent space surveillance stations, space resupply bases, permanent orbiting weapon delivery platforms, subsystems and components." On 21 September, General LeMay approved the plan and directed AFSC to initiate at once a design study and experimental investigation to select the configuration for a long duration MTSS.[19]

Soon after publication of the Space Plan, John Rubel, Deputy DDR&E, was briefed on it, the proposed space station, and other recommended USAF projects. The Air Force also discussed its space station requirement in a White Paper submitted to Secretary of Defense Robert S. McNamara on 17 November 1961, in connection with a USAF proposal to accelerate the Dyna-Soar (X-20) project.[§§] The paper pointed out that achievement of space rendezvous and developing docking and transfer techniques were already important aspects of NASA's program to land men on the moon. The ability to rendezvous, dock, and transfer men and supplies, the Air Force said, would lead directly to a capability to establish an orbital test station or laboratory which would be especially useful for evaluating systems in space.[20]

Figure 11. Robert S. McNamara
Source: CSNR Reference Collection

_____

§§ For a further discussion of Dyna-Soar and its relationship to MOL, see Chapter III.

While awaiting McNamara's comments on the White Paper and the recommended Air Force program, Lt Gen James Ferguson, USAF Deputy Chief of Staff, Research and Development, on 12 February 1962 discussed the space station proposal before a congressional committee. He said that much of DoD's space activities would require testing of subsystems in "the true space environment"

Figure 12. DynaSoar Space Glider
Source: CSNR Reference Collection

Figure 13. Early MOL Model
Source: CSNR Reference Collection

and that USAF officials were convinced that "a manned, military test station should be undertaken as early as possible." The Air Force, he added, was considering a coordinated effort with NASA, possibly using the Gemini vehicle as an initial transport for the orbiting station.[21]

On 22 February, in a lengthy memorandum to Secretary of the Air Force Eugene Zuckert, Secretary McNamara approved an accelerated Dyna-Soar program. Concerning the space station proposal, he agreed "that a space laboratory to conduct sustained tests of military men and equipment under actual environmental conditions impossible to duplicate fully on earth would be useful." He suggested the Air Force consider possible adaption of Gemini and Dyna-Soar technology and hardware for the initial development phase. McNamara's comments were taken as official guidance as the Air Force now turned its attention to intensive development planning. [22]

Beginning in March 1962 Air Staff and Air Force Systems Command representatives began working on space station planning documents for what was now designated a military orbital development system (MODS). On 26 March AFSC forwarded study data to Headquarters USAF which confirmed the technical feasibility of the concept and provided preliminary funding requirements. On 19 April, Dr. L. L. Kavanau, Special Assistant (Space), OSD, was briefed on the project and afterwards he suggested that the Air Force "quit emphasizing why it must have a space laboratory and get on with the design.[23]"

On 2 May 1962 Headquarters USAF issued an advanced development objective (ADO 37) for the MODS. Finally, in late May, after working closely with the Air Staff, AFSC submitted a proposed system package plan (PSPP) for a system, which it designated as Program 287. AFSC said MODS would consist of three basic elements: a station module (permanent test facility), a spacecraft (basic Gemini vehicle attached to the module), and the Titan III launch vehicle. The system would provide a shirt-sleeve working- environment for a four-man crew for 30 days. AFSC recommended a 15-month Phase I study effort be started at once in order to achieve an initial operational capability by mid-1966. It requested $14.7 million to begin studies during fiscal year 1963.[24]

Headquarters USAF subsequently directed AFSC to identify any internal funds which might be reprogrammed for MODS, pending project review and approval by the Secretary of Defense. On 8 June AFSC advised there were several programs (such as the mobile mid-range ballistic missile) which it believed would not be fully implemented and recommended reallocation of their funds. The Air Force, however, was still committed to the programs listed, whereupon USAF officials decided it would be necessary to submit a program change proposal (PCP) to OSD requesting support for a Phase I study. [25]

Meanwhile, Dr. Kavanau endorsed the proposed Phase I effort after hearing a new MODS presentation at the Space Systems Division on 19-20 June. He indicated that OSD would be receptive to receiving "a solid proposal" for a space test station and asked the Air Force to develop and submit its justification. Several weeks later the Air Staff completed the PCP which, together with a revised proposed system package plan, was submitted to the Chief of Staff. He approved the documents on 12 July 1962 and forwarded them to Dr. Brockway McMillan, Assistant Secretary of the Air Force (Research and Development).¶¶ Dr. McMillan later advised that he believed the $14.7 million requirement was too high and that half that amount appeared sufficient for program definition. The Air Staff subsequently revised the PCP in accordance with this guidance.[26]

MODS, BLUE GEMINI, AND THE FIVE-YEAR SPACE PROGRAM

During the summer of 1962 other important activities were underway which greatly affected USAF space station planning. One of the more important involved a special task force, headed by General Ferguson, which in July initiated a two-month effort to prepare a Five-Year USAF Space Program. In the final program document, the Ferguson task force described several man-in-space projects including the military orbital development system. The MODS proposal was given an especially strong endorsement by a Scientific Advisory Board sub-committee, which reported to General Ferguson on 25 September 1962:

> It is almost certain that as man's conquest of space proceeds, manned space stations with key military functions will assume strategic importance. It is therefore prudent for the Air Force to undertake R&D programs to explore the capabilities and limitations of man in space; to undertake exploratory development of special techniques to exercise military functions from manned orbital bases, and to program flight tests of primitive manned orbital bases with the capability of rudimentary military functions.

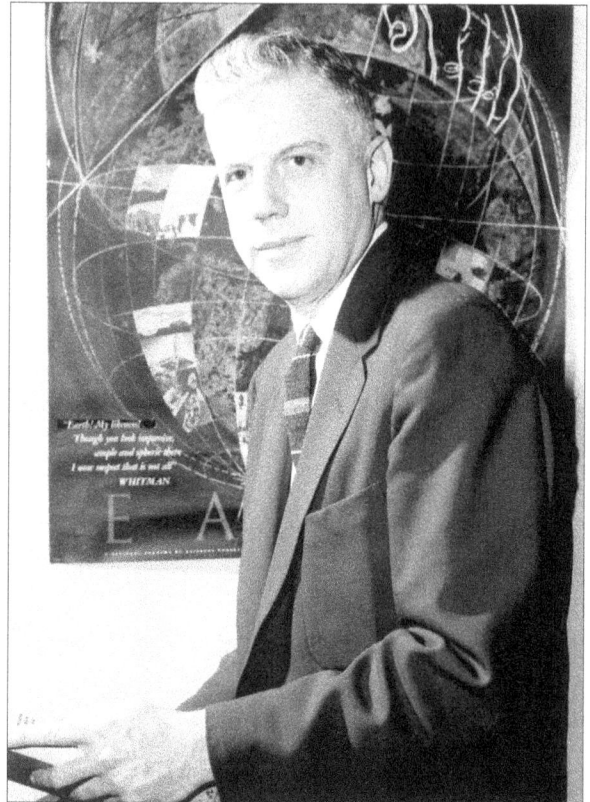

Figure 14. Brockway McMillan
Source: *CSNR Reference Collection*

The SAB recommended that the Air Force utilize NASA's Gemini vehicles as a means of initiating the military man-in-space program. [27]

The Five-Year Space Program document was reviewed and approved during September and October by the Air Council, major USAF commanders, the SAB, and a scientific advisory group headed by Dr. Clark Millikan. Prepared in loose-leaf format, it contained separate PCP's covering the USAF space projects. Total estimated costs to implement the program exceeded by far anything previously submitted to OSD by the Air Force. For fiscal year 1963 through 1967, it called for expenditures of more than $10 billion, about $6 billion more than the estimated costs contained in OSD's tentative guidelines for the same period. [28]

On 19 October 1962, the Chief of Staff forwarded the document to Secretary Zuckert and requested approval. He in turn dispatched it to OSD on 5 November with a general endorsement. Zuckert advised Secretary McNamara not to regard the PCP's in the program

¶¶ McMillan served as Assistant Secretary (R&D) until 12 June 1963, at which time he became Under Secretary of the Air Force, succeeding Dr. Joseph V. Charyk.

document as being submitted for approval in connection with the fiscal year 1964 budget. He said that specific recommendations would be forwarded separately.[29]

On 9 November Zuckert submitted his recommendations. He said he recognized the fiscal implications of the Five-Year Space Program, but explained that it had been deliberately prepared without regard to cost limitations. In fiscal year 1964 alone, the proposed projects would require $1 billion more than the amount tentatively approved by OSD. The Air Force Secretary said that, since such costs were unacceptable, he was limiting his recommendations to four specific programs—Midas, Saint, MODS, and Blue Gemini—with additional funds required totaling $363 million in fiscal year 1964. Of this amount, $75 million would be for MODS and $102 million for Blue Gemini. Previously, no funds had been provided for those projects.[30]

Concerning MODS, Zuckert argued that it possessed "distinct advantages beyond Dyna-Soar and the NASA Gemini program" and would provide a useful vehicle to help resolve some of the uncertainties concerning military space applications. As for Blue Gemini, in which the Air Force hoped to get some "stick time" in space, he said it would be available at an early date and could provide "an important and required steppingstone to MODS." While NASA's Gemini operations would be important for the general acquisition of information, Zuckert said it could not substitute "for actual Air Force experience with the vehicle."[31]

Figure 15. Zuckert and LeMay
Source: CSNR Reference Collection

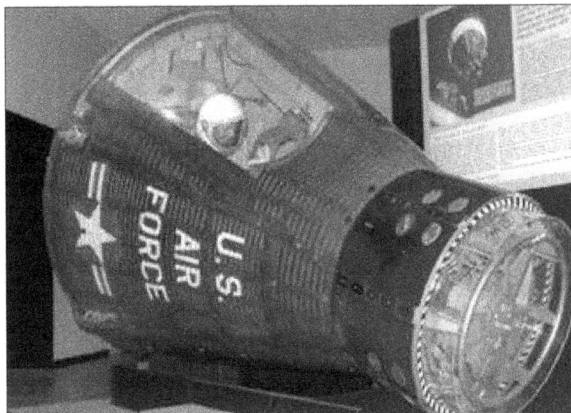

Figure 16. Gemini Capsule Used by USAF
Source: NMUSAF

Tentative USAF planning at this time called for six Blue Gemini launches beginning in May 1965. During the first four flights, the Air Force would investigate and evaluate manned space flight techniques and subsystems of particular interest for MODS and other space operations. There would be attempts to rendezvous and dock with an Agena vehicle, inspection of an Agena in orbit, post-docking maneuvers, and precise recovery. The final two flights would concentrate on mission subsystem testing. Each Blue Gemini pilot would first ride as a co-pilot on a NASA Gemini flight.

In summing up his fiscal year 1964 recommendation, Zuckert admitted that "certain items appear to be similar to activities included in the NASA program." However, he argued that while this might be considered in some quarters as "duplication," he felt it was essential to explore alternate approaches and to exploit different techniques to achieve effective, rapid progress in acquiring military space capabilities.[32]

THE NASA-DOD GEMINI AGREEMENT

Not unexpectedly, the Air Force's proposals were rejected in OSD. Defense officials objected not only to the price tag, but also to the duplication between USAF plans and projects already underway in NASA. Secretary McNamara told a congressional committee that the Air Force's recommendations posed "a real danger that two national programs will develop; one in the Defense Department and one in NASA." OSD's negative stand was discouraging to the Air Force, but an important change had in fact occurred. McNamara and his staff were now in general agreement that DoD—as the Air Force had repeatedly emphasized—had a "bona fide interest in manned space operations" in the near-earth environment. [33]

To help DoD pursue this interest, Secretary McNamara directed his staff to review the advantages, disadvantages and roles of Dyna-Soar versus Gemini while, at the same time, he approached NASA for an agreement to permit the Air Force to participate in the project. Concerning Gemini, DoD and NASA on 27 July 1962 had signed an agreement which called for Defense support of the project on a basis similar to that provided during Project Mercury. The 1962 agreement also confirmed management relationships between the space agency's Marshall Center and AFSC with regard to acquisition of the Agena vehicle, developed by the Air Force.[34]

NASA accepted McNamara's proposal for a new Gemini agreement since it was interested in preventing a duplicative DoD space program which might impinge on its mission. On 21 January 1963 McNamara and NASA Administrator James Webb signed the agreement "to insure the most effective utilization of the Gemini program in the national interest." It created a Gemini Program Planning Board (GPPB), one of whose aims was "to avoid duplication of effort in the field of manned space flight and to insure maximum attainment of objectives of value to both the NASA and DoD."[35]

The Gemini Board's functions were to include delineation of NASA and DoD requirements and to plan experiments to meet those needs. McNamara later remarked that this agreement not only would insure that there would be "one national space program instead of two," but that it would allow the Air Force, representing the Defense Department, "to participate fully in the manned earth orbit experimental and development work."[36]

On 8 February 1963 the Gemini Program Planning Board—composed of NASA, DDR&E, and USAF members—met for the first time. A month later it formed an ad hoc study group to compare NASA and DoD objectives and recommend possible DoD experiments which might be included in the Gemini flight program. Between 25 March and 26 April the ad hoc group met in almost continuous session at NASA's Manned Spacecraft Center (MSC) and, on 6 May, it reported to its parent body. The Board endorsed the work of the group and on 29 May it recommended to Webb and McNamara incorporation of a series of military experiments on Gemini flights that would cost approximately $16.1 million. It also recommended the Air Force establish a field office at the Manned Spacecraft Center to provide overall management of DoD participation.[37]

The Board rejected an ad hoc group proposal that the Gemini flight series be extended to include flights primarily of a military character. The Board felt that since military flights could not be performed within the scope of NASA's existing Gemini plans, they should be considered in a military follow-on program. Moreover, the Board felt that the degree of DoD participation in Gemini should be based on the long-term goals for military man in space and it urged DoD to expedite its decisions in that area.[38]

Secretary McNamara generally accepted the Board's recommendations. He authorized the Air Force to establish a field office at the Manned Spacecraft Center to provide overall management of the DoD portion of the Gemini program. With respect to the exhortation that DoD expedite decisions in the military man-in-space area, McNamara on 20 June 1963 advised Secretary Zuckert that—as a result of the plethora of USAF studies on military manned space flight (Dyna-Soar: Blue Gemini, MODS, Aerospace Plane, etc.)—"DoD will be faced with major new program decisions regarding manned space flight within the next year." Since space vehicle development was so expensive, he said it was necessary that DoD minimize the number of projects by multiple use of hardware and technology- within the entire national space program. He therefore directed Zuckert to submit a plan to assure integration of the several study efforts which might involve Gemini, *** thus providing him an additional basis for "comprehensive program decisions in the area of manned space flight as it relates to military missions."[39]

*** The Air Force submitted this plan to OSD on 23 August 1963. The Deputy for Technology, Space Systems Division (SSD), was assigned responsibility for the conduct of all Gemini-related studies and AFSC was to assure study integration.

(ENDNOTES)

1. Herman Oberth, Wege Zur Raumschiffahrt (Munich, 1923); also quoted in Willy Ley, Rockets, Missiles and Space Travel (New York, 1951), p 336 and 1961 ed., p 366.

2. Walter Dornberger, V-2 (New York, 1954), p 179.

3. James McGovern, Crossbow and Overcast (New York, 1964), pp 147-49.

4. "General Arnold's Third Report," in Reports of General of Army George C. Marshall, General of Army H. H. Arnold, and Fleet Admiral Ernest J. King (Philadelphia, 1947, p 463.

5. Wernher von Braun, "Multi-Stage Rockets and Artificial Satellites," in Space Medicine: The Human Factor in Flights Beyond the Earth, John P. Marbarger, ed. (Urbana, 1951), pp 1 -19.

6. Herman Oberth, Man in Space, trans. G. P. H. de Freville (New York, 1957); pp 69-70

7. Col. C. D. Gasser, WADC, "An Approach to Space Endeavor in Relationship to Current and Future Capabilities of the U. S. Air Force," 2 Jan 57.

8. WADC Technical Note 57-225(U), An Estimate of Future Space Vehicles Evolution Based Upon Projected Technical Capability, July 1957, p 27.

9. Ltr, Col. N. C. Appold, Asst to Dep Cmdr, Weapon Sys, ARDC to Dir R&D, Hq USAF, 26 Dec 57, subj: Initial Rprt on Unsolicited, Sputnik-Generated Contractor Proposals.

10. Memo (S), R. E. Horner, ·SAF (R&D) to Wm Holaday, D/Guided Msls, OSD, 24 Jan 58, subj: AF Astronautics Dev Prog; Project 7969, List of Advanced System and Space Vehicle Studies, Hq ARDC, 19 Mar 58.

11. Testimony of Brig. Gen. H. A. Boushey, 23 Apr 58, in House Hearings Before Select Cmte on Astronautics and Space Exploration, 85th Cong, 2d sess, p 523.

12. Chronology of Early Man-In-Space Activity, 1945-1958, prep by SSD and AFCHO (Feb 1965), pp 17ff.

13. Memo (S), Lt. Gen. D. L. Putt, Chairman, SAB to MIL Dir, SAB, 9 Apr 59, subj: Space Technology Problem Areas.

14. Development Plng Note 59-9 (S-RD), prep by D/Dev Plng, Oct 1959.

15. ASD, Military Test Space Station Evaluation (SR-17527), Tech Doc

16. Rprt, May 1962.

17. Ibid

18. Hist, D/Adv Tech, Jul-Dec 61, pp 38-39; Ltr, LeMay to SAFS, 12 Sep 62, subj: AF Space Prog Proposals, Tab L, in OSAF 55-67, vol 4.

19. ASD, Military Test Space Station Evaluation, p 6.

20. Air Force Space Plan, Sep 61, in OSAF 29-61; AFC 4/17C, 306 21 Sep 61; Hist, D/ Adv Tech, Jul-Dec 61, Jul-Dec 61, p 38-39.

21. White Paper on the AF Manned Mil Space Frog, 16 Nov 61, pp 5, 13.

22. Stmt by Gen. Ferguson, in House Hearings before Subcmte on Appn, 87th Cong, 2d Sess, 1963 Appropriations, Pt 2, pp 484-85.

23. Memo, McNamara to Zuckert, 22 Feb 62, subj: AF Manned Mil Space Prog.

24. Ltr, Col. C. Palfrey, Jr., Chmn, Adv Sys Wkg Gp to Chmn Space Panel, 20 Apr 62, subj: MODS.

25. Rprt of SRB Mtg 62-51, 4 Jun 62, subj: Mil Orbitl Dev Sys; Ltr, Gen. Keese to AFSC, 2 May 62, subj: ADO No. 37 for MODS; Rprt for Week Ending 8 Jun 62; Ltr S/Dev Plog to SP Dir AFSC, 13 Jun 62, subj: MODS Prog Actions.

26. Carl Berger, The Air Force in Space, Fiscal Year 1962 (AFCHO, 1966)

27. Chronology of Significant Actions Relating to the National Orbital Space Station (Sep 63); CSAF Decision, 12 Jul 62, subj: MODS.

28. Quoted in Ltr (S), LeMay to SAPS, 19 Oct 62, subj: Five-Year Space Prog.

29. Ibid.

30. Memo, Zuckert to SOD, 5 Nov 62, subj: Five Year Space Prog.

31. Memo, Zuckert to SOD, 9 Nov 62, subj: Five Year Space Prog.

32. Ibid; Gerald Cantwell, The Air Force in Space, Fiscal Year 1963 (AFCRO, 1966).

33. Memo, Zuckert to SOD, 9 Nov 62, subj: Five Year Space Prog.

34. Testimony of Secy Def Robert S. McNamara before House Subcmte on Approp, 88th Cong, 1st Sess, Pt 1, DoD Appropriations, p 376.

35. Hist, D/Sys Acquisition, Jan-Jun 62, pp 18-19.

36. Agreement Between the National Aeronautics and Space Administration and the Department of Defense Concerning.the Gemini Prog, 21 Jan 63.

37. McNamara Testimony, 11 Feb 63, before House Subcmte on Approp, 89th Cong, 1st Sess, Pt 1, DoD 1964 Appropriations, pp 257-258.

38. MFR, Lt. Col. John J. Anderson, 8 Feb 63, subj: Gemini Prog Plng Bd Mtg, 8 Feb 63; Minutes, Third Mtg GPPB, 7 Mar 63.

39. Memo, GPPB to SOD and Admin NASA, 29 May 63, subj: Recommendations by the Gemini Prog Plng Bd; Gerald T. Cantwell, The Air Force in Space, Fiscal ~ 1963 (AFCHO, 1967).

40. Memo (U), McNamara to Co-Chairman of GPPB, 20 Jun 63, subj: Recom of Gemini Prog Plng Bd.; Memo, McNamara to SAP, 20 Jun 63, same subj.

COU...
MATERIAL

PROPERTY OF THE
...ITED STATES GOVERNMENT

...F FOUND, DO NOT OPEN

BASELINE MOL MANNED MODE

USAF MOL/KH-10

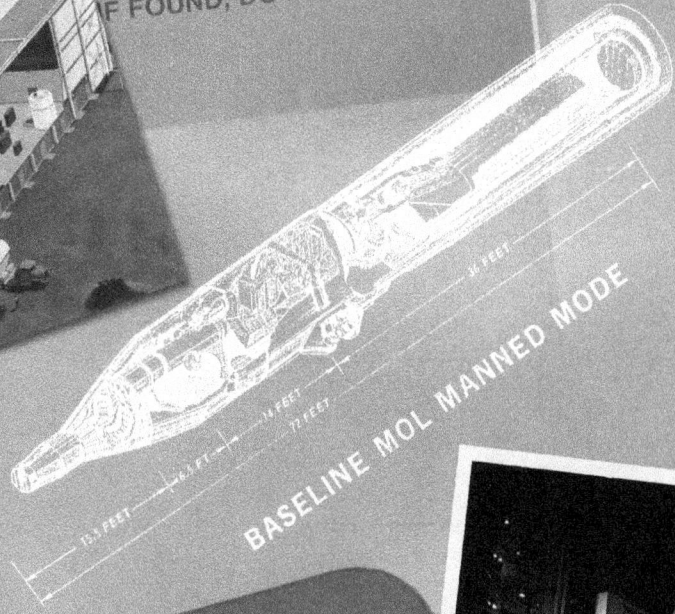

THE DORIAN FILES REVEALED:
A COMPENDIUM OF THE NRO'S MANNED ORBITING LABORATORY DOCUMENTS

CHAPTER II:
A NATIONAL SPACE STATION

A National Space Station

While the ad hoc committee of the Gemini Program Planning Board was working to identify the military experiments to be flown aboard the NASA vehicle, OSD in the spring of 1963 invoked a provision of the 21 January agreement to prevent the space agency from proceeding unilaterally with plans for a new space station study project. The provision was similar to one contained in a DoD-NASA agreement dated 23 February 1961, in which the two agencies agreed that neither would begin development "of a launch vehicle or booster for space without the written acknowledgement of the other." The January 1963 Gemini agreement stated that neither agency could initiate a major new manned space flight program in the near-earth environment without the other's consent.[1]

NASA was reminded of this restriction following a statement made to Congress by Dr. Hugh D. Dryden, Deputy Administrator of NASA, on 4 March 1963. Dryden reported that the space agency planned to award study contracts during fiscal year 1964 for "a manned orbiting laboratory orbiting the earth as a satellite." The completed studies, he said, would provide the information NASA required "to justify and support a decision (to proceed with a development) to be made in time for the fiscal year 1965 budget." USAF officials felt that these plans not only violated the NASA-DoD agreement but also constituted "a Phase I program definition of a MODS-

type manned space station." They further involved issuance of requests for proposals for demonstration of space station subsystem hardware.[2]

On 5 March Maj Gen O. J. Ritland, Deputy for Manned Space Flight, AFSC, advised Gen Bernard A. Schriever that—in light of NASA's proposals—he believed some kind of centralized management of planning for development of a space station was required. He reported to the AFSC commander that while the Air Force was pursuing its MODS studies, NASA had greatly intensified its contracting efforts and was planning to spend several million dollars for space station studies during fiscal year 1964.[3]

After this situation was brought to OSD's attention, on 15 March John Rubel, Deputy DDR&E, met with Dr. Robert C. Seamans, Jr., Associate Administrator of NASA, to discuss the issue. Several weeks later Dr. Harold Brown, DDR&E, also wrote to Administrator Webb about the subject. Secretary McNamara felt, Dr. Brown wrote, that it would be "contrary to existing NASA-DoD agreements... were NASA to initiate any of these projects without prior written concurrence from the Defense Department." He said that he and the Defense Secretary (then on an overseas tour) would be glad to discuss the subject with him.[4]

In a letter to McNamara on 24 April 1963 on the subject, Webb referred to NASA's "statutorily assigned functions" and its need to look constantly to the future "to insure U.S. leadership in the field of space science and technology." This was normally accomplished by letting contracts and doing some in-house work for advanced studies which, he said, seldom included hardware fabrication. According to Webb:

> ... such advanced exploratory studies do not fall within the purview of existing DoD-NASA agreements as they relate to the initiation of "major or new programs or projects"... While we would like nothing better than to have a two-way exchange of ideas and plans concerning the initiation of such advanced studies, we feel that a restriction which would require formal DoD concurrence as a precondition to the initiation of NASA

Figure 17. James E. Webb
Source: CSNR Reference Collection

Figure 18. Mercury Rocket
Source: CSNR Reference Collection

```
studies in this category, or vice
versa, would inevitably involve
an unduly complicated technical
monitorship and unwarranted delays.⁵
```

On 27 April McNamara and Webb met to discuss their differing interpretations of the DoD-NASA agreement and they reached a compromise of sorts. That is, the space agency head agreed that funded space station studies "should be jointly sponsored by the Department of Defense and NASA." Webb also accepted the argument that DoD and NASA would proceed with hardware development "only by mutual consent."⁶

AIR FORCE PROPOSES A NATIONAL SPACE STATION

Meanwhile, the Air Force recommended that a national space station project be initiated. In a memorandum to McNamara on 18 April, Secretary Zuckert suggested that—in view of NASA's "explicit interest" in an orbital station and USAF studies of the MODS concept—a near-earth space station project would involve an effort of major magnitude and consequently "should be undertaken as a national, rather than a departmental endeavor." He thought responsibility for such a program should be assigned to DoD "on behalf of all national interests." He said the assignment was logical "both because of the primary commitment of the NASA to the manned lunar landing program and because of the important military interests in near earth orbit.⁷

The USAF proposal struck a responsive chord in OSD. On 25 May, Secretary McNamara advised Zuckert that he considered "the Orbital Space Static Program as one requiring a new national mission to be assigned by the President on behalf of all national interests." He agreed that since the lunar landing assignment previously had been given to NASA, "the near-earth interests of the DoD might be considered a logical reason for assigning to the DoD this undertaking." However, he said the program needed careful consideration before the President and Vice President were approached on the subject. He expressed confidence that, if such an assignment were made to the DoD, "the Air Force could carry out its management responsibilities cooperatively with the NASA.⁸

To decide on an approach to the national space station, McNamara on 25 May proposed to NASA that the two agencies organize a "Manned Orbital Test Station Program Planning Group." He said its primary functions would be to monitor and, where necessary, to study potential manned orbital test static programs to insure that there was no duplication of effort. He suggested that the group report to the co-chairmen of the Aeronautics and Astronautics Coordinating Board (AACB).⁹ On 7 June, Webb noted that the AACB had already directed its Manned Space Flight Panel to study the best method of insuring DoD-NASA coordination of studies which might lead to a manned orbital space station development. He suggested they await the outcome of the AACB staff work already underway before considering creation of a new planning group.¹⁰

While these top-level discussions were underway, USAF and NASA representatives met and agreed informally to exchange information and requirements for their space station studies. NASA subsequently submitted to the Air Force descriptive material on all its space station studies and also provided certain of its requests for proposals. The Air Force in turn provided NASA data based on its MODS proposals. On 10 June 1963, referring to his agency's pending study contract negotiations, Dr. Joseph F. Shea, NASA Deputy Director for Manned Space Flight (Systems), also asked General Ritland to submit those "definitive requirements" which would meet the Air Force's space station needs in the near future.¹¹

On 18 June Col Donald Heaton, Director of Launch Vehicles and Propulsion, Headquarters AFSC, advised Dr. Shea that since there was general agreement one national program would serve the needs of both agencies, it appeared logical for each to sponsor separate pre-program definition studies. That is, he suggested NASA's studies concentrate on the configuration preferable to it but supporting DoD requirements to the maximum extent possible, and vice versa. "The product of either study", he said, "should be an adequate point of departure for a national program definition phase." Heaton also proposed that their pre-program definition studies be coordinated and that neither agency launch a program definition phase "without mutual agreement.¹²

Dr. Shea agreed to more direct contact between their study programs as well as a further exchange of space station data. He informed Heaton, NASA would continue to provide the Air Force copies of all significant documents related to those studies and he requested copies of USAF requirements documents and RFP's as they became available. In addition, he asked arrangements be made for NASA representatives to sit in on SSD briefings on its space station studies.¹³ Subsequently, NASA named Dr. Michael I. Yarymovych, Assistant Director of Manned Satellite Studies, Office of Manned Space Flight, to serve as its focal point for coordinating exchange of data with the Air Force.

Figure 19. Michael I. Yarymovych
Source: CSNR Reference Collection

Meanwhile, the AACB's Manned Space Flight Panel completed its review of NASA-DoD coordination and concluded that a formal exchange of information between the two agencies should be increased. On 27 June the panel suggested to the Board that data furnished include brief descriptions of projects (AF Forms 613 and NASA Task Descriptions), studies of supporting technology, significant in-house efforts, RFP's, work statements, contractor proposals, and final study reports. The panel agreed that significant meetings related to specific studies should be open to four observers from each agency.[14]

McNAMARA'S REPORT TO VICE PRESIDENT JOHNSON

The advantages of acquiring a national space station were being reviewed at this time not only in the government but also in various technical and professional journals. In July 1963, Air Force Magazine discussed in great detail the Air Force's MODS concept and NASA's proposed Manned Orbital Research Laboratory (MORL) in an article titled, "The Case for an American Manned Space Station." Perhaps coincidentally, Vice President Lyndon B. Johnson, chairman of the National Aeronautics and Space Council (NASC), on 22 July asked Secretary McNamara to submit a paper to him "expressing the possible uses of space stations in maintaining our national security." He also requested Administrator Webb to submit his ideas as to the manner in which a future space station development project should be approached.[15]

On 9 August McNamara forwarded a three-page letter to the Vice President which was, from the Air Force viewpoint, very gratifying indeed. The Secretary declared that an investigation of the role of military man in space was important to national security and that an orbital space station would help determine military utility. The station could serve, he said, as a laboratory and development facility to devise techniques for long duration life support and to test equipment for both manned and unmanned military missions. He said it was possible "to conceive of significant experiments and tests to improve our capability in every type of military operation where space technology has proven, or may prove, useful.[16]

In addition to its potential as a laboratory and development facility, Secretary McNamara stated:

```
... there is the probability that
it will evolve into a vehicle which
is directly used for military
purposes. It may provide a platform
for very sophisticated observation
and surveillance. Detailed study of
ground targets and surveillance of
space with a multiplicity of sensors
may prove possible. Surveillance
of ocean areas may aid our anti-
submarine warfare capabilities. An
orbital command and control station
has some attractive features. While
orbital bombardment does not appear
to be an effective technique at the
moment, new weapons now unknown may
cause it to evolve into a useful
strategic military tool as well as a
political asset.[17]
```

McNamara reported that the Defense Department had studied the space station concept for several years and "expected soon to approach industry with a Request for Proposal on a detailed pre-program definition study of an orbital space station." Data obtained from this study would permit OSD to determine the course of development and to start a program definition phase, "should a decision be made to proceed." Although he was not ready to make a recommendation, McNamara advised that the orbital space station program, if undertaken, would require a new national mission "to be assigned by the President on behalf of all national interests.[18]

Webb submitted his statement to the Vice President, also on 9 August. The space agency recognized, he said, "that an experimental Manned Orbital Laboratory (MOL) project, a mandatory forerunner of any long-duration manned space operational system, would be a major undertaking. "NASA and DoD", he said, were conducting exploratory studies of a MOL, both in-house and through contracts, and coordinating their efforts under the aegis of the AACB. After determining the national need for a MOL and evaluating various concepts, he proposed the two agencies submit to the President via the Space Council a recommendation "as to the management responsibility based on predominant interest and consideration of other pertinent factors, such as management competence, relation to other programs in progress, and international and political implications.[19]

A NEW NASA-DOD SPACE STATION AGREEMENT

Meanwhile, Secretary McNamara concluded that "a mere exchange of information" between DoD and NASA would be insufficient. Noting the Manned Space Flight Panel's report of 27 June to the AACB (cited above) had urged an increased exchange of data between the two agencies, McNamara suggested to Webb that they "now agree to a more formal coordination in this field." Toward this end, he submitted a draft agreement between DoD and NASA concerning studies of manned orbital test stations and requested Webb's signature.[20]

On 8 August Dr. Brown, in accordance with the above draft agreement, forwarded to NASA the Air Force's proposed study entitled "National Orbital Space Station (NOSS) Pre-Program Definition Study" and requested the space agency's concurrence. He reported approximately $1 million would be spent on the study in fiscal year 1964. On 23 August, NASA approved the study and expressed the opinion that it would be useful in defining the military potential of such a space station.[21]

On 23 August NASA also forwarded to DoD three proposed space station studies it planned to pursue and requested concurrence. They included a $3.5 million Phase II MORL study, a $200,000 Early Apollo Research Laboratory investigation, and a $200,000 Biomedical and Human Factors Measurement System study for a manned orbital laboratory. Asked by OSD to comment, the Air Force generally concurred in all the NASA study efforts up to final design and fabrication of mock-ups. It pointed out, however, that the final results would probably not be representative of national space station program candidates since DoD requirements were not being considered and would not be available until early in 1964 from the NOSS study.[22]

Webb, meanwhile, agreed with McNamara's view that the existing exchange of information was insufficient. However, he disagreed with the approach suggested in the DoD draft agreement. It did not take into account, he said, "some very important complexities which we both face in endeavoring to obtain the maximum of cooperation between the Department of Defense and NASA." He cited his previous agreement to submit studies which NASA proposed to make and to "fund in any area in which DoD was interested." Prior to such submission, he proposed a procedure under which officials of both agencies would coordinate to insure that the study, when submitted, included "those things which you and your colleagues regard as important and exclude those things which you may believe unnecessary." With the above in mind, Webb submitted an alternative agreement to cover the entire approach to a possible new manned orbital space station project.[23]

In brief, NASA's proposed agreement provided that: (1) advanced exploratory studies on a space station would be coordinated through the AACB; (2) upon joint evaluation of the studies, the Secretary of Defense and the NASA Administrator would make a joint recommendation to the President as to the need for such a project, including a recommendation as to which agency should direct the project; (3) if a decision were made to proceed with space station development, a joint NASA-DoD board would formulate the detailed objectives and specify the nature of the experiments to be conducted.

On 16 September 1963, McNamara accepted and signed the alternative agreement, although he still had "certain reservations." In a letter to Webb, he said his greatest concern was to insure that the requirements and design constraints of each agency could be fully incorporated from the beginning. As an example of the type of problem confronting them, McNamara cited NASA's proposed contractor effort for design of a Manned Orbital Research Laboratory at a cost of $3.5 million.

```
I believe that an effort of this
magnitude is premature by eight months
to a year since it will not be possible
prior to that time for us to provide
properly for the incorporation of
Defense Department judgments and
thoughts on military requirements
into the design. You must realize
that if on-going DoD studies provide
justifiable military objectives for a
space station development, there may
```

be the necessity for a significantly
different design approach which will
be responsive to agency's needs.[24]

In an effort to respond to this criticism of the MORL contract, the space agency revised its study task to "lay a broad foundation for a versatile space laboratory in such a way as to allow for later incorporation of a wide variety of experimental requirements." According to this revision, the MORL study would be carried out in parallel with DoD's space station studies and would make it possible "for a merging of the two with a minimum of delay." It would also cost less—$1.2 million instead of $3.5 million. NASA expressed the belief that this approach would facilitate the early initiation of a preliminary design phase that would accommodate the requirements of both agencies.*[25]

Meanwhile, in accordance with the September 1963 NASA-DoD agreement, Dr. McMillan directed AFSC to continue to provide Air Force documents to NASA. Later, on 8 November, Headquarters USAF provided AFSC formal guidance on the procedures to follow in transferring such information. A newly-established office, the Deputy Director of Development Planning, Space,[†] was designated the Air Staff focal point for conceptual planning of a possible new manned earth orbital and research project and for exercising "authoritative review" over exchange of space station data between the Air Force and NASA. AFSC was required to submit a monthly status report on all space station study activities, a list of new NASA space station reports, and other data.[26]

THE AIR FORCE OSS STUDY

For almost half a year—while the above top-level planning was proceeding—the Air Force's MOD study proposals had languished. However, in the spring of 1963, reacting to NASA's space station activities, the Air Force took steps to resurrect the project. On 16 May, Lt General Howell M. Estes, Jr., AFSC Vice Commander, wrote to General Ferguson suggesting that they initiate a "pre-Phase I" or "Zero Phase" MODS study at an estimated cost of $1 million. He said that since any space station program would probably be a joint DoD-NASA effort, it was imperative that DoD be in a position to evaluate the extent to which objectives of Program 287 (MOD)can be obtained by whatever program NASA proposes."[27]

After Dr. McMillan authorized Estes to proceed, on 22 May AFSC submitted a formal proposal to Headquarters USAF for a three-month pre-program definition MODS study effort. The study's basic purpose was "to enable DoD to evaluate the extent to which the objectives of MODS can be attained by the space station program expected to be proposed by NASA this fall. Specific study objectives were: (1) establish precisely the peculiar requirements for a military orbital development system, including the detailed design and schedule or the orbital tests which would be needed; and (2) to determine the minimum acceptable performance characteristics of the station module, shuttle vehicle, and associated equipment, as well as the criteria required to make the system militarily useful.[28]

While awaiting Headquarters USAF approval of the MODS study, AFSC on 27 May directed the Space Systems Division to begin preparation of RFP's, the MODS work statement, and other papers. On 12 June AFSC further advised that the Phase Zero study should consider "total DoD requirements" and that Army and Navy mission requirements would be sought for incorporation into an appendix to the work statements. AFSC said Army and Navy representatives also would be invited to participate in the review of contractor progress and final reports.[29]

Subsequently, on 24 June, AFSC invited the Army and Navy to participate in the Phase Zero study effort. It asked the two services to provide information on missions "which either potentially may most advantageously be conducted from space, or advanced through manned space experimentation and testing." To help prepare their submissions, AFSC forwarded background information on the study and advised that their designated representatives would be able to attend contractor progress briefings. If a development program followed, Army and Navy personnel also would be included in the AFSC project office.[30]

Early in July 1963 AFSC prepared a revised Form 613c for the National Orbital Space Station, the new designation superseding MODS. However, this name was short-lived, as the study project was finally identified as the Orbital Space Station (OSS), adopting terminology used by the Secretary of Defense. On 17 July Secretary Zuckert forwarded the OSS descriptive task to the Secretary of Defense and advised he intended to initiate the study at a cost of $1 million. Its primary purpose would be to examine "on an overall parametric basis" the contributions such a vehicle could make to enhancing military objectives. Zuckert reported the Air Force intended to exchange data and maintain close coordination with the Army, Navy, and NASA.[31]

---

* OSD approved the revised study as well as the two other studies planned by NASA.
† Established 23 September 1953. See also Chapter V, "Evolution of the MOL Management Structure."

A week later Dr. McMillan signed a new program change proposal for the space static study and forwarded it to OSD. It listed the following tentative system development milestones: first contract award, September 1964; logistic support vehicle qualification test flight, March 1968; achievement of an initial operational capability, December 1968. The Air Force requested OSD's approval of expenditures of $75 million in fiscal year 1965 to implement a program definition phase, and expenditures of $324 million in fiscal year 1966. Total R&D costs through fiscal year 1969 were estimated at $786 million, but the Air Force said that a more accurate estimate would be made during program definition.[32]

Deputy Secretary of Defense Roswell Gilpatric subsequently advised the Air Force that the PCP was under review and its details being discussed by Dr. Brown with Dr. Alexander Flax, the new Assistant Secretary of the Air Force (R&D).[‡] Gilpatric said Brown would coordinate the USAF proposal with NASA prior to issuance of a request for proposals. As noted earlier, Dr. Brown did submit the proposed study to NASA and received the latter's formal concurrence in late August. A few days later Brown authorized the Air Force to proceed with the $1 million study.[33]

Figure 20. Alexander H. Flax
Source: CSNR Reference Collection

In authorizing the OSS study, the defense research director advised the Air Force that "the Secretary of Defense and I will have a more detailed interest than usual "in the outcome of the work because of the national importance attached to it. To insure program objectives were being met, Brown asked to review results of the source selection before the contracts were awarded. He said the immediate objective of the study should be directed toward "the building of a space station to demonstrate and assess quantitatively the utility of man for military purposes in space." He continued:

The space station so contemplated would be a military laboratory, and its characteristics must be established with some specific mission in mind if its function is to be a genuine military one. The principal missions to be considered are those that can be included in a broad interpretation of reconnaissance: surveillance, warning, and detection can be considered in this context. Other missions such as those assuming the use of offensive and defensive weapons shall not be considered unless it can be explained in detail how such missions might be done better from a space station than any other way.[34]

On the basis of this additional DDR&E guidance, an ad hoc team which included representatives from AFSC, SSD, and the Air Staff assisted by ASER[§], revised the various documents—the DD 613c, RFP, and work statement. On 13 September 1963, Dr. Flax approved the revisions and authorized General Ferguson to proceed with the study. He asked that it be completed by March 1964 and the final USAF report be available by April 1964. He further requested submission of monthly progress and status reports to himself and Dr. Brown, following completion of each task. Dr. Flax's approval was promptly forwarded to AFSC.[35]

On 18 September SSD sent RFP's to 45 firms which had responded to a formal advertisement the previous July which announced the Air Force's intention to contract for a space station study. It established a tentative schedule calling for contractor selection by 15 November, receipt of OSD approval by 22 November, and completion of contractor negotiations by 5 December.[36]

---

‡   Dr. Flax was sworn in as Assistant Secretary of 8 July 1963, succeeding Dr. McMillan, the new Under Secretary.

§   Analytical Services, Inc., a non-profit research organization.

Subsequently, OSD issued a Decision Guidance paper on the USAF program change proposal, approving establishment of the Orbital Space Station as a DoD program element. However, it limited Air Force expenditures in fiscal year 1965 to $5 million (versus a requested $374 million). In the three succeeding fiscal years OSD indicated tentative annual expenditures of $100 million. In establishing the OSS as a program element, however, OSD advised that the decision did not constitute approval of any specific program or study effort.[37]

As it turned out, the award of the OSS study contracts was delayed, due to major new decisions made in late 1963, which significantly altered direction of the study effort.

ENDNOTES

1. Agreement Between NASA and DoD Concerning the Gemini Prog, 21 Jan 63; Ltr (U) Dep SOD Roswell Gilpatric to Webb, 23 Feb 61.

2. Ltr (S-SAR), Ritland to Schriever, 5 Mar 63, subj: Centralization of Plng Activities for a Natl Space Station.

3. Ibid.

4. Ltr (U), Brown to Webb, 11 Apr 63, w/incl.

5. Ltr (U), Webb to McNamara, 24 Apr 63.

6. Ltr (U), McNamara to Zuckert, 25 May 63, subj: A Natl Orbital Space Station Frog.

7. Memo (S), Zuckert to SOD, 18 Apr 63, subj: Orbital Space Station Frog.

8. Ltr (S), McNamara to Zuckert, 25 May 63, subj: A Natl Orbital Space Station.

9. Ltr (U), McNamara to Webb, 25 May 63.

10. Ltr (U), Webb to McNamara, 10 Jun 63.

11. Ltr (U), Dr. J. F. Shea, NASA, to Gen. Ritland, AFSC, 10 Jun 63.

12. Ltr (U), Col. Donald Heaton, D/Vehicies, AFSC, to Dr. Shea, NASA, Jun 63, subj: Suggestions for Coordination of Space Station Plng.

13. Memo (U), Shea to Heaton, 20 Jun 63.

14. Rprt (U), Manned Space Flight Panel to AA.CB, 27 Jun 63, signed by D. Brainerd Holmes, NASA, and Dr. McMillan, Air Force.

15. Ltr (U), Vice President Johnson to McNamara, 22 Jul 63, n.s.

16. Ltr (C), McNamara to Vice President Johnson, 9 Aug 63.

17. Ibid.

18. Ibid.

19. Ltr (U), Webb to Vice President Johnson, 9 Aug 63.

20. Ltr (U), McNamara to Webb, 8 Aug 63.

21. Ltrs (U), Brown to Seamans, 8 Aug 63; Seamans to Brown, 23 Aug 63.

22. Activity Rprt (S), D/Adv Plng, Space, Week Ending 13 Sep 63.

23. Ltr (U), Webb to McNamara, 17 Aug 63.

24. Ltr (U), McNamara to Webb, 16 Sep 63.

25. Ltr (U), Dr. George E. Mueller, Dept Asst Administrator for Manned Space Flight, to Flax, 11 Oct 63.

26. Memo (C), McMillan to Vice c/s USAF, 27 Sep 63, subj: The AF and the Natl Space Frog; Ltr (S); Lt. Gen. James Ferguson, Hq USAF, to AFSC (Gen.Ritland), 8 Nov 63, subj: Implementation of DoD/ NASA Agreement.

27. Ltr (S), Gen.Estes to Hq USAF (Ferguson), 16 May 63, subj: Initiation of a Phase Zero Study for Prog. 287.

28. Ltrs (C), Col. D. M. Alexander, DCS/Plans, AFSC to Gen. Keese, 22 May 63, subj: Initiation of a Phase Zero Study for Prog. 287; Col. Alexander to Hq USAF, 22 May 63, subj: FY 64 Integrated Space Plng Studies; Msg (C) SCLDS 27-51-11, AFSC to SSD, 27 May 63.

29. Msg (C) MSFA 14-6-61, AFSC to SSD, 14 Jun 63.

30. Ltr (C), Col.Heaton to Capt. W. T. O'Bryant, BuWeps, 24 Jun 63, subj: Request for Navy Rqmts for Manned Space Station Testing. A similar letter was sent to the Arm.y's Chief of Research and Development.

31. Ltr (C), Zuckert to SOD, 17 Jul 63, subj: Space Technology Study.

32. DD 1355-1 Program Change (S), R&D, 24 Jul 63, subj: NOSS, signed by Dr. McMillan.

33. Memo (U), Gilpatric to SAF, r..d•., subj: Memo (C), Brown to SAF, 30 Au5 63, subj:

34. Memo (C), Brown to SAF, 30 A 63, subj: Space Technology Study; Mil Orbiting Space Station. Mil Orbiting SFace Station.

35. Activity Rprts (S), D/Adv Pl::::.,Space, Weeks Ending 13 Sep and 20 Sep 63; Meo (C), Flax to DCS/ R&D, 13 Sep 63, subj: OSS Pre-Program Definition Study; Ltr (S), Ge. Kinney to AFSC, 13 Sep 63, subj: Approval of OSS Pre-Program Jefinition Study.

36. AFRDPC Activity Rprt (S), D/J..iv Plng, Space, Week Ending 4 Oct 63.

37. Sec/Def Decision/Guidance (S), signed by Gilpatric, 10 Oct 63, subj: Orbital Space Station, counter-signed by Zuckert on 4 Nov 63.

MOL ASSEM...
INTEGRATION BUILD...

MATERIAL

PROPERTY OF THE
...NITED STATES GOVERNMENT

...F FOUND, DO NOT OPEN

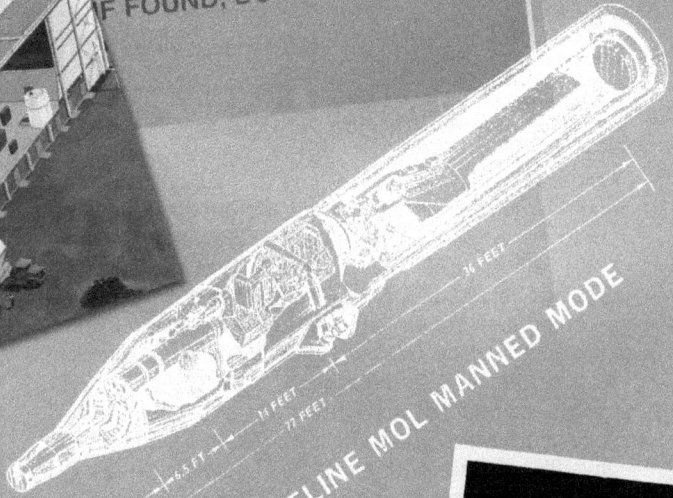

BASELINE MOL MANNED MODE

15.5 FEET   6.5 FT   11 FEET   72 FEET   36 FEET

USAF MOL/KH-10

THE DORIAN FILES REVEALED:
A COMPENDIUM OF THE NRO'S MANNED ORBITING LABORATORY DOCUMENTS

CHAPTER III:
DYNA-SOAR KILLED,
MOL APPROVED

DYNA-SOAR KILLED, MOL APPROVED

As noted earlier, about the time of the signing of the January 1963 Gemini agreement, McNamara asked his staff to review and compare the Air Force's Dyna-Soar (X-20) with Gemini. This unexpected review troubled USAF officials since only a year before the Secretary had authorized the Air Force to drop its suborbital flight plan (approved by OSD in April 1959)and go directly to an orbital flight test program. The Air Force was strongly committed to Dyna-Soar—a piloted orbital space glider which could effect a controlled landing in a conventional manner at a selected landing site—as its best hope for achieving an operational space capability by the late 1960s.

But despite the earlier approval, the Secretary of Defense apparently retained many doubts about Dyna-Soar, as he made clear in remarks before a congressional committee in February 1963:

    I personally believe that rather
    substantial changes lie ahead of
    us in the Dyna-Soar program, but we
    are not prepared to recommend them
    to you yet. I say this, in part,
    because of the Gemini development.
    Gemini is a competitive development
    with Dyna-Soar in the sense that
    each of them are designed to provide
    low earth orbit manned flight with
    controlled re-entry. Dyna-Soar does
    it one way, and with flexibility, and
    Gemini another.

    We are very much interested...in
    the Gemini project. When we become
    more familiar with it and understand
    better its potential I suspect it
    will have a great influence on the
    future of Dyna-Soar...

    The real question is: What do we have
    when we finish (Dyna-Soar)? It will
    cost to complete, in total, including
    funds spent to date, something on
    the order of $800 million to $1
    billion. The question is: Do we
    meet a rather ill-defined military
    requirement better by proceeding

down that track, or do we meet it better by modifying Gemini in some joint project with NASA?"[1]

THE DYNA-SOAR/GEMINI REVIEW

With these questions in mind, on 18 January 1963, McNamara requested DDR&E to undertake the review. Specifically, he asked to be provided information on the extent to which Dyna-Soar would provide "a valuable military capability" not provided by Gemini, as well as the extent to which NASA's spacecraft "as then conceived, could meet military requirements." A few days later the Air Force was directed to submit a paper detailing its position.[2]

The task of preparing the USAF position paper was assigned to Maj Gen Richard D. Curtin, Director of Development Plans, who was assisted by AFSC and Air Staff representatives. Completed on 26 February, the paper proclaimed Dyna-Soar as "the single most important USAF development project," and "fundamental to the preservation of the image of the Air Force for the future." The project was fully justified on the grounds it was expanding the nation's reservoir of scientific and technological knowledge. The Air Force argued that Dyna-Soar was not competitive with Gemini and was a logical extension of the X-15 type of research vehicle.[3]

While the Curtin paper was being coordinated within the Headquarters, General LeMay voiced concern over the latest "crisis" and he suggested that it might have resulted from the Air Force's enthusiasm and efforts to obtain a role in the Gemini project. On 2 March he urged Secretary Zuckert to clarify the USAF viewpoint with OSD. He said the Air Force might have inadvertently given the Secretary of Defense the impression that

Figure 21. X-15 Research Aircraft
Source: CSNR Reference Collection

Figure 22. Eugene Zuckert, Secretary of USAF, with X-20 Model in Las Vegas, 1962
Source: CSNR Reference Collection

it was more interested in the Gemini approach to a manned military space capability than in Dyna-Soar. "Our interest in Gemini," he said, "is strictly on the basis of an effort in addition to the X-20 program and to the extent supportable by available and approved funding."[4]

In seeking to defend Dyna-Soar, the Air Force at this time received welcome support from NASA. On 9 March Dr. Raymond L. Bisplinghoff, Director of NASA's Office of Advanced Research and Technology, and Dr. McMillan, prepared a joint USAF-NASA review of the space agency's technical interest in the X-20. Essentially, NASA took the position that if the Air Force did not develop the X-20, someone else would have to pursue it or something similar. NASA's prime interest in the X-20 was that it would provide a valuable tool for advancing the technology of highly maneuverable re-entry systems.[5]

The USAF position paper on Dyna-Soar was submitted to OSD on 11 March. Two days later John Rubel, Deputy DDR&E—who had been conducting the Gemini/X-20 review for Dr. Brown—submitted a paper to Secretary McNamara. Rubel posed a series of questions indicative

of the doubts about Dyna-Soar. "How important, really," he asked, "are the X-20 objectives; more particularly, how much is it worth to try to attain these objectives? What would be lost if the project were cancelled and its principal objectives not attained on the current schedule, or at all?" In his paper, Rubel discussed the differences between the X-20 and the ballistic-type Mercury and Gemini capsules, examined the advantages and disadvantages of each, and concluded that flexible re-entry and landing was not "immediately important." He listed for the Secretary of Defense four options which might be considered in connection with Dyna-Soar's future, including project termination.[6]

With the Air Force and DDR&E papers in hand, McNamara in mid March undertook an on-the-scene review of the X-20 program. Accompanied by Dr. McMillan, he visited the Martin and Boeing plants at Denver and Seattle. Afterwards, on the flight back to Washington, he expressed to McMillan his concern that the Air Force was putting too great an emphasis on controlled re-entry when it didn't know what the X-20 would do once in orbit. First emphasis, he said, should be on what missions could be performed in orbit and how to perform them; then the Air Force could worry about re-entry.[7] On his return to the Pentagon, McNamara asked Secretary Zuckert to review USAF Space projects to determine their applicability to the following four missions: (1) Inspection and destruction of hostile satellites; (2) protection of U.S. satellites from destruction; (3) space reconnaissance; and (4) use of near-earth orbit offensive weapons.[8]

The Air Force's response was forwarded to OSD on 5 June 1963. In it Dr. McMillan admitted that neither the X-20 nor Gemini, as then defined, would produce on-orbital operational capabilities of military significance. Each system, he said, possessed growth possibilities, but they would require major additional expenditures. With respect to what was being done on the four missions listed by McNamara, Dr. McMillan said there had been few real USAF accomplishments*, even though the Air Force's R&D program was directed toward their ultimate achievement. As for Gemini and Dyna-Soar, McMillan again restated the Air Force's view that there was no fundamental or unwarranted duplication and urged that the X-20 development be energetically pursued.[9]

BROWN RECOMMENDS A DEFENSE SPACE STATION AND NASA MAKES A COUNTER-PROPOSAL

By the fall of 1963, after considering these reviews and other factors such as costs, Secretary McNamara concluded that Dyna-Soar should be terminated and that advantage be taken of the Gemini vehicle used in conjunction with a DoD space station project[†]. A final decision was not immediately announced, pending Dr. Brown's analysis and study of possible approaches to the space station development. On 14 November, Brown completed this work and submitted an 11-page memorandum to McNamara, listing six alternative configurations for a space station using differing combinations of booster and vehicles (including Titan II, Titan III, Saturn IB, Gemini and Apollo). He estimated the costs of the individual projects would range from $4709.5 million to $1,286.0 million.[10]

Brown expressed his preference for Alternative 3, which called for using the Titan III booster to orbit a four-room, four man station. According to this plan, the station itself would be launched unmanned, with the

crew subsequently rendezvousing with it in a Gemini spacecraft or other similar ferry. The estimated cost of Alternative 3 was $983.0 million.

The Director of Defense Research acknowledged that, in all of the proposed space stations described, the method for returning the astronauts to earth was "primitive." That is, it involved essentially the ballistic trajectory and parachute descent with surface impact on the ocean. Brown believed it desirable to have an improved ferry vehicle—a low lift/drag maneuvering re-entry system— for a conventional ground landing. He suggested that the Air Force's ASSET (aerothermodynamic/elastic structural system environmental test) program[‡] be augmented using small-scale models and that it might eventually lead to development and launch of a full-scale ferry vehicle able to perform the first rendezvous with the proposed station in 1968.

As for resources, Dr. Brown thought enough funds would be available in fiscal year 1965 for the new project if OSD used the X-20 resources plus other national funds related to manned earth orbit programs. He recognized, however, there was a danger that inadequate funding in subsequent years might leave the United States without any manned military space program. OSD's decision should, therefore, be made with the determination to "see the program through the orbital test phase of the space station." If his proposal were accepted, it would enable the Air Force to undertake a series of manned earth orbit launches about nine months earlier than could be expected with Dyna-Soar. In conclusion, Brown recommended:

> That a military space station program
> be initiated, taking advantage of
> the Gemini developments, based upon
> a package plan which cancels the X-20
> program and assigns responsibility
> for Gemini and the new space
> station program to the Air Force,
> the effective date for transfer of
> management responsibility for Gemini
> being October 1, 1965.[11]

In accordance with the NASA-DoD space station agreement of September 1963, Dr. Brown submitted a copy of his 14 November memorandum to the space agency for review and approval of the proposed development. After studying Brown's recommendation,

---

* McMillan's memorandum did not touch on the unmanned satellite reconnaissance project, which was not considered an Air Force program. See pp 58-59.

† These events took place against the backdrop of new Soviet manned space achievements. Between 11 August 1962 and 16 June 1963, the Russians launched three more male and one female cosmonauts, bringing their total orbital time in space to 381 hours versus America's approximately 51 hours. The last two ships, Vostok 5 and 6, came within three miles of each other.

‡ The first 1,140-pound ASSET vehicle was launched from Cape Kennedy on 10 Sep 63 and reached an altitude of 201,000 feet and a velocity of 16,106 feet per second. Although a malfunction in the recovery system resulted in failure to recover the ASSET, the AF obtained most of the desired data from 130 temperature, pressure, and acceleration pickup points.

NASA made a "counter-proposal." It asked the Defense Department to consider a program which would not extend "quite so far as the establishment of a space station, at least as its first objective." On 30 November, after representatives of both agencies discussed the subject further, Dr. Brown submitted to Secretary McNamara an additional memorandum which described the NASA counter-proposal. He said it appeared likely that the NASA staff would advise Webb "to agree, in principle, to a manned military space program" which was separate from but coordinated with, the NASA activity. However, they would not agree to assigning DoD "the responsibility for a space station" since they remained uncertain of their own needs for such a vehicle. Consequently, they suggested DoD fulfill its needs with an orbiting military laboratory "which does not involve the complications of personnel ferry, docking, and resupply."[12]

NASA's proposal, it should be noted, was quite similar to the Air Force's 1962 plan for MODS. Like the USAF plan, it also would use the Titan III to launch a Gemini capsule and a cylindrical test module capable of supporting a crew of from two to four men for 30 days. The Gemini spacecraft would weigh 7,000 pounds and the module, 15,000 pounds. Cost of the system was estimated at $593.0 million.

After comparing NASA's counter-proposal with his Alternative 3, Dr. Brown agreed that the space agency's plan was "an entirely reasonable and orderly development approach which might well be followed whether or not the final objective is the establishment of a space station." However, he thought that while much valuable military testing could be accomplished using NASA's approach, it was not fully equivalent to a space station because it lacked "the operations of rendezvous, docking, resupply and crew rotation." If Secretary McNamara accepted NASA's counter-proposal, it would have the effect of delaying a Presidential decision on which organization would have management responsibility for a space station, "since their proposal would not be defined as a station.§" Dr. Brown said that while NASA's alternative was acceptable for "a near-term manned military space station," he felt it inferior to his own recommendation, which he now reconfirmed.[13]

After reviewing DDR&E's memorandum, the same day—30 November—McNamara met with Webb and the two agency chiefs reached an oral agreement that DoD would proceed to build a military space station. As for the approach, McNamara decided to accept NASA's

§   NASA did not want DoD to even use the term "space station." An alternate name, suggested by Dr. Yarymovych, was "manned orbiting laboratory." As was noted earlier, the Air Force in a planning document published in 1959 had referred to a "manned orbital laboratory" as one of its requirements.

alternate proposal rather than the more complicated and expensive system development described in DDR&E's Alternative 3.

THE AIR FORCE'S LAST EFFORT TO SAVE DYNA-SOAR

The Air Staff had been waiting, somewhat nervously, for OSD's decisions on the Dyna-Soar/Gemini review, but received no official word until 12 November when Dr. Brown informally advised Dr. Flax of the impending decision. However, it was not until 2 December that the Air Force received copies of Dr. Brown's two lengthy memoranda to McNamara (of 14 and 30 November) with a request for comments. To help Dr. Flax prepare a detailed response to the Brown memoranda, a technical team from the Space Systems Division, headed by Brig Gens Joseph Bleymaier and Joseph J. Cody, was flown to Washington. Meanwhile, within Headquarters USAF there ensued a last-minute effort to save the X-20. The Air Staff prepared a 14-page paper which proposed three alternate approaches for development of a space station using the X-20 as a small two-man station or as a ferry vehicle. On 4 December the Assistant Vice Chief of Staff, Maj Gen John K. Hester, forwarded it to Secretary Zuckert. He said:

> I completely support the objective as specified in these documents (the 14 and 30 November memoranda) of achieving a manned military space capability in the most practical and least expensive manner. However, I cannot agree with the conclusion that the Gemini route coupled with cancellation of the X-20 program and initiation of a low L/D maneuvering re-entry system will achieve this objective, nor do I agree that the approach will lead to a militarily meaningful space program, either operationally or from an economic or timely point of view. Instead, I believe that a reoriented X-20 program offers a highly promising way to achieve a low cost, effective manned military capability at an early date.[14]

General Hester recommended that the X-20 be considered for use in the proposed military manned space station. The same day he submitted these views, the Air Force Secretary—after discussing the issue with Deputy

Figure 23. MOL Model
Source: CSNR Reference Collection

Defense Secretary Gilpatric—wrote to McNamara: "I hate to see us getting into a position of abandoning a program such as Dyna-Soar and start a new program which is based upon program data and costs that could be quite optimistic. In addition, I think it is going to be very hard to make clear to Congress and the public the basis for the action that is proposed." Zuckert asked for an opportunity to discuss the subject with McNamara, "if' the final decision has not been made.[15]

Dr. Flax, meanwhile, prepared and forwarded to Zuckert his views on the "unwise" OSD proposal to cancel Dyna-Soar. He also noted that DoD had given no serious consideration to using the X-20 as a possible element of any space station program, and he commented on (among other things) the possible additional costs involved in using the Gemini:

With the Gemini vehicle, even with a large number of recovery areas, it may be necessary to provide backup systems for considerably longer

periods of up to a day or more. Also, in order to assure the capability for emergency sea recovery, it would be necessary to keep naval and air elements deployed on a continual basis over the entire period of manned space station flight. Costs of such deployments over the period of a year could easily negate any savings associated with any cost differential between Gemini and X-20 type vehicles. Even if emergency situations are ignored, the cost of regular monthly recoveries at sea for Gemini re-entry vehicles would substantially exceed the cost of land recovery of X-20 vehicles and this may well offset any payload advantages of the Gemini.[16]

On 5 December Secretary Zuckert forwarded Flax's comments to McNamara and again expressed his opinion that the X-20 deserved to be given serious consideration for a role in the manned military space experiment. The same day Flax also dispatched to DDR&E the Air Force's quick-reaction study and evaluation of the recommendations contained in Brown's 14 and 30 November memoranda. Among other things, the USAF

Figure 24. MOL Concept
Source: CSNR Reference Collection

technical team challenged the cost figures for Alternative 3. According to its calculations, the cost would be about $1.5 billion rather than $983.0 million. Commenting on the new estimate, Dr. Flax remarked that he believed the team's figure was low and should be increased about 30 percent, bringing probable costs to $1.9 billion.[17]

Unfortunately, these attempts to save Dyna-Soar were wasted. On 5 December, even as the USAF papers were being received in OSD, Gilpatric advised Zuckert that the Defense budget for fiscal year 1965 would reflect "several changes" in the military space program. One involved initiation, effective 1 January 1964, of a "Manned Orbital Program" with the simultaneous cancellation of the X-20 program. The other change required the Air Force to redirect and augment its "Advanced Re-entry and Precision Recovery" program (ASSET)¶. Gilpatric requested the Air Force to submit plans reflecting these changes by 31 December 1963.[18]

¶ Under this program, a series of studies were undertaken and several prototype re-entry test vehicles were developed and test flown for possible application to future USAF manned space project.

ANNOUNCEMENT OF THE MOL PROJECT

At a press conference on 10 December, Secretary McNamara formally announced that the Defense Department intended to build and launch a two-man orbital laboratory into space in late 1967 or early 1968 "to determine military usefulness of man in space." At the same time he announced cancellation of Dyna-Soar, stating that the substitution of MOL for it would save $100 million in the budget scheduled to be sent to Congress in January.

The Defense Secretary described MOL as "an experimental program, not related to a specific military mission." He recalled that he had stated many times in the past "that the potential requirements for manned operations in space for military purposes are not clear." Despite this, he said DoD would undertake "a carefully controlled program of developing the techniques which would be required were we to ever suddenly be confronted with a military mission in space."

McNamara emphasized that the entire program would be "Air Force managed." Both NASA and DoD, he reported, had agreed that the MOL project was "a wise move from the point of view of the nation." The two agencies also had agreed that, although it was under USAF management, NASA's requests for participation in MOL would be recognized "to the extent that does not compromise the Air Force mission, in the same way that the Gemini has recognized the Air Force request for piggyback payloads... to the extent it doesn't compromise the lunar landing priority and requirement."[19]

THE DOD-NASA MOL AGREEMENT

On 27 December 1963 Dr. Albert C. Hall, representing DDR&E, and Dr. George Mueller of NASA summarized in a joint paper their agencies' views and agreements on MOL and "the minimum elements of manned earth orbit programs." They agreed that DoD requirements—"the early effective demonstration of man's utility in performing military functions (for example, earth surveillance) from orbit"—would not be aimed at an operational "space station" in the context usually attributed to that term. They also agreed that NASA's requirements would properly emphasize scientific and research aspects of orbital flight.[20]

In the paper, Drs. Hall and Mueller reviewed various possible system candidates for manned orbital flight and the OSD decision to select the Gemini/MOL approach. They agreed that continuing space agency studies might confirm NASA's need "for a space station of proportions which will permit a centrifuge and/or require crew sizes of four or more." They recognized that a national requirement might develop for a large orbiting station and agreed that both agencies would continue to coordinate their studies in that direction. Concerning management interfaces for the Gemini/MOL, they concurred that "if the Gemini B capsule is procured from the Gemini contractor that it should be procured through NASA" and that a coordinating board should be established to define the relationships and execute the necessary agreements.

In summary, Drs. Hall and Mueller listed the following NASA- DoD agreements and conclusions:

1. The Gemini B/MOL was a single military project within "the National Space Program" and was being implemented by DoD in response to military test requirements in preparation for possible requirements.
2. DoD would initiate, under USAF management, a MOL program directed toward determining the military utility of man in orbit.
3. DoD would make use of the NASA-developed Gemini, modified as required to be the passenger vehicle for the laboratory**.
4. Titan III would be employed as the MOL booster.
5. NASA experimental requirements would be incorporated in the MOL Program.
6. A Coordinating Board would be constituted to resolve Gemini B/MOL interface between DoD and NASA.
7. The X-20 program would be cancelled in favor of the MOL Program.
8. The ASSET program would be augmented by DoD.
9. DoD and NASA would coordinate on an accelerated test program to determine the characteristics and suitability of various forms of maneuverable recovery vehicles.
10. Both agencies would continue their study of requirements for large or operational type space stations and would utilize the AACB and its panels to coordinate these studies.[21]
11. In late January 1964 Drs. Brown and Seamans signed a DoD/NASA agreement authorizing the Air Force to negotiate a Gemini B design contract with McDonnell, provided that the arrangement did not set a pattern for any follow-on engineering and procurement contracts. A new contract would require NASA's specific concurrence so as not to interfere with its Gemini program.††

** In late January 1964 Drs. Brown and Seamans signed a DoD/NASA agreement authorizing the Air Force to negotiate a Gemini B design contract with McDonnell, provided that the arrangement did not set a pattern for any follow-on engineering and procurement contracts. A new contract would require NASA's specific concurrence so as not to interfere with its Gemini program. See Chapter XIII, History of MOL, Plans and Policies, Vol. II.
†† See Chapter XIII, History of MOL, Plans and Policies, Vol. II.

ENDNOTES

1. McNamara Testimony, 2 Feb 63, Before House Cmte on Armed Services, 88th Cong, 1st Sess, Hearings on Mil Posture, pp 465-67.

2. Memo (S), McNamara to DDR&E, 18 Jan 63; DDR&E to ASAF (R&D), 28 Jan 63, subj: Gemini and Dyna-Soar Frogs; Cantwell, pp

3. Rprt (S), Air Staff Board Mtg 63-1, 1- Mar 63.

4. Ltr (S), LeMay to SAF, 2 Mar 63, subj: DJD Review of USAF Space Prog.

5. Ltr (C), R. L. Bisplinghoff and B. McMillan to Zuckert, 9 Mar 63, subj: Review of Technical Interests in. X-20 Prog.

6. Memo (C), Rubel to SOD, 13 Mar 63, subj: X-20 Prog. Review.

7. Memo (S), McMillan to SAF, 15 Mar 63.

8. Memo (C), McNamara to SAF, 15 Mar 63, subj: Review of AF Space R&D Progs.

9. Memo (S), McMillan to SOD, 5 Jun 63, subj: Review of AF Space R&D Prog.

10. Memo (S), Brown to McNamara, 14 Nov 63, subj: Approaches to a Manned Mil Space Prog.

11. Ibid.

12. Memo (S), Brown to McNamara, 30 Nov 63, subj: Evaluation of an Orbital Mil Test Module.

13. Ibid.

14. Ltr (S), Maj. Gen. J. K. Hester, Asst Vice C/S USAF to SAF-OS, 4 Dec 63, subj: Approaches to a Manned Mil Space Prog.

15. Memo (S), Zuckert to SOD, 4 Dec 63.

16. Memo (S), Flax to SAF, 4 Dec 63, subj: Manned Mil Space Prog.

17. Memos (S), Zuckert to McNamara, 5 Dec 63; Flax to Brown, 5 Dec 63.

18. Memo (S), Gilpatric to SAF, 5 Dec 63, subj: Change in DoD Space Prog.

19. DoD News Release No. 1556-62, 10 Dec 63; Times, 11 Dec 63.

20. Memo (S), Drs. Hall and Mueller to Drs. Brown and Seamans, 27 Dec 63, subj: Joint Review of Manned Orbital Progs.

21. Ibid.

MOL ASSE
INTEGRATION BUILD

MATERIAL

PROPERTY OF THE
ITED STATES GOVERNMENT

F FOUND, DO NOT OPEN

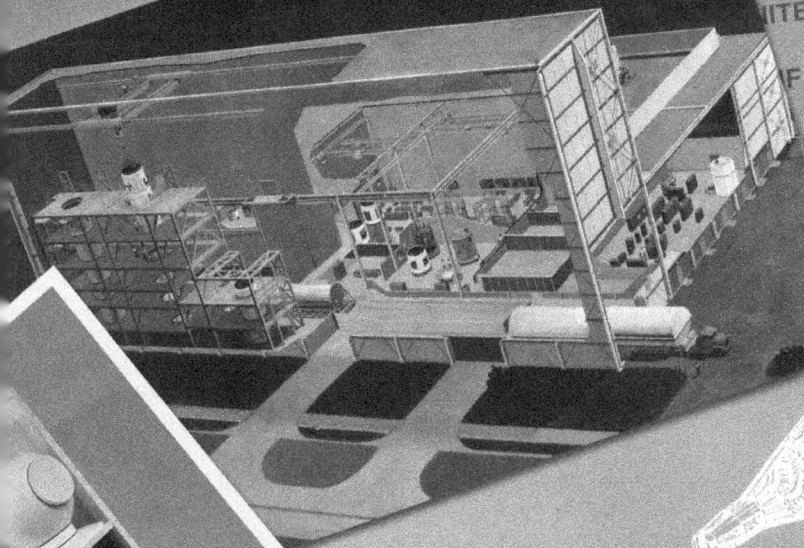

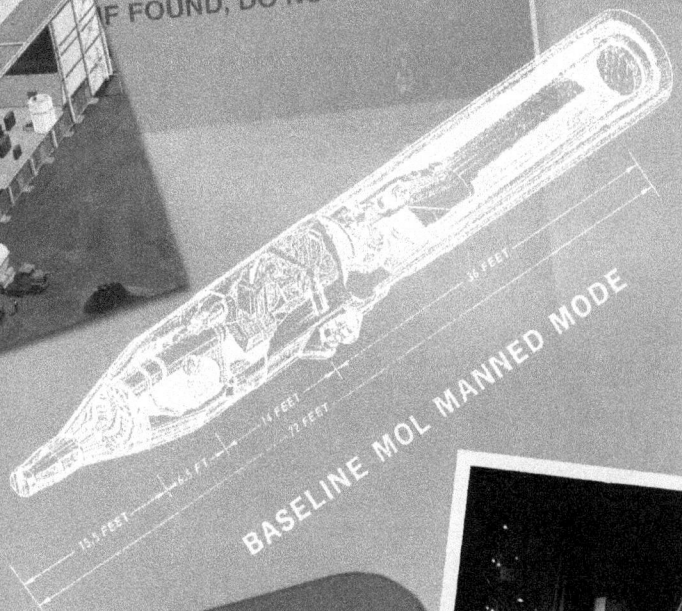

BASELINE MOL MANNED MODE

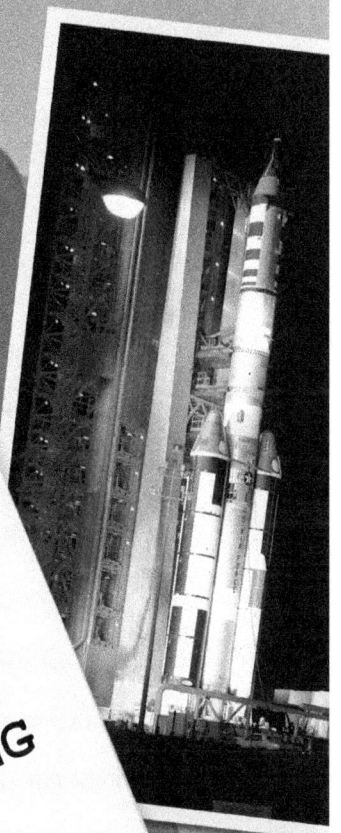

USAF MOL/KH

THE DORIAN FILES REVEALED:
A COMPENDIUM OF THE NRO'S MANNED ORBITING LABORATORY DOCUMENTS

CHAPTER IV:
PLANNING THE MANNED ORBITNG
LABORATORY
DECEMBER 1963-JUNE 1964

Planning the Manned Orbiting Laboratory
DECEMBER 1963-JUNE 1964

The MOL decision made, the Air Force in December 1963 began an arduous effort stretching over many months to define its primary objectives, identify the military experiments the astronauts would perform, and study the kinds of equipment and subsystems needed. This work began after Dr. Brown forwarded detailed program guidance to the Air Force on 11 December. According to his instructions, the basic purpose of the Manned Military Orbiting Laboratory (MMOL)* was to assess the military utility of man in space. Since man was not considered useful unless he performed a variety of tasks in space, MOL equipment was to be chosen both to support the astronaut and challenge his flexibility and judgment. DDR&E visualized the following type of manned orbital activity:

> (The) astronaut will carry out scientific observations of both space and earth. He will adjust equipment to ensure its maximum performance. He will maintain the repair equipment. He will be measured to see if he is capable of coping with the unusual— either in his observation or in his equipment operation. Indeed, it is planned that he will be challenged so severely that room in the laboratory must be planned to provide minimum elements of personnel comfort such as rest, exercise, and freedom from the confinement of a space suit.[1]

Accordingly, Dr. Brown suggested the Air Force design military test equipment and adopt procedures to measure the degree of improvement that could be achieved by employing man in space. Since reconnaissance was considered a logical mission, he proposed the Air Force install camera equipment in the laboratory module to provide "threat warning intelligence." To test the astronauts' capabilities, he suggested that targets simulating key enemy localities be transmitted to them and they be required to respond to emergency requests for information by locating and photographing these points, performing on-board processing and photo-interpretation, and transmitting the data to earth. Much of the above proposed test activities, he said, could be simulated on the ground or in aircraft.

In addition to using photographic equipment, he suggested the Air Force install an optical viewer in the MOL having a sufficient field of vision to allow an astronaut to search for targets of opportunity, identify them, and report in real time. He thought that several sets of high quality direct optical equipment would be needed. He suggested that tests also might be performed with an infrared system using a variety of wave lengths to help determine the operator's ability to interpret data, optimize the signal and identify the greatest sources of noise, and report receipts of signals from ballistic missiles, ships, ground installations, and other sources. He believed it might be demonstrated that infrared systems became operationally feasible only by using "the discriminatory intelligence of man in the loop.[2]

Besides possible experiments and equipment to be used, Brown's instructions also covered a number of other program areas. He directed the Air Force to make the maximum use of NASA's control facilities at Houston† and the worldwide tracking network built for the Mercury, Gemini, and Apollo programs. Concerning procurement of Gemini B, he asked the Air Force to work through NASA officials if the changes required to adapt the capsules to MOL were of a minor nature. If major modifications were required, he said the Air Force would be authorized to deal directly with McDonnell. Funding to support this preliminary study and development planning effort was set at $10 million for fiscal year 1964 and $75 million for fiscal year 1965.

In his covering memorandum to the above program guidance, Dr. Brown asked the Air Force to submit a development plan to OSD by year's end and he further stated:

> The degree of success of the MOL Program is certain to have national importance. The nature of the cooperative effort with NASA will require decisions that must be made by the Secretary of Defense. In recognition of these conditions, the DDR&E will expect to have a larger measure of direct involvement than is the usual case. In working out a plan for Air Force/DoD relations, the Air Force should recommend a

---

* At the request of President Johnson, the word "military" was later dropped.

† This proposal was discussed but never implemented.

method of management control which will satisfy this requirement and at the same time be acceptable to the Air Force. An arrangement similar to that which now exists in the Titan III program should be considered.[3]

THE AIR FORCE RESPONSE

On 16 December 1963, Headquarters USAF sent copies of the program guidelines to General Schriever and directed him to submit a MOL development plan. He in turn contacted Maj Gen Benjamin I. Funk, head of the Space Systems Division, and requested preparation of both a MOL development plan and one covering an accelerated ASSET development program. He suggested the MOL document be in the form of an illustrated briefing using a preliminary system package plan (PSPP) format. He said that, in view of the "narrowing" of the MOL program's original scope by OSD and SSD's extensive experience conducting space station studies, the Division should be able to immediately focus on the task.[4]

Figure 25. Bernard A. Schriever
Source: CSNR Reference Collection

Schriever further suggested that the development plan provide for six MOL launches with the first manned shot to take place in the last quarter of calendar year 1967. He also provided guidance on other actions required. He said SSD should seek detailed knowledge about the Gemini system—about its launch and flight operations and control, spacecraft engineering, etc. It should reassess the Air Force's participation in the NASA Gemini program

with particular reference to the relationship between MOL and the military experiments being planned for piggyback flight aboard space agency vehicles. Finally, he said it should make a careful study of proposed MOL experiments and support its recommendations with data obtained from analysis, ground tests, and aircraft simulations where possible.

The AFSC commander concluded:

We must move out immediately and aggressively on the MOL Program for which we have waited and prepared for so long. I cannot overemphasize the national importance of this military manned space undertaking and am confident that we can rise to meet the difficult challenge it presents.[5]

On receipt of this guidance, General Funk instructed General Bleymaier, who had been in charge of Titan III development, to take on the job of full-time director of a MOL task force. Bleymaier's task force quickly organized itself and began work on a preliminary plan which was completed by the end of December 1963. As the task force visualized it, the MOL would be used primarily for the surveillance-

Figure 26. Ben I. Funk
Source: CSNR Reference Collection

reconnaissance mission. To get the project underway, it recommended the Air Force deal directly with McDonnell rather than through NASA on acquisition of the Gemini B spacecraft. It also proposed that: (1) the OSS studies be cancelled or deferred, having been "somewhat overtaken by the MOL decision;" (2) the Air Force continue to fly piggy back experiments aboard NASA's Gemini spacecraft; (3) the Martin-Marietta Corporation be selected as the booster-payload integration contractor; and (4) contract definition activities be started at once.[6]

On 2-3 January 1964 Generals Funk and Bleymaier led an SSD briefing team to Washington to present their proposed MOL implementation plan to Headquarters AFSC and Pentagon officials. After hearing the briefings, General Schriever approved their submission to higher headquarters together with a paper on MOL management drafted by his staff. This paper recommended that the Air Force establish a high-level management office, with Schriever as its director, to serve as the primary agency between the Secretary of the Air Force and a SPO (system program office) to be established at SSD‡. On 4 and 6 January, Bleymaier presented the plan to the Air Staff Board (ASB)and members of the Designated Systems Management Group (DSMG), including Secretary Zuckert and Drs. McMillan and Flax.[7]

Figure 27. Joseph F. Bleymaier
Source: CSNR Reference Collection

At the formal DSMG briefing, Maj Gen William W. Momyer, chairman of the ASB, reported the Board's opinion that the AFSC plan was responsive to OSD's guidelines. He commented that while those guidelines were not ones the Air Force would have adopted if it controlled MOL decision-making, the important consideration was "to take advantage of an environment which will allow us to proceed... recognizing that in the future we may be allowed to expand it to accommodate other valid Air Force needs and aims." General Momyer said the Board has a number of questions about the proposed plan. One involved the launch schedule, which the Board members felt should be moved up in view of NASA's plans to launch its three-man Apollo spacecraft in early 1967§. The Board also was concerned about "putting all the Air Force man-in-space eggs in the reconnaissance basket" and recommended reexamining the mission area.[8]

On his part, Secretary Zuckert concluded that, as the plan appeared to be responsive to top-level guidance, the Air Force should submit it to OSD. He concurred in an AFSC recommendation that the ASSET program plan be withheld pending completion of a study of the scope of that project.[9]

With Zuckert's approval, on 7 January AFSC briefed members of Dr. Brown's staff including Dr. Hall. Two days later Maj Gen A. J. Kinney, Assistant Deputy Chief of Staff, Research and Development, and his staff met informally with Dr. Hall to discuss the proposed

development plan. Hall remarked that, while AFSC had made a commendable effort, he felt the plan was unresponsive in certain areas and needed considerably more work in others, particularly concerning preprogram definition activities.

He said further that he disagreed with the AFSC recommendation that the OSS studies be cancelled or deferred. Also, he thought both McDonnell and Martin-Marietta should do studies to determine how the MOL systems integration job should be accomplished, while the Air Force investigated "the nature of the experiments and attendant equipment which would go into the laboratory cannister." As for the proposed launch schedule, he also took a position similar to the Air Staff Board's, that is, he felt that the schedule should be moved up to provide for unmanned launches in calendar year 1966 and early 1967 and for a first firing of a manned vehicle in the second quarter of calendar year 1967.[10]

Dr. McMillan also had a somewhat negative reaction to the AFSC presentations which related to his highly secret activities as Director of the National Reconnaissance Office (NRO). Created in 1961, this covert organization was responsible for conducting unmanned satellite reconnaissance of the Soviet Union. Because of Moscow's special sensitivity to overhead reconnaissance of its territory, the President promulgated a stringent security policy which stated that the United States should not in any way officially acknowledge or confirm or deny the operational employment of a satellite reconnaissance system¶. All information relating to reconnaissance was to be rigidly controlled to avoid provoking the Russians and the word itself was not to be used. For example, in response to United Nations queries, the American delegation would use the word "observation."

In the course of supervising this highly successful, OSD-managed satellite reconnaissance program, Dr. McMillan decided it might be worthwhile to investigate "possible manned reconnaissance tasks." On 7 June 1963 he instructed the Directorate of Special Projects (SAFSP), which developed and operated the unmanned reconnaissance systems, to undertake a study and simulations to determine man's ability to recognize "high priority targets" and to point "high resolution cameras so as to obtain coverage of these targets.[11]

The proposed investigations, which were given the project designator MS-285, were subsequently initiated on 2 December 1963 by the Eastman Kodak Company, an SAFSP "black" contractor involved in developing the

---

‡   See Chapter V for a further discussion of the evolution of the MOL management structure.

§   The unvoiced fear was that an operational Apollo in 1967 might undercut support of MOL.

¶   Moscow's sensitivity gained worldwide attention in the spring of 1960 after a U-2 reconnaissance aircraft was shot down over its "territory. A fuming Premier Nikita Khrushchev torpedoed the Big Four summit conference in May after President Eisenhower refused to apologize for the U-2 missions.

optics for the unmanned reconnaissance program. In the area of prime photographic functions, Eastman Kodak undertook to consider man's ability to: (1) search, detect, and recognize targets; (2) select alternate targets; (3) aim cameras; (4) detect motion and control exposure for unusual lighting conditions; and (7) record and report target data. Funded by a $351,201 SAFSP contract, the Eastman Kodak studies were to be completed by July 1964.[12]

Figure 28. John L. Martin
Source: CSNR Reference Collection

Such was the situation when Secretary McNamara announced the MOL project and discussions ensued about its surveillance- reconnaissance mission. Concerned about the security aspects of the new program, the military Director of the NRO Staff**, Brig Gen John L. Martin, Jr., on 14 January 1964 reminded McMillan that the entire U.S. satellite reconnaissance effort was being conducted in the "black" and had been a forbidden subject within the Air Force since late 1960. Also, he noted that camera contractors involved had been restrained from making any public disclosures and he suggested contact be made with the "black" contractors who had been active either in the OSS studies or preliminary MOL planning "to reestablish satellite reconnaissance discipline which existed prior to the exceptions which were made for these programs." He also urged that any MOL flights be made from Cape Kennedy since launches from the West Coast would lead to the obvious assumption of immediate

reconnaissance employment. "There is," he said, "no other credible reason for low altitude polar launches for such a vehicle."[13]

It was with General Martin's strictures in mind that Dr. McMillan, following the AFSC presentation, notified General Ferguson that the proposed development plan had placed too much emphasis on an operational reconnaissance system. Development of such a system, he said, was "not an approved objective" and he warned that it was "absolutely crucial" to MOL's survival that it be directed toward specified and approved objectives. He said that before program definition could begin, the Air Force would have to establish a specific set of MOL objectives and requirements and define the criteria to be used in evaluating "trade-offs among objectives." Until this was done, program go ahead would not be authorized. He urged General Ferguson to draw up a specific list of candidate experiments or experimental areas to be analyzed and studied during program definition.[14]

Figure 29. James Ferguson
Source: CSNR Reference Collection

Responding to McMillan's guidance, General Ferguson directed his staff to take steps to insure that project goals, requirements, and criteria were clearly defined. Following a series of meetings during February, Schriever's and Ferguson's planners agreed that "the objectives of the MOL should not be based on a single set of experiments aimed only at one mission, such as reconnaissance." AFSC was directed to prepare a unified document which identified a minimum number of experiments to help assess the utility of man in space.[15]

** Known otherwise as the Office of Space Systems (SAFSS).

Meanwhile, Dr. Flax submitted a memorandum to Dr. Brown on the preliminary USAF approach to initiating MOL development. He reported that, as a start toward program management, a system office would be established at SSD with responsibility for MOL, Titan IIIC, and Gemini while studies of higher echelon management proceed. He said the Air Force would proceed with the OSS studies, revising the original work statements to drop the preliminary vehicle design requirement and emphasize identification of technical requirements, experiments, equipment, etc. It also would let contracts to McDonnell and Martin-Marietta for the Gemini B and Titan IIIC studies. To support these investigations, Dr. Flax asked Dr. Brown to release $10 million in emergency fiscal year 1954 funds. He estimated the cost of all pre-program definition studies—including the $1 million previously earmarked for the OSS studies—at $18.60 million.[16]

On 29 January 1964 the Defense Research Director authorized Flax to proceed with negotiations for the OSS study contracts only. Concerning the other proposed studies, he said "a convincing account of experiments to be done in the MOL Program must first be provided to the Secretary of Defense" before they would be authorized. His staff, he noted, was "working with the Air Force on such a document." Dr. Brown was referring to Dr. Hall and Bruno W. Augenstein, his Special Assistant (Intelligence and Reconnaissance), who were meeting with SSD and Aerospace Corporation officials on the West Coast to discuss not only MOL experiments, but mission and equipment justifications.[17]

DDR&E's direct involvement in the MOL planning process—which Dr. Brown had indicated would occur—troubled General Schriever. In a message to General Funk on 21 January, he said it was "imperative that the results of the SSD/Aerospace contribution to the DoD personnel presently working with you be provided me for joint discussion with DDR&E." Recalling the AFSC presentations made to Headquarters USAF and OSD personnel earlier in the month, he expressed concern that "the many approaches and alternatives to mission assignments and equipment definition in the MOL program may be in conflict and thus jeopardize approval.[18]

Despite McMillan's efforts and those of NRO to stop references to MOL as a manned reconnaissance system, the main emphasis in various "white" papers prepared by SSD and the Aerospace Corporation was on the surveillance mission. For example, an SSD scientific advisory group headed by Dr. McMillan—after reviewing the proposed MOL implementation plan—concluded that "reconnaissance-surveillance is a most practical and acceptable military mission for experimentation and that

other missions such as satellite inspection should remain secondary mission possibilities." The advisory group[††] felt that: adding a man to the reconnaissance system "could most dramatically reduce the complexity, expense and unreliability which would be inherent in an unmanned, automated system to accomplish the equivalent amount of militarily important information gathering.[19]

Reconnaissance also was considered a prime MOL mission in an Aerospace document on "MOL Experiments and Testing Philosophy," dated 13 February 1954, which propose d a number of experiments using optics "for daily sampling of enemy reactions during tense international situations." Reviewing the requirement for an effective optical system, Aerospace noted that:

```
Such a system is a 60-inch diameter
cassegrainian type telescope with
diffraction limit optics over the
useful field of view. This caliber
of optics using high resolution
film such as S0132 (Eastman Kodak
4404) and the man to adjust the
image motion compensation to better
than- 0.1 percent, will yield ground
resolutions of {better than one foot}
from 100 nautical miles altitudes
with 20 degree sun angle light
conditions, neglecting degradations
caused by atmospheric seeing. Under
low light levels associated with
5-degree sun angles such as would be
useful over the Soviet Union during
the winter months, ground resolutions
of {better than one foot} could be
realized.[20]
```

The Aerospace paper went on to discuss in same detail the advantages and disadvantages of using optical systems with larger or smaller diameters than 60 inches, and outlined the work sequence by which an astronaut might point a camera and compensate for image motion using an auxiliary pointing and tracking telescope.[21]

Hall and Augenstein similarly concluded that observation experiments should be given "careful and predominant attention" in the MOL program. On 5 March, in a lengthy memorandum to Dr. Brown, they reported on what they termed were "vigorous and productive discussions" of MOL missions and experiments extending over a period of many weeks. They advised that sufficient agreement and

---

†† The members were Drs. Gerald M. McDonnel, Homer J. Stewart, and Ernst H. Plesset. Special advisors were Drs. Nicholas J. Hoff, Laurnor F. Carter, Arthur E. Raymond, and Prof. Cornelius T. Leonde.

understandings had been reached so that the Secretary of Defense should "provide authoritative guidance to the USAF to pursue the next phase of effort."[22]

On 9 March Dr. Eugene Fubini, Deputy DDR&E, forwarded a copy of the Hall-Augenstein report to McMillan. He reminded the Under Secretary that a list of proposed MOL experiments, together with a brief statement of their military and/or scientific value, was required by OSD before it would approve the project.[23]

In March 1964 there was still another group which recommended that manned space reconnaissance be pursued. A panel of Project Forecast, established the previous spring by Secretary Zuckert and General LeMay with General Schriever as its Director, declared that the areas of most promise for manned reconnaissance were "those of high resolution photography, infrared imagery, and the all-weather capabilities of the synthetic array side-looking radar." The panel estimated that high resolution camera systems could be built within a few years that would "yield ground resolutions of less than {one foot}. It believed the systems could be enhanced by using man to point at the proper targets and adjust for image motion compensation.[24]

With the consensus being that the reconnaissance mission should be given the main emphasis in the MOL program, Dr. McMillan in early March met with General Schriever to clarify future approaches to the proposed pre-Phase I studies. The two men agreed that certain overt experiments related to reconnaissance would attempt "to determine man's capability, with appropriate aids, to point an instrument with accuracy better than ½ mile, to adjust for image motion to better than 0.2 percent, and to focus precisely (if this is necessary)." McMillan agreed that these activities could be classified under the normal security system (as "secret") and simply stated as objectives without indicating to contractors or others how they might compare with existing or projected unmanned satellite reconnaissance projects.

He recommended use of a telescopic system for the pointing, image motion compensation, and focus experiments. Pointing accuracy could be recorded with a simple collimated camera of resolution easily available from unclassified equipment. IMC performance could be recorded by photographing stars or by use of long exposures. He agreed that photography of the quality approaching that needed for reconnaissance might be undertaken on some orbital flights. An experimental camera held under special security (an F/16 camera with 240" focal length) might be made available as government-furnished equipment.[25]

In a letter to Schriever sent under the NRO BYEMAN security system, McMillan further advised that the NRO had initiated separate studies which would compare carefully the potential cost and performance of very high resolution systems, both manned and unmanned. He said that these studies would be kept current with the overt MOL program and that NRO's objective would be to insure that, "at such time as the evidence from MOL experiments warrants the decision, the basis for a timely development of a manned system will be at hand." Concerning the experimental camera, he advised that activities related to it would be handled exclusively within NRO channels and that special clearances would be given selected AFSC personnel who would be kept regularly informed of results.[26]

Meanwhile, the NRO announced formal guidelines for its covert studies, being performed under the code name "DORIAN." It stated that the studies and any subsequent hardware activities which were directed toward development of "an actual reconnaissance capability for the Department of Defense's manned orbiting laboratory are under the sole direction and control of the (S) National Reconnaissance Office and are part of the (TS) National Reconnaissance Program." Normal military security would apply to other MOL study activities conducted outside Project DORIAN. The fact that certain actual reconnaissance studies were under way for application to MOL, and the existence and participation of NRO, etc., were to be considered extremely sensitive and required handling under the BYEMAN security system.[27]

POLICY FOR THE CONDUCT OF THE MOL PROGRAM

At the end of March 1964, Dr. McMillan issued a statement of policy to govern the conduct of the early phases of the MOL program. Once again he emphasized that the primary objectives of the program were experimental: "To obtain authoritative data, in an economical way, on the possible contributions of man to the performance of military missions in space, and to obtain data on man's performance sufficient to form a basis for design and evaluation of manned systems." He further directed that:

> No requirement to develop an operational system will interfere with the requirements imposed by the experiments to be performed; cost and schedules will be defined by the needs of the experimental program.
>
> Experiments will be performed on orbit only after prior tests on the ground and, if necessary, in aircraft, adequately define and justify orbital tests.
>
> Granted that an orbital flight is justified by its primary experimental purposes, such secondary experiments as are desirably and conveniently carried along may also be included.[28]

Among experimental areas of military interest, McMillan listed "observations of the earth and earthbound events, and detection of an interaction with other space vehicles, both cooperative and uncooperative." The basic function of man was to search for and select targets or subjects for observation, to navigate precisely, adjust and maintain equipment, and summarize and report data. It was expected that man would facilitate various mission-related experiments including detection, classification, identification, and tracking of such targets as fixed installations at known locations; fixed installations having varying degrees of ambiguity as to location; and ground vehicles, ships, space vehicles, missile launches, explosions including nuclear, etc.[29]

Like Dr. McMillan, the Director of Defense Research and Engineering also emphasized at this time that ground simulation is and thorough advance study would have to precede any experimental MOL effort in space and that orbital experiments would be designed to test man and

determine just what he could do. Dr. Brown explained this approach during an appearance before the Senate Committee on Aeronautics and Space Sciences:

> If you just send a man up there without knowing what experiments he is going to do when he gets there, what you are likely to find is that everything he can do you have a machine that can do just as well.
>
> I am gradually becoming convinced that there are some things he can do better, but I want the experiment specified first so when he goes up there he will actually be able to show he can do better.
>
> I think I can give you one specific example: I think a man can probably point a telescope more accurately than automatic equipment can. However, unless you design the equipment to measure that before you send a man up, and unless you give him a piece of equipment that will answer that question...you are not going to get the answer.[30]

Headquarters AFSC, meanwhile, had reconciled itself to the fact that the MOL development plan would not be approved until it had presented to the Secretary of Defense "a convincing account of MOL program experiments which will satisfy the objectives of demonstrating qualitatively and quantitatively the military usefulness of man in space." On 9 March General Funk was instructed to submit a preliminary technical development plan (PTDP)—to include descriptions of proposed experiments—that could serve as the single authoritative MOL reference document.[31] In a separate letter sent to General Funk under the BYEMAN security system, the SSD commander was advised that the reconnaissance mission remained extremely sensitive and that the PTDP should avoid any reference to it.[32]

Several weeks before receiving this guidance, SSD had set up a working group under General Bleymaier to identify the proposed MOL experiments. Designated the MOL Experiments Working Group and headed by Col William Brady, the SSD System Program Director for MOL, it had a membership of several dozen military and industrial representatives. During February and March 1964 the group examined more than 400 proposed experiments submitted by various defense and industrial agencies and categorized them into a number of technical areas. Committees of experts were then formed in such specialties as optics, infrared, radar, communications, etc., to analyze the proposals to determine whether common objectives and equipment might satisfy mere than one experimental objective.[33]

After eliminations and consolidations, 59 experiments were identified. These were further scrutinized, evaluated, and finally reduced to 12 primary and 18 secondary MOL experiments which were incorporated into the preliminary technical development plan submitted to AFSC on 1 April. The 12 original primary experiments were:

P-1—ACQUISITION AND TRACKING OF GROUND TARGETS. To evaluate man's performance in acquiring pre-assigned targets and precisely tracking them to an accuracy compatible with the requirements for precise Image Motion Compensation (IMC) determination.

P-2— {remains classified.}

P-3—DIRECT VIEWING FOR GROUND AND SEA TARGETS. To evaluate man's-ability to scan and acquire land targets of -opportunity, to scan and detect ships and surfaced submarines, and to examine ships and surfaced submarines for classification purposes.

P-4—ELECTROMAGNETIC SIGNAL DETECTION. To evaluate man's capability for making semi-analytical decisions and control adjustments to optimize the orbital collection of intercept data from advanced electromagnetic emitters.

P-5—IN SPACE MAINTENANCE. To evaluate man's capability to perform-malfunction detection, repair, and maintenance of complex military peculiar equipment.

P6—EXTRAVEHICULAR ACTIVITY. To evaluate man's ability in the performance of extravehicular operations peculiar to future military operations, including external spacecraft maintenance.

P-7—REMOTE MANEUVERING UNIT. To evaluate the astronaut's ability to control the Remote Maneuvering Unit (RMU).

P-8—AUTONOMOUS SPACECRAFT POSITION FIXING AND NAVIGATION. To evaluate the-capability of a man-using various combinations of equipment to act as a spacecraft navigator and provide autonomous navigation.

P-9—NEGATION AND DAMAGE ASSESSMENT. To evaluate man's ability to carry out a negation and damage assessment function.

P-10—MULTIBAND SPECTRAL OBSERVATIONS. To evaluate man's ability to detect high radiance gradient background events and missile signatures using multiband spectral sensors and to provide additional measurement data on backgrounds and missile signatures.

P-11—GENERAL PERFORMANCE IN MILITARY SPACE OPERATIONS. To obtain reliable and valid measurements of more basic performance as it relates to applied mission functions and physiological changes occurring during the stresses of the MOL flights.

P-12—BIOMEDICAL AND PHYSICAL EVALUATION. To evaluate those effects of weightlessness which can potentially compromise mission success. Sufficient data are required to validate supporting measures employed, devise improved methods, if necessary, and afford plausible estimates of biomedical status for missions longer than 30 days.

Three other experiments, later added, were: P-13, ocean surveillance; p-14, manned assembly and service of large antennas; and F-15, manned assembly and service of large telescopes.

The SSD development plan described the pre-Phase I MOL activities which would precede issuance of a request for proposals to industry for project definition. It discussed the US approach to program management, procurement philosophy (i.e., adopting the "associate contractor" concept to procure major elements of the system), organizational responsibilities, and steps leading to a first unmanned launch, which the plan tentatively scheduled for June 1968, and the first manned flight sometime in calendar year 1969.[34]

After General Bleymaier and Colonel Brady briefed Dr. Flax on the plan on 6 April, it was officially submitted to Headquarters USAF on the 8th. AFSC requested authority to proceed with the pre-Phase I effort and estimated the cost at $5.5 million (another $500,000 was later added for the Navy's ocean surveillance studies). Concerning the OSS contracts, whose cost was included in the above total,.AFSC brought to the attention of the Headquarters that it had never received authority to award them. Dr. Flax subsequently signed a "Determination and Findings" (D&F) on 13 April 1964 authorizing AFSC to negotiate the final OSS study contracts.[35]

MOL PRE-PHASE I GO-AHEAD IS APPROVED

On 10 April Dr. Flax forwarded to Dr. Brown two copies of the AFSC plan and requested funds and authority to proceed with pre- Phase I MOL activities. The Defense Research Director, after reviewing the document, reported to Secretary McNamara on 21 April that OSD-Air Force discussions had clarified the MOL experiments approach and that the Air Force had requested go-ahead authority for the pre-Phase I studies only, with funding listed as follows:[36]

	MILLIONS
6 Experiment Study Contracts (USAF to Industry)	$1.0
3 Experiment Studies (Navy in-house)	0.5
6 Support Studies of Laboratory Subsystems	1.2
Gemini B. Detailed Study (McDonnell)	1.0
Titan III Interface Study (Martin)	1.0
Apollo Applications Study (North American)	0.2
One-Man Gemini Applications Study (McDonnell)	0.1
Aerospace Corporation Support	1.0
Total	$6.0

So that McNamara might know the spectrum and detail of the experiments already selected by the Air Force, Dr. Brown listed the 12 primary and 18 secondary experiments. He explained that the advantages of having a man in space vehicle were in his ability to recognize patterns, interpret them in real time, and report the results, and his ability to point a sensor (telescope-camera) and provide image motion compensation. He said the proposed MOL experiments should provide answers to the question whether better results could be obtained by using a man as compared to an unmanned system of the same weight. He advised he planned to release $6 million of deferred fiscal year 1964 funds for the Air Force to begin the studies-if the Defense Secretary did not object.[37]

On 27 April, Dr. McMillan also reported to McNamara on USAF plans for reconnaissance studies, experiments, and possible developments connected with the MOL project, and NRO actions. He advised that the "black" effort was being handled within the BYEMAN control system, while certain other studies were carried out openly as part of the MOL program under normal classification. He said:

> Should the MOL experiments demonstrate satisfactorily that a man may be able to make important contributions to the effectiveness of satellite reconnaissance missions, it will be necessary to compare carefully the potential cost and performance of very high resolution systems, manned and unmanned. Such comparisons will require complete access to the present unmanned satellite reconnaissance program. They will be carried out exclusively by the (S) NRO as Project DORIAN.[38]

Based upon this information from two of his top scientific and technical advisors, Secretary McNamara authorized the start of pre-Phase I activities. Whereupon, on 29 April Dr. Brown advised Zuckert of the release of the $6 million for the MOL studies. However, he laid down certain conditions by requiring the Air Force to: (1) delay contract negotiations for Apollo and one-man Gemini studies until his office had approved the proposed work statements; (2) delete experiment P-9 unless the -Air Force could show its compelling importance; and (3) give special emphasis to ground simulation testing during all experimental studies.[39]

On 4 May, Dr. Flax forwarded Brown's instructions to the Vice Chief of Staff for "action as directed," noting that AFSC would have to obtain advance approval of all work statements. AFSC and SSD, however, found the latter requirement irksome since the procedure was time-consuming and would delay the letting of contracts. The OSS study contracts were an example. The work statements, first submitted to Headquarters USAF on 19 February, were not approved until 12 March, the D&F was not signed until 13 April, and authority to proceed with the contracts was not given until 20 May. Actual letting of three contracts—to Douglas, Martin-Marietta, and General Electric—was not accomplished until 27 May 1964.[40]

As expected, there were long delays which served to extend the contracting process through the spring and summer months. Thirteen additional contracts were awarded on the following dates: 8 June, the Gemini B spacecraft study (McDonnell); 15 June, attitude control and stabilization (Minneapolis-Honeywell); 17 June, Titan III interface (Martin-Marietta); 7-16 July, electrical power subsystems (Allis-Chalmers, North American Aviation, and General Electric); 13 July, environment control subsystems (Garrett Corp. and Hamilton Standard); 22 July, autonomous navigation (Hughes); 23 July, multiband spectral observation definition (Aerojet-General); 10 August, image velocity sensor subsystem (IBM) ; 1 September, short-arm centrifuge (Douglas); and 24 September, manned electro-magnetic (EM) signal detection (Airborne Instrumentation Laboratories). The most expensive contracts were for the Gemini B study ($1,189,500) and the Titan III interface investigation ($910,000). Total costs of 13 pre-phase I studies came to $3,237,716.‡‡

Dr. McMillan, meanwhile, had again reemphasized to General Shriever the importance of using simulations during the various experiments and studies. On 15 May, he informed the AFSC commander that he thought the contractors might be misled by the preliminary technical development plan's emphasis, especially as it related to the proposed image velocity sensor subsystem study. "The objective," he said, "is to determine what man can do in acquiring and tracking and compensating for image motions. Design of orbital gear is incidental to a third phase of the task. The first two phases are simulation and aircraft tests, which are prerequisite to, rather than concurrent with, the third phase." Consequently, he asked that the plan be rewritten to clearly establish the main objective. This additional guidance was dispatched to SSD on 28 May and a revised development plan was published on 20 June.[41]

---

‡‡ In addition to these contracts, SSD was responsible for a cost plus-incentive fee (CPIF) contract previously negotiated with LingTemco-Vought Astronautics for development of a Modular Maneuvering Vehicle unit for the NASA Gemini program. The cost was $5,890,183.

ENDNOTES

1. See TS Atch, "Manned Orbital Program, Department of Defense," to Memo (S), Brown to Flax, 11 Dec 63, subj: Manned Orbital Program.

2. Ibid.

3. Memo (S), Brown to Flax, 11 Dec 63, subj as above.

4. Ltrs (S), Maj Gen H. C. Donnelly, Asst DCS/R&D to AFSC, 16 Dec 63, subj: MOL; Schriever to Funk, 16 Dec 63, Subj: MOL Program.

5. Ibid.

6. Memo for Record, Maj B. J. Loret (MSF-1), Hq AFSC, 19 Jun 64; Mins (S), 78th DSMG Mtg, 6 Jan 64; SSD Hist Rprt (S), MOL Program, 1st Half CY 1964, p2.

7. Rprt of ASB Mtg 64-1; Mins (S), 78th DSMG Mtg, 6 Jan 64; Ltr (S/SAR), Lt Gen Howell M. Estes, Jr., Vice Cmdr, AFSC to Hg USAF (Ferguson), 8 Jan 64, Subj: Prelim Plan for Implem of MOL Program.

8. Mins (S), 78th DSMG, 6 Jan 64.

9. Ibid.

10. Memo (S), Kinney to Ferguson, 10 Jan 64, Subj: MOL.

11. Msg SAFSS-1-M-0136 (S-BYEMAN), Martin to Greer, 7Jun 63.

12. SAFSP-F/30-7-863 (S-BYEMAN), Greer to Martin, 30 Jun 63; SAFSS-1-M-0215 (S-BYEMAN), Martin to Greer, 14 Oct 63; 0319022 Martin to Greer, 4 Jan 64.

13. Memo (TS/DORIAN), Gen Martin, Dir/NRO Staff to McMillan, 14 Jan 64, Subj: MMOL.

14. Memo (S), McMillan to Ferguson, 15 Jan 65, Subj: Rqmts & Objs of MOL.

15. Ltr (U), Kinney to McMillan, 19 Feb 64, Subj: Rqmts & Objs of MOL; Activity Rprts (S), D/Adv Plng, Space, Wks Ending 7, 14 Feb 64.

16. Ltr (S), Flax to DDR&E, 18 Jan 64, subj: Manned Orbital Program.

17. Memo (C), Brown to Flax, 29 Jan 64.

18. Msg (S) SCG 29-1-34, AFSC to SSD (Schriever to Funk), 29 Jan 64.

19. Ltr (C), Lt Col B. C. Gray, Secy to SSD Advisory Gp to distr, 31 Jan 64, Subj: Comments on SSD Division Advisory Gp on the MOL Program.

20. Memo (S), MOL Experiments and Testing Philosophy (AS64-000-00575), 13 Feb 64; Ltr (S) Gen Hester, Vice CSAF to SAFUS, 24 Feb 64, Subj: OSS Study and MOL Experiments.

21. Ibid.

22. Memo (C), Hall and Augenstein to Drs. Brown and Fubini, 5 Mar 64, subj: Experiments for MOL and Purposes fo the MOL Program.

23. Memo (C), Fubini to Under SAF, 9 Mar 64, Msg 60208 (U), USAF to AFSC, 12 Mar 64.

24. Rprt (S/SAR), Support Panel, Project Forecast, "Space," Vol I, pV-3, 4, Mar 64.

25. Ltr (S), McMillan to Schriever, 3 Mar 64; Msg (S/DORIAN), WHIG-0974, Martin to Gen Greer, 3 Mar 64.

26. Quoted in Msg (S/DORIAN) WHIG-0974, Martin to Greer, 3 Mar 64.

27. Msg (TS/DORIAN), WHIG-0984, Mazza to Lt Col Ford, 5 Mar 64.

28. Memo for Record (C), Dr. McMillan, 26 Mar 64, subj: Policy for Conduct of MOL Program; Ltr (U), AFSC (Col Jacobsen) to SSD, 9 Apr 64, subj as above.

29. Ibid.

30. Dr. Brown's testimony, 9 Mar 64, before Senate Cmte on Aeronautical and Space Sciences, 89th Cong, 2nd Sess, 1965 NASA Authorization, pt 2, p 461.

31. Msg (S) 9-B-13, AFSC to SSD, 9 Mar 64.

32. Ltr (TS-BYEMAN), Col R. K. Jacobsen, Hq AFSC to Funk, 10 Mar 64, subj: MOL Program.

33. Ltr (C), Funk to Blemaier, 17 Jan 64, Subj: Determination of Experiments for the MOL Program, Rprt (S), Candidate Experiments for MOL, prepared by MOL Experiments Wkg Gp, 2 vols.

34. Prelim Techn Dev Plan for MOL (S/NOFORN), Apr 64 (revised as of 30 Jun 64).

35. Ltr (S), AFSC to Hq USAF (Ferguson), 8 Apr 64, subj: MOL; D&F 64-IIc84, 13 Apr 64, signed by Flax.

36. Memos (S), Flax to DDR&E, 10 Apr 64, subj: M)L; Brown to McNamara, 21 Apr 64, subj: Initiation of MOL Pre-Phase I.

37. Memo (S), Brown to McNamara, 21 Apr 64, subj: Initiation of MOL Pre-Phase I.

38. Memo (S-BYEMAN), McMillan to SecDef, 27 Apr 64, subj: Recon Aspects of the MOL.

39. Memo (C), Brown to zuckert, 29 Apr 64, subj: Approval of USAF FY RDT&E MOL Program.

40. Memo (U), Flax to Vice CSAF, 4 May 64, subj: Approval of USAF FY 64 RDT&E MOL Prog; Ltr (U), USAF to AFSC, 5 May 64 ; Memo (S), Flax to Ferguson, 20 May 64, subj: OSS Studies; Ltrs (S), SSD to Douglas, Martin, GE, 27 May 64, subj: OSS Study.

41. Ltr (S), McMillan to Schriever, 15 May 64; Ltr (S), Ritland to McMillan, 28 May 64.

MOL ASSEM...
INTEGRATION BUILD...

MATERIAL

PROPERTY OF THE
...ITED STATES GOVERNMENT

...F FOUND, DO NOT OPEN

BASELINE MOL MANNED MODE

USAF MOL/KH-10

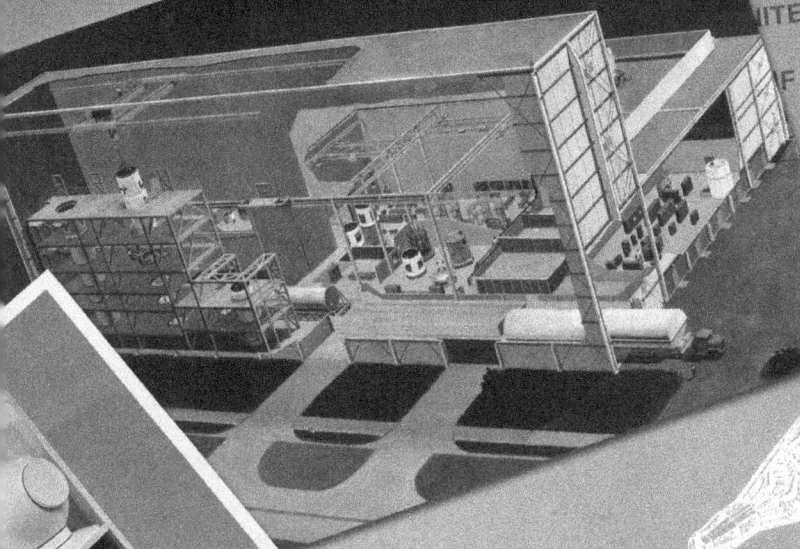

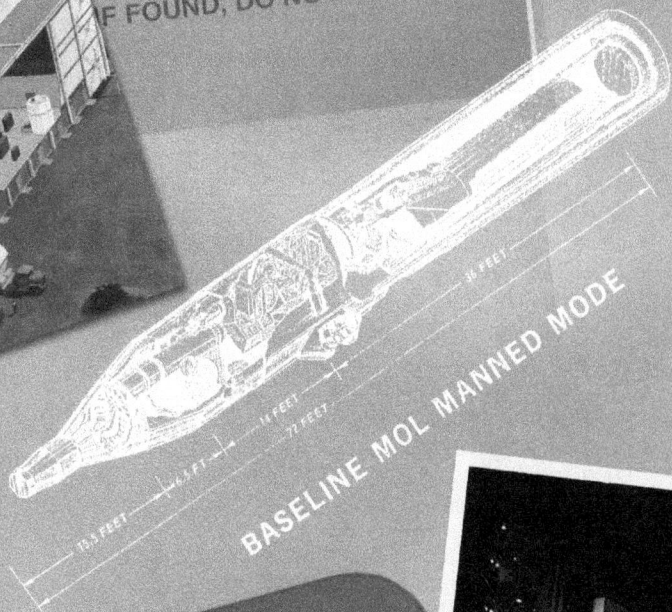

THE DORIAN FILES REVEALED:
A COMPENDIUM OF THE NRO'S MANNED ORBITING LABORATORY DOCUMENTS

CHAPTER V: EVOLUTION OF THE MOL MANAGEMENT STRUCTURE

Evolution of the MOL Management Structure

Several months before the pre-Phase I studies got under way, Headquarters USAF took steps which eventually led to establishment of the unique MOL management structure. This special organization had its origins in events which took place during the summer of 1963, at a time when DoD and NASA were investigating the proposed national space station concept discussed in Chapter III. USAF officials believed that development of a national space station would require an effort comparable to "the Manhattan project, our ICBM program, and the Lunar program" and they felt it essential that the Air Force be chosen executive manager.[1]

It was with this goal in mind that General Ferguson on 7 August 1963 reported to Gen William F. McKee, the Vice Chief of Staff, that NASA was already organized "in depth" to thoroughly define and establish a space station project. If the Air Force was to succeed in becoming executive manager, he said it would be necessary to adjust the Air Staff organization since USAF field agencies would not be in a position to cope with a project requiring top level coordination with such groups as Secretary McNamara's staff, Congress, etc. Accordingly, he requested permission to set up an office within his Directorate of Development Planning "to plan, define, and establish a national space station program under the executive management of the Air Force."[2]

Figure 30. Osmond J. Ritland
Source: CSNR Reference Collection

General McKee approved the request and on 23 September the office of "Deputy Director of Development Planning, Space" was formally established as the Air Staff focal point to coordinate with NASA on plans for development of the national space station. Col Kenneth W. Schultz was named to head the new office, which was seen as paralleling the arrangement within DDR&E's office, where Dr. Hall was serving as Deputy Director for Space. Besides working with NASA, Colonel Schultz was given responsibility for managing USAF space planning studies and coordinating with the Army, Navy, and other governmental agencies.[3]

Dr. McMillan lauded the above action as "a timely organizational step." On 27 September, he wrote to General McKee that it might also be appropriate to give the new Deputy Director the job of reviewing proposed agenda items for the monthly space station reporting meetings with NASA, controlling USAF attendance, and reviewing all Air Force space briefings intended for the space agency. Advising that while he did not intend to downgrade General Ritland's role as AFSC Deputy Commander for Manned Space Flight,* he thought Colonel Schultz should be responsible for keeping him informed of all significant space station data exchanges between NASA and the Air Force.[4]

Subsequently, General Ferguson advised General Ritland that the new Air Staff office would exercise "authoritative review over exchanges of space station data between the USAF and NASA." It also would be responsible for all correspondence which promulgated or altered official USAF positions or policies, and handle coordination of NASA studies. He directed General Ritland to submit to the new office a monthly status report on all AFSC Space station study activities.[5]

When in December 1963 Secretary McNamara announced the plan to kill Dyna-Soar and initiate MOL, an entirely new factor was introduced into the management picture. Dr. Brown shortly afterwards asked the Air Force for recommendations on a "method of (MOL) management control" and, as was noted earlier, suggested adopting the arrangement followed by SSD in the Titan II development project. While General Schriever and his staff thoroughly agreed on the need for

---

* Established in the spring of 1962 to coordinate certain Air Force activities with NASA in support of the lunar landing program.

a strong, centralized field organization, they also believed there should be strong and clear lines of authority to the highest levels of the Air Force.

In January 1964, at the request of headquarters USAF, AFSC prepared a MOL management paper which General Schriever submitted to Dr. McMillan on the 20th. The stated objective was to provide "continuing positive direction and control of the program by the Secretary of the Air Force while assuring the necessary flexibility at the operating management level." To achieve that objective, AFSC recommended placing General Schriever at the head of a "MOL Special Program Office" to be located in the Washington area, preferably in the Pentagon. It would be responsible for overall review and program control, report directly to the Secretary in directing the project and implementing his decisions.[6]

To insure Headquarters USAF participation, AFSC suggested assigning Air Staff representatives on a full-time basis to the SID at the Space Systems Division. They would be responsible for keeping their home offices informed while they worked for and represented the system office in their respective functional areas. In addition, AFSC proposed that NASA be requested to provide one or two people to work full-time with the SPO on the West Coast. On major program and policy matters, it visualized the MOL office as coordinating with the Chief, Vice Chief, or appropriate Deputies, and with Dr. Seamans or other NASA officials. The system office would report directly to the "MOL Special Program Office."[7]

On 1 February 1964, while the Air Staff was mulling over AFSC's proposal, General Schriever moved within his own headquarters to establish a new office of "assistant Deputy Commander for Space for MOL" under Col R. K. Jacobsen. Shortly thereafter he met with Dr. McMillan to discuss these management changes and the AFSC plan for a Pentagon-level program office. Initially, McMillan thought well of the proposal. On 6 February he advised Schriever he agreed that there should be a MOL office in the Washington area responsible for developing and maintaining "an experimental plan binding on the program, and in particular on SSD, after approval by the Air Staff, SAF-CS, and DDR&E." As he saw it, the MOL office would be responsible for coordinating with NASA, insuring support from all AFSC elements, programming and managing resources, and monitoring progress and providing timely information to the Secretary, Air Staff, and DDR&E.[8]

However, McMillan also favored appointing a special assistant to the Secretary to help him review MOL program progress. He said that while the assistantship would be a full-time job, the director of the MOL office might wear the "second hat" if an individual with appropriate qualifications could be found. He said further that he believed the MOL office should be headed by a general officer, with his appointment and that of the Secretary's special assistant being considered together.[9]

While awaiting a firm decision on the top-level organization, General Schriever on 10 March directed General Funk to establish an SSD Deputy Commander for Manned Space Systems and to delegate it full authority for MOL development. Subsequently, he named General Bleymaier to head the new office. Colonel Brady was designated. System Program Director under Bleymaier, and 18 other officers were initially assigned to him. Later, on 7 May, a permanent Navy MOL field office was established as an integral part of the SPO.[10]

These organizational actions completed the basic field-level organization, but in the meantime little had been done about the Washington office. On 12 March, Schriever and Ferguson met with Drs. McMillan and Flax to discuss the matter, at which time the Air Force Under Secretary expressed "certain reservations" about his earlier agreement of 6 February. He now indicated that AFSC's proposed top-level management proposal was not acceptable.[†] The next day a dissatisfied General Schriever wrote to Ferguson and reviewed the steps he had taken within his headquarters and at SSD to establish "effective internal AFSC management" of MOL. He said these steps were valid and adequate for AFSC operations but, in his opinion, as he later advised McMillan, the Air Staff arrangement was "inadequate for the task that lies ahead."[11]

Dr. McMillan, however, subsequently decided on a different type of arrangement. In late April, after receiving authority to go ahead with the pre-Phase I studies, he wrote to the Chief of Staff about his previous approval of establishment of Colonel Schultz's office and said:

> The project of developing a Manned
> Orbital Laboratory as directed by
> the Secretary of Defense will involve
> an extraordinary degree of intra-
> governmental and inter-service
> relationships, particularly during
> the early phases. The Office of the
> Secretary of Defense, as well as that
> of the Secretary of the Air Force,
> will be continuously concerned with
> details of the policies governing
> the MOL development...

---

†   The NRO had already raised questions about management of the "black" aspects of MOL development activities.

```
I now believe that the MOL Project
Office should assume the responsibility
to meet the requirements implicit
in the decision by the Secretary
of Defense to proceed with further
studies relative to this project.
For the reasons outlined...above,
this office must be specifically
and directly responsive to the
requirements of the Secretary of the
Air Force, as well as to the Chief
of Staff.¹²
```

McMillan's memorandum set off an Air Staff organizational study aimed at creating the needed coordination office. The recommendation that emerged from this study was to redesignate the Deputy Director of Development Planning, Space, as the "Assistant for the MOL Program" and to expand the office. This recommendation was accepted and on 9 June 1964 the Vice Chief formally announced redesignation of the office, which he said would assume "the normal Air Staff functions involving MOL activities as directed by the Chief of Staff and the Secretary of the Air Force."¹³

As it turned out, this proved to be an interim arrangement in the evolution' of the MOL management structure.‡

ENDNOTES

1. Ltr (C), Ferguson to McKee, 7 Aug 63, subj: Nat'l Space Station Program.

2. Ibid.

3. Ltr (U), Gen McKee, Vice USAF to Deputies, et al, 23 Sep 63, Subj: The AF and the Natl Space Program.

4. Memo (U), McMillan to McKee, 27 Sep 63, subj as above.

5. Ltr (S), Ferguson to Ritland, 8 Nov 63, subj: Implementation of DoD/NASA Agreement.

6. AFSC Draft Proposal (U), MOL Mgt, 20 Jan 64.

7. Ibid.

8. Ltr (U), Schriever to Ferguson, 1 Feb 64; Ltr (U), McMillan to Schriever, 6 Feb 64.

9. Ltr (U), McMillan to Schriever, 6 Feb 64.

10. Ltr (S), Schriever to Funk, 10 Mar 64, subj: Field Level Mgt of MOL; AFSC News Release No. 43-R-30, 13 Mar 64; Historical Rprt, MOL Prog, 1st Half 1964, p 2.

11. Ltr (U), Schriever to Ferguson, 13 Mar 64; Ltr (C), Schriever to McMillan, 18 Aug 64.

12. Memo (U), McMillan to CSAF, 29 Apr 64, subj: MOL Coordination Office; AFCCS Draft Numbered AF Ltr, subj: MOL Coordination Office.

13. Ltr No. 28 (U), McKee to Deputies, et al, 9 Jun 64, subj: as above.

---

‡ These further changes will be discussed in Chapter VII.

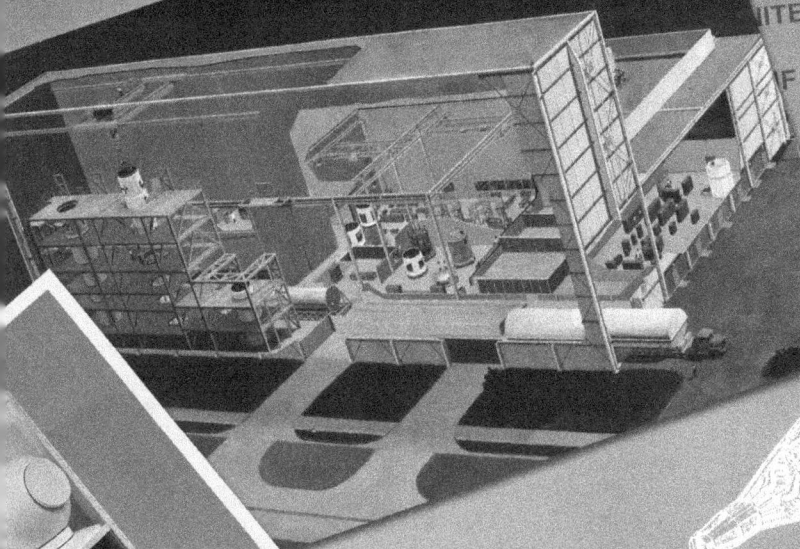

MOL ASSEM
INTEGRATION BUILD

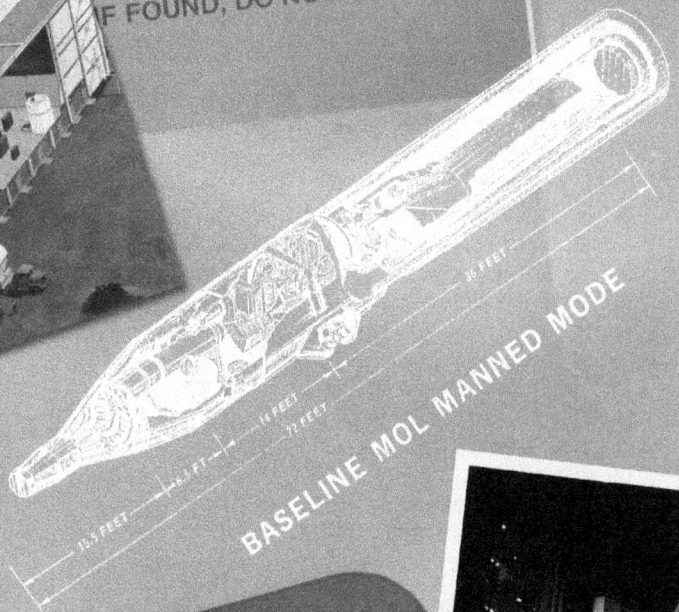

MATERIAL

PROPERTY OF THE
UNITED STATES GOVERNMENT

IF FOUND, DO NOT OPEN

BASELINE MOL MANNED MODE

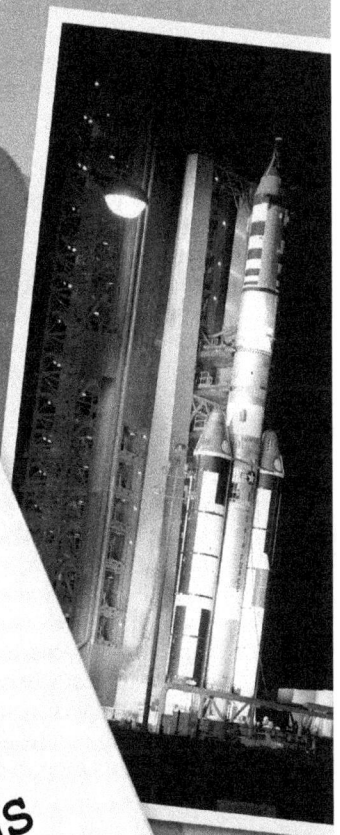

USAF MOL/KH-10

THE DORIAN FILES REVEALED:
A COMPENDIUM OF THE NRO'S MANNED ORBITING LABORATORY DOCUMENTS

CHAPTER VI:
RESULTS OF THE
PRE-PHASE I INVESTIGATIONS

RESULTS OF THE PRE-PHASE I INVESTIGATIONS

AFSC's original preliminary MOL development proposal of January 1964 called for about 20 months of pre-Phase I and Phase I study activity* leading to full-scale hardware development beginning in September 1965. It projected the first unmanned MOL launch in June 1968 and the first manned flight sometime in calendar year 1969. This schedule was criticized, as noted earlier, by Dr. Hall of OSD and General Momyer, chairman of the Air Staff Board. Both men recommended that the Systems Command review the proposed schedule to see if it could be accelerated to insure that MOL remained competitive with NASA's Apollo and Apollo Extended projects. On 23 April, during a meeting with USAF officials, Dr. Hall again urged that they try to achieve an earlier MOL launch.

Secretary Zuckert subsequently requested General Schriever to take another look at the MOL schedule. The AFSC commander in turn directed General Bleymaier to undertake a preliminary review of possible alternative schedules and to submit a report. Bleymaier completed this task on 1 June 1964; he concluded that if the time allotted for project definition could be sharply curtailed, MOL experimental test flights could begin 18 months after contractor go-ahead, a MOL with limited subsystems suitable for manned flight could be made available within 24 months, and one with complete subsystems in about 32 months.[1]

General Bleymaier noted that SSD's experience with other major development programs, such as Titan III, indicated that the study phase prior to receipt of OSD go-ahead authority for hardware development ran as long as 17 months to 2 years. Since many months of study had already been devoted to the MOL concept, he thought it "logical and feasible to reduce the definition phase to six months and the contractor selection to four months with program go-ahead at that point." If this was done, he felt that the first MOL flight could be achieved "approximately 30 months from initiation of the current (pre-Phase 1) studies which were approved last month."[2]

ALTERNATIVE MOL SCHEDULES

Unfortunately, Bleymaier's report was based on the assumption that Dr. Brown's 29 April approval of the start of pre-Phase I studies would be followed by the prompt award of the various contract. But, as we have seen, SSD and AFSC were forced into the time-consuming procedure of obtaining higher headquarters approval of work statements beforehand. Consequently, the awarding of the contracts dragged out through the entire summer. In July, however, Schriever asked Bleymaier to prepare a briefing on alternative MOL schedules based on his preliminary study.[†] Also, in advance of Phase I approval, the AFSC commander directed SSD to establish a MOL source selection board; he later named Brig Gen Jewell C. Maxwell its chairman. The board, officially organized on 27 August 1964, did not, however, begin its work for many months.[3]

Meanwhile, the possibility of adopting alternative MOL schedules to help shorten the development cycle was brought to the attention of top Pentagon officials. Dr. Flax thought the subject worth pursuing (as did, later, Dr. Brown) and on 3 August he asked General Ferguson to undertake a formal study of alternative schedules. In this connection he said that, among other things, an accelerated MOL program "could generate additional meaningful payloads for the Titan IIIC research and development launches.[4] Headquarters USAF dispatched Dr. Flax's recommendation to AFSC on 7 August. SSD subsequently was asked to do the formal study. Besides considering the possible use of Titan III R&D launches, the Division was to identify any program elements—hardware, facilities, etc.—for which funding or procurement actions might be initiated in advance of overall MOL project approval.[5]

After this new study of alternative MOL schedules was completed, Colonel Brady, MOL System Program Director, briefed General Schriever and Drs. McMillan and Hall on 15, 16, and 18 September respectively. He stated that, if authorized to go ahead with certain contracting efforts, the Air Force would be able to launch "a 2-man 1,500-cubic foot laboratory as early as mid-1967, and a 4-man, 3,000-cubic foot laboratory by late 1968." To achieve this accelerated schedule, SSD requested authority to prepare and release to industry

---

*   On 26 February 1964 OSD issued a new directive (No. 3200.9) which formalized what it called the "Project Definition Phase," which was previously termed Phase I or Program Definition. PDP was defined as a period of time set aside for precise planning of engineering, management, schedules and cost factors, prior to commitment to a full-scale development project.

†   This briefing was given to Schriever on 5 August 1964.

by 2 November 1964 a complete RFP package for the MOL laboratory vehicle. It also, prior to completion of the pre-phase I studies, wished to negotiate new "level of effort" contracts with McDonnell for the detailed design and initiation of Gemini B spacecraft development and with Martin-Marietta for work on Titan III integration and structural engineering tests.[6]

Dr. McMillan approved SSD's proposals end forwarded them to DDR&E on 18 September with a recommendation that SSD be authorized to prepare the RFP package. Meanwhile, SSD pushed ahead with the major task of planning, outlining, and drafting the formal request for proposals, including a work statement and annexes which covered such subjects as system engineering, PERT/time/cost factors, configuration control, etc. On 1 October, still awaiting OSD approval, it requested permission to publicize a synopsis for the laboratory vehicle procurement so as to insure that all qualified sources were aware of the impending competition and could be included in the bidders list. Dr. McMillan forwarded this request to DDR&E on 6 October.[7]

In his delayed response, Dr. Brown advised McMillan that the 18 September request for release of the RFP to industry was still under discussion in his office. As for SSD's proposed synopsis, he said that since DoD and NASA were currently engaged in discussions of space station projects (see pages 88-9) its publication "at this time" might have an adverse effect. Further, he remarked, the status of MOL planning was still "too premature to warrant the interpretation which contractors are likely to place upon the act of publication." Therefore, he planned to withhold "a decision on the synopsis" until his staff completed its review of the RFP proposal.[8]

These delays were exceedingly frustrating to USAF officials. On 13 November General Schriever complained to Dr. McMillan that AFSC had been left "without current direction or intention on which to base the allocation of command manpower and resources to meet MOL milestones. If it was OSD's intention to revert to the original development schedule, "program ramifications" must be recognized, he said. He referred specifically to the launch dates in the late 1960s, which would place the Air Force "in a poor competitive posture with NASA's current and extended Apollo programs." He reiterated his strong desire "to undertake a more progressive MOL program."[9]

On 23 November, Under Secretary McMillan sought to get a decision from OSD. Forwarding a copy of SSD's completed laboratory RFP, he pointed out to DDR&E that it had been so structured that it would be applicable whether or not the accelerated MOL schedule

was approved. He said the Air Force was prepared to proceed immediately with a two-shot pre-MOL program integrated with the Titan IIIC R&D schedule and he requested permission to negotiate sole source (level of effort) contracts with Martin-Marietta for the Titan III and McDonnell for the Gemini B.[10]

Dr. Brown, however, could not act at this time pending a decision on the fiscal year 1966 budget, being reviewed by top officials. Instead, he advised the Air Force that it would be necessary to stretch out Titan III development to insure the booster would be available for use in the MOL and defense communication satellite programs. He requested a new study be made of various alternatives, including one which would delay completion of Titan III development as much as 6 to 12 months. He suggested a test program consisting of 15 Titan III flights, with two additional vehicles being produced and assigned to MOL.[11]

In forwarding this request to AFSC, Headquarters USAF said it was aware that "requisite to any final determination of firm recommendations" on Titan III was the MOL decision. However, it felt the delay in obtaining the decision did not preclude initiating a study of various Titan III program adjustments. SSD shortly thereafter began the requested study, although—as General Funk wrote to General Schriever on 7 December 1964—the Division found it "extremely difficult to plan a worthwhile program in a vacuum."[12] But the very day he sounded this pessimistic note, an important budgetary meeting got under way in Washington which produced a decision to proceed with the program.

DOD/NASA "DUPLICATION, "CONGRESSIONAL CRITICISM AND THE BUDGET CONFERENCE OF 7-8 DECEMBER 1964

Some 18 months before, it will be recalled, Secretary McNamara and NASA Administrator Webb submitted statements to the Vice President on manned space stations and on 16 September 1963 agreed to coordinate each agency's advanced exploratory studies of such a vehicle. They also agreed they would eventually submit a joint recommendation to the President on the need for a National Space Station, including which agency should management the development if the requirement was accepted.

Meanwhile, NASA had already embarked on an extensive study program of earth orbital stations. In fiscal year 1963 it contracted for 23 studies totaling $4.049 million; in fiscal year 1964 it planned to spend an additional $5.750 million for follow-on investigations, and it scheduled still other studies for fiscal year 1965.[13] As the results of the early studies came in, NASA planners worked up three possible approaches to a manned space station. One involved a plan to extend the Apollo spacecraft "stay time" in space; another called for development of a 4 to 6-man laboratory; the third proposed a 10 to 20-man vehicle. NASA gave special attention to the extended Apollo concept, known as Apollo X. It was seen as remaining permanently in orbit, with crew rotation and resupply being provided by Gemini or Apollo-type ferry vehicles. Besides being used for observations of the behavior of men in space, such a vehicle would permit NASA to conduct scientific and engineering experiments.[14]

By the summer of 1964, as the space agency moved vigorously ahead with these studies, DoD officials concluded that some sort of joint management arrangements were needed to prevent program duplication. On 25 September Secretary McNamara wrote to Webb about the matter. He reported that DoD had obligated $5.5 million for the MOL program, budgeted $33 million in fiscal year 1965, and planned to make a much larger commitment in 1966. Then referring to NASA's studies of Apollo as a possible forerunner of a national space station, and noting its plans to spend additional sums in fiscal year 1965, the Defense Secretary said:

> I know we both feel, because of the important leverage they exert on subsequent programs of potentially enormous size, that studies in the area of manned earth orbit research and development should be carefully controlled, and that the purpose of our joint agreement of... 1963 is to achieve the necessary control. Of course, as studies progress and become more closely associated with particular conditions, joint control may become more difficult to achieve.[15]

McNamara saw little reason for having "two separate large programs" because of the great expenditures they would entail and he proposed they adopt a management plan to consolidate the work of both agencies. Under terms of this DoD plan, NASA would agree that MOL was the flight forerunner to a scientific or military operational space station. It would accept responsibility for the scientific program to be carried out using the MOL, while the Air Force continued as operating manager of

the development effort. A DoD-NASA board would be established to carry out the coordination. After completion of MOL flights, DoD and NASA would decide on the need for a new large military operational or scientific space station, the extent to which their individual requirements could be met by a single program, and which agency should have development responsibility.[16]

On 14 October 1964, in a lengthy reply to McNamara's letter, Webb politely rejected the DoD proposal. He agreed it was timely that they reassess their efforts and said that NASA was ready to help in the detailed planning of the MOL module configuration and in other areas. However, he pointed out that the Apollo system represented a capability for earth, orbital operations "that will be in being before 1969" (i.e., long before MOL) and that much could be gained" by exercising this capability." In doing so, he said, the space agency "would, of course, be most desirous of continuing to support the military needs." Webb said further:

> In a new and rapidly developing field such as astronautics wherein new opportunities as well as perhaps constraints and limitations are revealed almost from day to day, it seems to me that we should not attempt rigidly to interpret or classify programs in terms of possible undertakings in the future. In the area of manned spaceflight, both in potential scientific and military applications, I view Gemini, Apollo, and the DoD MOL all as important contributors to the ultimate justification and definition of a national space station.[17]

While Webb's support of the MOL program was welcomed, the fact remained that a cost-conscious administration or Congress might balk at funding two "duplicative" space station developments. Indeed, the Air Force's fears were aroused at year's end when Senator Clinton P. Anderson, the powerful chairman of the Senate Space Committee, urged President Johnson to merge the two projects. In a letter to the White House, the Senator argued that $1 billion could be saved over a five-year period if MOL were cancelled and USAF funds applied to the Apollo-based space station. In support of his argument, he noted that MOL was "dead-ended" since it could not grow beyond its two-man, 30-day mission without developing a resupply system. He said that while he agreed the military should be given a

Figure 32. Donald F. Hornig
Source: CSNR Reference Collection

chance to exploit the potential value of a manned space program, the Apollo X would provide the Air Force a broader-based capability on which to build.[18]

The Senator, a close friend of the President, subsequently reported he had received a "sympathetic response" from the White House. He said an agreement had been worked out between the military and civilian programs which had "gone a long way toward answering the questions I raised." As he understood it, under its provisions the Air Force and NASA would take advantage of each other's technology and hardware development "with all efforts directed at achievement of a true space laboratory as an end goal."[19]

The agreement cited by Senator Anderson was reached during a two-day budget conference held 7-8 December 1964 between McNamara, Webb, Kermit Gordon (Director of the Bureau of the Budget), and. Dr. Donald F. Hornig, the President's Scientific Advisor. After discussing the status of MOL, the four men agreed that the primary objectives of the Defense Department project, in order of priority, would be:

1. The development of technology-contributing to improved military observational capability for manned or unmanned operation. They saw this as possibly including intermediate steps toward operational systems.

2. Development and demonstration of manned assembly and service of structures in orbit with potential military applications, such as a telescope or radio antenna.

3.Other manned military experimentation,including the programs studied during the past year.[20]

They further agreed that DoD would emphasize the first two primary objectives and undertake to determine the vehicle characteristics that would be required. Vehicle studies and investigations military experiments would be coordinated with NASA, which also would undertake to identify "specific configurations of the Apollo which may have the capability of accommodating experiments" relating to the two primary Defense Department objectives. Results of the ASA studies were to be made available to the Pentagon by 30 April 1965, at which time DoD would attempt to determine whether any of the Apollo configurations could meet its objectives "in a more efficient, less costly, or more timely fashion."[21]

The conferees also agreed that the President's fiscal year 1966 budget would include $150 million for the MOL program. However, these funds would not be released until all studies had been completed and the results were reviewed by McNamara, Hornig, and Gordon. They anticipated that these reviews would take place about May 1965. Subsequently, in a memorandum to the President on 12 December, Secretary McNamara advised that his future recommendations on the MOL would be based on agreement between the three which took into account costs and the issue of NASA-DoD program duplications.

Pre-Phase I Study Conclusions

The agreement at the 7-8 December meeting to proceed with MOL was based in large measure on the results obtained from the "black" MS-285 investigations undertaken by the Directorate of Special Projects. Eastman Kodak, one of its "black" contractors, on 22 July 1964 had reported its preliminary conclusion that man could indeed "make substantial contributions to a satellite reconnaissance mission." Reviewing a plan to adapt a strip versus frame camera approach to a manned reconnaissance satellite, Eastman Kodak stated that:

... the frame camera makes optimum use of man's ability to point accurately at a ground target. The primary object of a manned mission should be very high resolution coverage (both photographic and visual) of limited target areas of major importance. The man's ability to put the camera on the target reduces the need for a large field, thereby making the frame camera feasible. The astronaut looks through the primary optical system at the target and adjusts tracking rates of the tracking mirror until the image is stationary in the field. Visual target inspection can be performed concurrently with photography. As the target is tracked the number of photographs is limited only by the frame rate of the camera.

Such a system should provide at least 20 separate looks at a given target, each from a different angle and under different seeing and tracking conditions. This number of photographs would also provide the opportunity for subsequent image enhancement... The resolution improvement with the manned frame camera approach can be matched in a strip camera only by using a lens of applicable greater aperture and weight.[22]

As part of Project MS-285 studies of a manned reconnaissance system, the Directorate of Special Projects also contracted with Lockheed to obtain use of its manned reconnaissance simulator, developed by the company with its own funds. This "black" contract initially covered two major experimental sessions on the simulator and extended from 28 May to 7 September 1964[‡].

Used to determine man's ability to aim sensing devices to acquire and track ground targets, the Lockheed simulator provided the astronaut with a televised simulated view of the earth as it appears from a satellite. A gimbaled television camera which transmitted the image to the operator could be programmed in pitch and roll to simulate vehicle rates and pointing error rates as they would actually occur. The pilot, through a two axis control stick, was able to center the target and perform a "rate killing" tracking operations and also change magnification over a 9 to 1 range.

On the basis of numerous runs on the Lockheed simulator, the Directorate of Special Projects reported on 22 September:

All data to date has confirmed our original assumptions that man could correct the line of sight to an accuracy of less than 0.1 degree and that he can reduce image motion or smear rate to a value of less than 0.1 percent of V/H (velocity over height). The data indicates that the target acquisition and centering task ranges from approximately 3 to 8 seconds with a mean resultant displacement error in the line of sight of 0.06 degree when performing at five times magnification and 0.02 degree when performing at 45 times. We have consistently demonstrated the ability to reduce the residual tracking rate error to values of .025 percent of V/H. This level of rate performance is accomplished within the first 2 seconds of tracking time. We have performed the tracking task in the presence of varying levels of tracking mirror vibration. Overall tracking performance is at least as good as stated above. We are still analyzing the results in detail to determine the precise effects of magnification and vibration...[23]

Figure 33. Edward M. Purcell
Source: CSNR Reference Collection

Figure 34. Sidney D. Drell
Source: CSNR Reference Collection

[‡]   The original contract cost $150,000. Two further extensions of the contract through 19 October 1964 brought the cost to $175,067.

The results of the above simulations were briefed to the Reconnaissance Panel of the President's Science Advisory Committee (PSAC) on 21 October 1964. Among those in attendance, besides Prof. E. M. Purcell, Panel Chairman, were: Drs. Hornig, E. H. Land, H. F. York (formerly DDR&E), N.F. Golovin, D. H. Steininger, Prof. Sidney Drell, and Mr. Willis Shapley, a representative of the Bureau of the Budget.

After this briefing, Professor Drell was dispatched to Lockheed on 11 November to get a first-hand look at the equipment used for the simulations. Over a four-hour period, he was given a complete briefing and demonstration together with some operational time in the simulator. He emerged from the session apparently convinced that the numbers which had been reported to the Panel as representing man's ability were indeed valid and that the pilot was consistently performing the IMC task to an order of magnitude better than results obtained from existing unmanned systems.[24]

As the "black" studies and simulations contributed to the decision to proceed with MOL, so did the conclusions of the various "white" study contractors, whose pre-Phase I reports began flowing into the Air Force late in the year. Copies also were made available, at DDR&E request, to the Institute of Defense Analysis (IDA) for review and evaluation.[25]

An analysis of the studies and reports of simulation test it received led SSD to conclude that the basic MOL concept and the value of employing man to perform specific military tasks in space had been confirmed. It reported that the results of studies of MOL experiments P-1, -2, and -3 and contractor simulation tests demonstrated "that man can accomplish IMC to better than .2% consistently and was limited only by the quality and magnification of the optics and the inherent stability of the vehicle." Extensive B-47 flights conducted with a modified bombsight and using two cameras also had verified that man had the ability "to acquire unknown targets as small as trucks and trains and make an accurate count of the total present."[26]

In the area of electromagnetic signal detection, a test employing a KC-135 confirmed that man was able to discriminate false alarm signals, select signal bands of interest, and assess and classify the signals within seconds after receipt. Tests to determine in-space maintenance (Experiment P-5) capabilities, using the Air Force's zero "g" KC-155 and submersible tests, proved that man had "the inherent potential to accomplish any level of maintenance and repair conceivable, being constrained only by the time available, the fineness of the task, and the presence of a pressurized suit."

The Division reported other test and study results were equally encouraging. During exercises in the zero "g" KC-135, it was demonstrated that man could stabilize and maneuver himself in AMU and similarly could effectively operate an RMU by TV or direct viewing (Experiments P-6 and -7). In the autonomous navigation and geodesy area (Experiment P-8), simulations were performed which indicated that, with small fields of view, man could acquire and point at identifiable landmarks within 15 arc seconds. In another experiment (P-10) involving multiband spectral observations, four men operating radiometric and calibration instruments and automatic trackers installed in a KC-135 demonstrated their ability to calibrate, point, monitor displays, change plans, and assist in data interpretation.[27]

The IDA review and assessment of the pre-Phase I studies also tended to support the basic validity of the MOL concept. IDA reported, among other things, that there appeared to be no known insurmountable problems to providing life support and environment control systems for 30-45 days and that attitude control systems for specified attitude holding and slewing of the MOL were within the technological state of the art. It also concluded that the ground support network for MOL appeared adequate but only for initial flights (other facilities would be needed for MOL follow-on systems) and that the three OSS studies contained "general operational concepts of value" to the MOL project.[28]

In a separate analysis of the simulation aspects of the MOL experiments, IDA reported that many aerospace firms had installed equipment to simulate the critical photo reconnaissance mission. Their preliminary results, IDA said, indicated that "man will be able to contribute to the task of pointing a high resolution camera and tracking a target with the rate accuracy of -3 arc/sec necessary to achieve very high resolution photography (-1ft)." However, IDA cautioned that important inputs such as stabilization and attitude control parameters and realistic navigation errors needed to be included in the simulations before final conclusions could be reached about system tracking accuracy.[29]

ENDNOTE

1. Ltr (C) Bleymaier to Schriever, 1 Jun 64, subj: Alternative MOL Schedules.

2. Ibid.

3. Msgs (C), MSF-1-9-6-1, AFSC to SSD, 9 Jun 64; SCG 15-6-29, Schriever to Funk, 15 Jun 64; MSF-1-8-7-4, AFSC to SSD (Bleymaier), 8 Jun 64; Msg-1-23-7-8, AFSC to SSD, 23 Jul 64.

4. Memos (C), Flax to Ferguson, 3 Aug 64, subj: MOL Schedule Alterna tives; Brown to Flax, 31 Aug 64, subj: Unmanned MOL Tests.

5. Msg (C), MSG-1-14-8-6, AFSC to SSD (for Bleymaier), 14 Aug 64.

6. Ltr (C), Ritland to McMillan, 17 Sep 64, subj: MOL Program.

7. Memo (S), McMillan to DDR&E, 18 Sep 64, subj: MOL Program; Msgs (C), SSHKM 000028, SSD to AFSC, 1 Oct 64; MSF-1-10132, 2 Oct 64; Memo (C), McMillan to DDR&E, 6 Oct 64, subj: MOL Program.

8. Memo (C), Brown to McMillan, 14 Oct 64, subj: MOL Synopsis.

9. Ltr (C), Schriever to McMillan, 13 Nov 64.

10. Memo (S), McMillan to DDR&E, 23 Nov 64, subj: MOL Program.

11. Memo (S), DDR&E to SAF, 17 Nov 64, subj: Add'l Release of Funds for FY 65 RDT&E on Titan III and Request for Program Stretchout.

12. Ltr (S), Hq USAF to AFSC, 25 Nov 64, subj: Alternative Adjustment Study of Titan III R&D Program; Ltr (S), Funk to Schriever, 7 Dec 64, subj: Brown Memo for SAF, 17 Nov 64.

13. Memo (U), Flax to Zuckert, 12 Jun 64, subj: DoD/ NASA Space Station Coordination.

14. Statement by James Webb, 4 Mar 64, before Senate Cmte on Aeronautical and Space Sciences, 89th Cong, 2nd Sess, Pt 2, *1965 Nasa Authorization*, p 297.

15. Ltr (U), McNamara to Webb, 25 Sep 64.

16. Ibid.

17. Ltr (C), Webb to McNamara, 14 Oct 64.

18. Senator Anderson's remarks are quoted in *Aviation Week*, 7 Dec 64, p. 16.

19. Quoted in the *N.Y. Times*, 20 Dec 64.

20. Memo for Record (S), Kermit Gordon, Director, BOB, 10 Dec 64, subj: Manned Orbiting Laboratory (MOL Program 1966 Budget).

21. Ibid.

22. Msg 5931 (S-DORIAN), Lt Col Knolle to Lt Col Howard, 30 Sep 64.

23. Ibid.

24. Msgs Whig 1828 (TS-DORIAN), Stewart to Greer, 12 Oct 64; 6411 (S-DORIAN), Berg to Carter, 24 Nov 64.

25. Memos (U), Brown to Flax, 22 Oct 64, subj: MOL Pre-Program Definition Phase Studies; Flax to Brown, 5 Nov 64, same subj; Ross to DDR&E, 25 Nov 64, same subj.

26. Atch 1 (S) to SSD Report, "MOL Growth Potential," Jan 65.

27. Ibid.

28. IDA Study S-185 (S), Review and Assessment of USAF/SSD MOL Pre-Program Definition Phase Studies, Mar 65.

29. IDA Study S-179 (S), Simulation Aspects of the MOL Program, Jan 65.

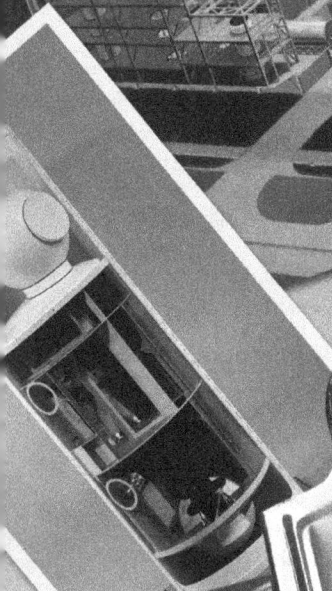

MOL ASSEM...
INTEGRATION BUILD...

MATERIAL

PROPERTY OF THE
...NITED STATES GOVERNMENT

...F FOUND, DO NOT OPEN

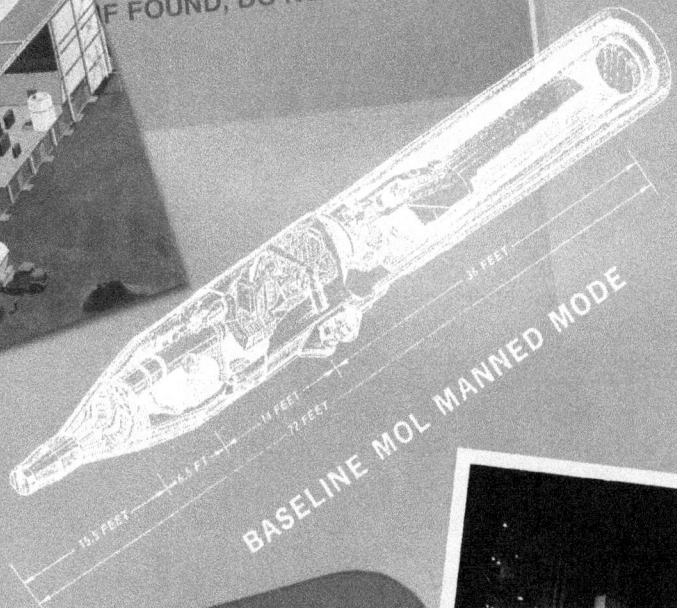

BASELINE MOL MANNED MODE

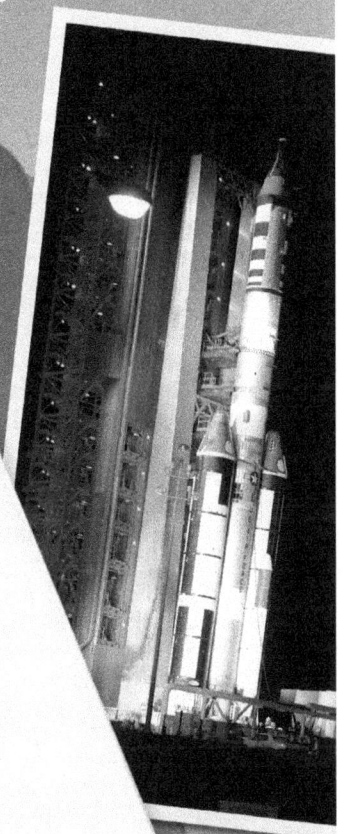

USAF MOL/KH-10...

THE DORIAN FILES REVEALED:
A COMPENDIUM OF THE NRO'S MANNED ORBITING LABORATORY DOCUMENTS

CHAPTER VII:
THE LABORATORY VEHICLE
DESIGN COMPETITION
JANUARY–JUNE 1965

THE LABORATORY VEHICLE DESIGN COMPETITION
JANUARY - JUNE 1965

In accordance with the agreements reached at the budget meetings on 7-8 December 1964, DDR&E submitted new instructions to the Air Force which formally changed MOL program objectives. On 4 January 1965 he directed Dr. McMillan to initiate additional studies for an experimental military program which would contribute "to improved military observational capability for manned or unmanned operation" and to development and demonstration of manned assembly and servicing of structures in orbit with potential military applications such as a telescope or radio antenna.[1]

In ordering the new studies, Dr. Brown asked the Air Force to carefully assess whether any of NASA's Apollo configurations could be used in place of the Gemini B/MOL. To help make this determination, he said NASA would be requested to submit data on the Apollo system to the Air Force by 30 April 1965. The USAF evaluation of both configurations was to be submitted to him by 15 May.

Dr. Brown authorized the Air Force to award three contracts to industry for preliminary design studies of the MOL laboratory vehicle, based on the Titan IIIC/Gemini B combination. He asked that the proposed lab configurations provide for assembly and servicing of large optical devices and radio telescopes in space, for testing high resolution surveillance radar concepts, and be capable of being used as a manned experimental facility. He required that the three contractors be qualified to build the laboratory module, whether the final Titan IIIC/Gemini or NASA's Saturn IB/Apollo combination was chosen. It was OSD's intention that the final contractor would be selected from the above three firms without further competition from industry.

The Defense Research Director also asked the Air Force to re-examine its proposed MOL unmanned flight schedule to take advantage of planned Titan III R&D test launchings in order to provide for "qualification of components of the MOL system." To preserve the option for proceeding with development, he advised Dr. McMillan that certain fiscal year 1965 funds would be released for studies and work on the Titan booster and Gemini B.[2]

On 8 January Dr. McMillan forwarded these DDR&E instructions to General Schriever and directed their implementation. Concerning the lab vehicle preliminary design studies, he advised General Schriever to consider only the Titan IIIC/Gemini B configuration, but with the understanding that it was being used "solely for illustrative purposes and is not intended to prejudice the final decision on booster or personnel carrier subsystems." The three contractors selected to do the studies should be able to develop end build their proposed laboratory for integration with either Saturn IB/Apollo or Titan IIIC/Gemini B. He also asked the AFSC commander to prepare a work statement for NASA defining MOL requirements, to enable the space agency to determine whether any of its proposed Apollo configurations could accommodate the planned equipment and experiments.[3]

Almost simultaneously, Dr. McMillan sent instructions to Maj Gen Robert E. Greer which were somewhat similar to those dispatched to General Schriever. That is, he directed General Greer to initiate certain "black" studies to define the technical characteristics of large optical system payloads and large antennas for use in achieving "improved military observational capability for manned or unmanned operations." Study results were to be submitted by 15 May.[4]

Figure 35. Robert E. Greer
Source: CSNR Reference Collection

WORRIES OVER MOL SECURITY

When NRO officials in Washington and Special Projects personnel on the West Coast read Dr. Brown's 4 January memorandum, they were startled by its reference to acquiring a MOL reconnaissance capability. Twelve months before, in January 1964, the Director of the NRO Staff had expressed concern over the breakdown of security discipline resulting from widespread MOL discussions at that time. Subsequently, the NRO devised and established Project DORIAN as a means of controlling all information relating to satellite reconnaissance activities.

Now, in January 1965, General Martin* bluntly informed Dr. McMillan that Brown's memorandum constituted a violation of NRO security. He said its implication, although it did not explicitly use the word "reconnaissance," was obvious. "The overall impression created in the minds of unwitting people involved," he said, "has been that MOL has finally been assigned a reconnaissance mission." He emphasized the need for prompt security decisions before any further MOL correspondence was issued, pointing to the following dangers if no action was taken:

> The security of MOL reconnaissance aspects is inescapably tied to the security of the unmanned satellite reconnaissance program. Exposure of MOL reconnaissance capability to anyone outside the BYEMAN system automatically will provoke pressure for disclosure of unmanned reconnaissance data. Such personnel will want to know how the MOL capability compares with the unmanned satellite reconnaissance capability.[5]

As a consequence of this situation, a working group which included General Greer's special assistant, Col Ralph J. Ford, and Col Paul E. Worthman of the NRO Staff, was formed to prepare recommendations on the security aspects of the MOL program. The group proposed a basic approach which would provide for "an absolutely clear and separate division between reconnaissance oriented tasks (DORIAN/BYEMAN) and non-reconnaissance related tasks in MOL." General Greer endorsed this approach and on 30 January 1965 recommended to Dr. McMillan that all payload studies and. development of passive-in-nature Sigint and terrestrial (of the earth) image forming sensors, "having

practical intelligence collection application," be controlled by the Director of NRO under Project DORIAN and within the BYEMAN security system. Studies and work not involving reconnaissance payloads, i.e., such as those concerned with general experiments to determine man's usefulness in space, would be subject to normal security restrictions, such as those contained in DoD Directive 520012 and AFR 205-3.[6]

Dr. McMillan approved these recommendations and, on 5 February, he issued a paper titled "Special Security Procedures for the Department of Defense Manned Orbiting Laboratory," a copy of which he sent to Dr. Brown. He advised the Defense Research Director that it had become clear from detailed analysis of work statements, procedures and methodology, "that the only practical recourse is to keep any reconnaissance, including active sensor work, black." He said he thought this could be handled in a manner "which does not detract from the efficiency of the current activities nor will it hamper the DoD-NASA exchange of information..."[7]

Seven weeks later McMillan also issued additional guidance and security policies to govern MOL study and developmental activity. He emphasized that all payload study and other work would have to be cleared by him to insure appropriate security controls.[8]

A REVISED MOL MANAGEMENT STRUCTURE

Dr. Brown's redirection of the program on 4 January was followed by important changes in the MOL management structure. As noted earlier, General Schriever during 1964 had urged that a strong central management office be set up in the Washington area. McMillan, however, decided that an Air Staff coordination office, which in mid-1964 was organized as the Office of the Assistant for MOL†, would be sufficient for the time being. General Schriever objected to this arrangement as inadequate. He said the new office simply could not provide the leadership, channels, and direction needed for the program. On 18 August 1964, he once more strongly urged the Air Force Under Secretary to establish "a single integrated office."[9]

In response to General Schriever's criticism, Dr. McMillan on 3 September met with Colonel Schultz, the Assistant for MOL, and Mr. Frank Ross, of the Office of the Assistant Secretary for Research and Development, to discuss the management question. Afterwards, he asked them to prepare working papers on MOL management

---

* At this time General Martin, former Director of the NRO Staff, was understudying General Greer, who was scheduled to retire on 1 July 1965.

† Redesignated on 9 November 1964 as the Assistant for Manned Flight, to take into account its other responsibilities including the NASA Gemini experiments project and certain coordination activities involving the space agency.

alternatives and advised he was thinking of establishing a "MOL Policy Management Committee" to help oversee the program. The committee, which would consist of himself, Dr. Flax, Generals Schriever and Ferguson, and a Secretariat, would enable "the principals in the decision-making chain" to meet at regular intervals "to facilitate agreement on major policy matters." Colonel Schultz and Mr. Ross subsequently submitted several alternative management proposals to the Under Secretary which, in general, incorporated some of the ideas contained in AFSC's original 1964 plan.[10]

While these activities were under way, General Schriever undertook to strengthen his own management structure. He designated Brig Gen Harry L. Evans, who was nearing the end of a two-year tour of duty with the Joint Chiefs of Staff (JCS), as his Assistant Deputy Commander for Space for MOL. Evans, who had previously worked under General Schriever at the Ballistic Systems Division, had had major responsibilities for a number of early USAF satellite systems. On 30 October 1964 the AFSC Commander informed Dr. McMillan that he planned to bring General Evans into his headquarters; he again urged him to provide the top-level management needed to insure program success.[11]

In early January 1965 the AFSC commander's year-long campaign for better MOL management began to produce some results. Dr. McMillan decided that General Evans could be of great help in the: Office of the Secretary of the Air Force in overseeing the new MOL studies. He directed Dr. Flax to seek General Evans' immediate release from the JCS. Since his tour was to end 1 February 1965, this proved to be no problem. General Evans promptly reported in and was provided temporary office space in the conference room of the NRO staff, then under the direction of Brig Gen James T. Stewart‡.[12]

Meanwhile, McMillan discussed his MOL management plan— the establishment of a management committee and the post of Special Assistant for MOL (Evans' title)— with Cyrus Vance, the Deputy Secretary of Defense. Vance agreed that management would be a "dominant factor" in ensuring successful implementation of studies leading to a MOL decision. "The objective, of course," Vance wrote McMillan on 7 January, "is the creation of a system which will allow the exercise of firm control which will unquestionably be needed to prevent the program from becoming prohibitively complex and costly, and at the same time to deal effectively with the many governmental elements that are involved in such a large program, particularly during the early stages."[13]

On 18 January Secretary Zuckert, approving McMillan's management plan, issued a formal order establishing a "Special Assistant for the Manned Orbiting Laboratory." He was to report directly to the Under Secretary and "be primarily responsible for assisting the Office of the Secretary in managing the MOL Program." In addition, he was made responsible for maintaining liaison with and providing MOL program status information to OSD and other interested government agencies, in particular to NASA.[14]

McMillan described General Evans' new assignment as being "in addition" to his assigned duty as Schriever's Assistant Deputy Commander for MOL. In the latter capacity, Dr. McMillan said, "General Evans will be responsible, under General Schriever, for field-level management of the program. His straddling of both Secretarial and working-level positions in the management structure provides him with an ideal vantage point from which to effect the important exchange of program information" with NASA. This arrangement was considered an interim organizational structure "for the study phase conducted between January and June 1965."[15]

Simultaneous with the announcement of establishment of the Office of Special Assistant, Secretary Zuckert approved formation of a MOL Policy Committee. Designated as official members of this "key policy body" were the: Secretary of the Air Force, Chairman; Under Secretary; Chief of Staff; Commander of AFSC; Assistant Secretary for Research and Development; and Deputy Chief of Staff, Research and Development. The committee was responsible for reviewing and making recommendations on all MOL matters, including program objectives, plans, programs, schedules, and milestones. The Special Assistant was to provide the committee secretariat.[16]

In notifying the Air Staff of these new management arrangements, Secretary Zuckert stressed that the success of the MOL program would depend "on how well we execute our mandate in the next few months... how rapidly we can implement this unique management concept." He expressed belief that the MOL Policy Committee would permit "most rapid application of the broadest level of Air Force support to the program, and will insure that we have applied our best judgment and experience to MOL policies and guidance."[17]

After reviewing the new management arrangements, Secretary Vance informed McMillan that OSD had no objections to them and he advised that DDR&E was prepared to participate to the extent the Air Force considered desirable as problems arose. (Endnote 18)

‡   Gen Stewart succeeded Gen Martin, who was at this time understudying Gen Greer at the Directorate of Special Projects.

Within Headquarters USAF, however, some questions were raised over "the limited degree of Air Staff participation" in the program. On 9 February 1965 the Space Panel expressed the opinion that while the Chief of Staff had concurred with the management organization, "he did so as an initial means of providing necessary response to OSD, and did not necessarily envision it as a continuing method of program management."[19]

In any event, by early 1965 the MOL management structure consisted of the MOL Policy Committee and the Special Assistant in the Pentagon. Within AFSC there was Schriever's Assistant Deputy Commander for Space for MOL and a system project office on the West Coast. Finally, the Directorate of Special Projects on the Coast also had major responsibilities in the "black" area.

BASELINE MOL MANNED MODE

Figure 36. MOL Drawing
Source: CSNR Reference Collection

NASA'S MOL/APOLLO STUDY

Even as MOL management was being strengthened, Drs. Brown and McMillan were initiating discussions with NASA to obtain space agency contributions to the study program. In a letter to Dr. Seamans in early January in which he solicited NASA's cooperation, McMillan remarked:

> As I see it, from the point of view of the Department of Defense, the central question relative to the Manned Orbiting Laboratory is one of existence: the question whether or not to proceed with a major program of manned military space flight. This is a question to which the Secretary of Defense must develop an answer. Furthermore, before any such program is undertaken... he must reach agreement with the President's Science Advisor and the Director of the Bureau of the Budget that the program of military, engineering and scientific experiments and steps toward operational capability is worth the cost and does not duplicate approved programs in any other agency.[20]

McMillan said that if a decision was made to proceed with the military project, many contingent decisions would follow. Those that would directly affect NASA would involve the manner in which MOL might support space agency objectives and whether or not NASA hardware and resources would be used. To clarify the issue of "program duplication," he said information was needed on whether Apollo could be used for MOL. He referred to a recent suggestion made by Adm W. F. Boone of NASA that they form an ad hoc board within the Aeronautics and Astronautics Coordination Board to consider the results of DoD and NASA studies and arrive at findings. McMillan agreed an ad hoc group would be helpful but he opposed involving the AACB because of security.[21]

Dr. Brown also wrote to Seamans about space agency inputs. On 11 January he proposed that NASA submit a briefing and supporting documents to DoD by 1 March giving its best estimate of Apollo capabilities to serve as a military facility for earth orbit operations. He also solicited information on: (1) any Apollo improvements which it was likely NASA would undertake as part of its program; (2) the times at which Apollo equipment could be procured and operated by the Air Force for orbital operations without interfering with the national lunar landing program; and (3) the cost history of Apollo and the Saturn booster. In addition, he asked Dr. Seamans to submit a description of NASA's planned scientific earth orbit experiments which MOL might be able to perform.[22]

Recognizing the importance of the DoD study and the implications it might have on the space agency's program (i.e., it would be a great coup should Saturn/Apollo be selected for the MOL program), Dr. Seamans promised

NASA's full cooperation. Thus the space agency acted promptly when—advised that Evans would be in charge of coordinating DoD studies—it designated Mr. Robert F. Garbarini as his counterpart, responsible for exchanging pertinent data and guiding preparation of NASA reports. Evans and Garbarini met on 13 January and formed a six-member DoD/NASA ad hoc study group which discussed exchange of data and submission of USAF descriptions of proposed MOL experiments.[23]

Figure 37. Saturn Booster
Source: CSNR Reference Collection

During the next several months the Evans-Garbarini group conferred on at least eight more occasions. The two officials also engaged in an extensive correspondence, agreeing on guidelines and ground rules for mutually acceptable formats for submission of cost estimates and a study plan. On 1 March General Evans delivered to NASA two SSD reports on proposed MOL primary and secondary experiments and two other documents on MOL performance and design requirements. Other data requested by Garbarini—on the proper ordering of MOL experiments with respect to priority, interdependence, number of flights, orbit altitudes and inclinations, and flight duration—also were provided.[24]

Even before most of this data was in hand, NASA organized a MOL-Apollo task tea to prepare the space agency report. It also contacted its various centers for assistance and let three contracts (to Grumman, Boeing, and North American Aviation) to help identify and define proposed scientific experiments which might be conducted aboard the MOL. A total of 84 NASA

experiments were identified and a report describing them was sent to OSD on 17 March 1965. Included were a number of "earth viewing" experiments which NASA proposed to conduct using various high performance optics, infrared, or radar sensors.

In evaluating the various material provided it by the Air Force, the MOL-Apollo task team quickly noted that not all information on planned USAF experiments had been made available. For example, it had been given no information on two primary experiments—P-14 (manned assembly and service of large antennas) and P-15 (assembly and service of a large telescope). On 17 March Seamans brought this matter to Brown's attention. He reported that NASA had only six weeks remaining to complete its study and that it was "imperative" that experiment descriptions on P-1 and P-15 and other USAF experiments be forwarded.[25]

But, of course, OSD was unable to comply with this request because of security. In view of the decision that all references to sensor or reconnaissance payloads would be controlled under Project DORIAN and the BYEMAN security system, NASA was officially informed that the P-15 experiment had been deleted[§]. Dr. McMillan, however, did agree to release information on Experiment P-14 and a description of it was forwarded to NASA on 22 March. Thus, with the primary MOL experiment, P-15, being withheld, the MOL-Apollo task team could only reach the erroneous conclusion that all USAF experiments could be accommodated by certain Apollo configurations which it shortly proposed.[26]

The fact that NASA had been going through a somewhat unreal exercise became apparent to Dr. Michael Yarymovych in mid-March, when he was detailed to General Evans as a technical advisor from the space agency. After he was briefed on P-15 as the primary MOL experiment, Dr. Yarymovych strongly urged that Mr. Garbarini and other NASA staff members be informed since they were "just wasting their time." In April Garbarini and several other members of the MOL-Apollo committee were given a DORIAN briefing on the "black" aspects of the program by Maj Harvey Cohen, the MOL security officer. This new information—while it enlightened them—could not be considered in the NASA report, which was to be based entirely on the "white" data submitted earlier.[27]

Another major handicap NASA faced in promoting the Saturn/Apollo configuration involved schedules and costs. Because of the priority commitment to the lunar landing program, the space agency found it would be

§ NASA's top officials, Dr. Seamans and Mr. Webb—given DORIAN clearances in the fall of 1964—apparently were briefed on the reasons why.

unable to make Saturn IB boosters available to DoD until mid-1969 and Saturn V until 1970—too late for the proposed Air Force launch schedule. Also, Saturn/Apollo costs were substantially higher than the Air Force estimates for Titan I /Gemini B¶. But, in its formal report to OSD dated 5 May 1965 and titled, "Utilization of Apollo Systems for NASA and DoD Experiments in Earth Orbit," NASA declared that with slight modifications, Saturn/Apollo could meet all the requirements of the MOL Program.

However, several weeks earlier—during a meeting of the MOL- Apollo study committee—Dr. Seamans had hinted to the members that "constraints of pressurized volume and early flight schedules required in the interest of national security" would tend to prejudge the hardware selection "in favor of Gemini B/Titan IIIC." He expressed his confidence that "military earth orbital operations" would not have an adverse effect on NASA's plans to build a space laboratory. Justification of such a laboratory, he said, would be based "upon the quality of its experimental program and the values of extended lunar exploration."[28]

APPROACHING A MOL DECISION

On 23 January 1965, Secretary of Defense McNamara told a press conference that an aerospace industry competition would soon get under way for design studies of an orbiting laboratory system. He said the purpose was to develop technology "to improve the capabilities for manned and unmanned operations of military significance." McNamara made a special point of emphasizing the phrase, "orbiting laboratory system," explaining to the newsmen that OSD had "not eliminated the manned phase of the program" but had broadened its concept "to include unmanned activities as well as manned." The reason, he said, is that "manned and unmanned systems are always competitive."[29]

Several days later he and NASA Administrator Webb issued joint statement pledging close cooperation and coordination of each other's space projects. The primary purpose of the statement was to provide a basis for the impending budget message "and Congressional testimony and public remarks of all officials concerned." The two agency chiefs said that they intended to avoid "duplicative programs" and that any manned space flights undertaken in the years ahead by DoD or NASA would "utilize spacecraft, launch vehicles, and facilities already available or now under development to the maximum degree possible.[30]

¶ Estimated 10-year developmental and operational costs for Apollo (90-day flights) were $5837 million, for Apollo (30-day flights), $6948 million. Titan III/Gemini B costs (30-day flights) over 10 years were estimated at $4999 million.

McNamara discussed the use of certain NASA hardware in the MOL program sometime later in a letter to Vice President Hubert Humphrey, Chairman of the Space Council. The Vice President had asked for comments received from Congressman Olin Teague (D-Tex), the ranking member of the House Committee on Science and Astronautics, who had urged President Johnson "to take a look as soon as possible and make a decision" whether or not Gemini would be used by DoD. Representative Teague was particularly concerned that the valuable Gemini industrial team at McDonnell Aircraft would be disbanded if a MOL decision was not made—since all of NASA's Gemini spacecraft were already in production.[31]

Figure 38. Hubert H. Humphrey
Source: CSNR Reference Collection

In his letter to the Vice President, McNamara cited the various agreements and steps taken by DoD and NASA to insure maximum benefits were obtained from "the national investment" in the Gemini program. Concerning MOL, he said his decision might take one of three forms:

First, it may be determined that the cost of the MOL program is too high to be commensurate with its military value. While I do not expect this conclusion, it is a possibility, and in that case I will not proceed. Or, second, it may be determined that the MOL program is worth the cost but the use of Apollo hardware is the more effective approach. Or, third, it may be determined that we will proceed with the Gemini B approach to MOL. In the third alternative, we will, of course; take advantage of Gemini capability.[32]

Meanwhile, in accordance with his announcement of the laboratory vehicle design competition, AFSC on 25 January released its request for proposals to 23 aerospace contractors. By mid-February seven contractors submitted proposals to the Air Force, which were promptly reviewed and evaluated by the MOL Source Selection Board headed by General Maxwell. On 25 February the newly-created MOL Policy Committee met for the first time to hear the Board's recommendations. Attending this initial session were Secretary Zuckert, Gen John F. McConnell, Drs. McMillan and Flax, Mr. Leonard Marks, Jr., Assistant Secretary for Financial Management, and Generals Schriever, Ritland, and Ferguson**. After General Maxwell's briefing, the committee decided— as the third and fourth contractors in the competitive evaluation were very close in ranking—that AFSC should award four rather than three study contracts.[33]

Several days later, on 1 March, the Air Force announced that Boeing, Douglas, General Electric, and Lockheed were the successful bidders. Each was awarded 60-day contracts totaling about $400,000 per firm. They were directed to submit their final study reports by 30 April. One of them, the Air Force said, would be selected to begin MOL project definition.[34]

Even as these "white" activities got under way, the Directorate of Special Projects was pursuing certain "black" studies. In mid-January 1965 it redirected the existing Eastman Kodak DORIAN effort, organized initial technical meetings to discuss the latest guidance, and established milestones to meet the 15 May reporting date. The Directorate also contracted with two other firms— Perkin-Elmer and ITEK—for studies of large lightweight optical elements and, in addition, it assembled a team of highly experienced personnel in the fields of optics and satellite reconnaissance. This team was charged with investigating and considering large lightweight "optical element together with their application in manned satellite reconnaissance" systems.[35]

With the "white" and "black" studies proceeding nicely, Dr. McMillan suggested to Dr. Brown that it might be desirable— from both the Air Force's and industry management standpoints—to announce the successful laboratory vehicle contractor immediately after review

and approval of the Source Selection Board's findings and recommendation. He pointed out that, even with an early decision, the four contractors would have to be supported at their current contractual levels between the time they completed the studies and announcement of the project definition phase winner. In addition, he thought it might be necessary to support the winning contractor for a period after the announcement.

Consequently, McMillan proposed that OSD provide $0.8 million to sustain the four contractors for 30 days after completion of their studies and $1 million to sustain the selected contractor after announcement of the winner. Dr. Brown agreed and he subsequently released $1.8 million in fiscal year 1965 funds[††]. Supplemental agreements of approximately $200,000 each were negotiated with the four laboratory vehicle study contractors, which extended them through May 1965. [36]

Meanwhile, during March and April laboratory contractor briefing teams made mid-term presentations to the Air Force and the results pointed to the final contractor selection. Thus, a NASA representative who attended the briefings thought that Douglas had made a "very strong presentation that indicated large corporate support behind the study." DORIAN security, it might be noted, proved ineffective as far as these presentations were concerned. According to this NASA official: "Experiment P-15 was discussed by all contractors, although it has been dropped by the Air Force. Designs ranged from 55" to 100" aperture optical telescopes."[37] On 1 May the contractors' final documented reports were completed and submitted to the Air Force.

The Source Selection Board promptly began its evaluation of the MOL laboratory vehicle proposals. Simultaneously, AFSC began drafting its program recommendations and, on 15-16 May, gave a preliminary briefing to Dr. McMillan. Afterwards, he suggested that certain additional data be incorporated into the presentation and he asked for another briefing.

Meanwhile, on the basis of progress reported to him on the "black" payload investigations, McMillan issued new program guidance to the Directorate of Special Projects. In a message on 20 May, he advised:

The development of optical technology leading to optical systems capable of improved resolution is the primary objective of the MOL program. The initial objective is to develop and demonstrate at the earliest time an operationally useful high resolution

---

** Also in attendance were members of the Secretariat, including General Evans and Col David L. Carter and Maj D. S. Floyd.

†† This brought total approved FY 1965 MOL expenditures to $20,300,000.

manned optical reconnaissance system capable of achieving at least {better than 1 foot} ground resolution. Other mission applications of the MOL program such as sea surveillance, COMINT and ELINT are secondary and may be accommodated if no appreciable compromise to the orbital vehicle which meets the primary objective is required.[38]

Reviewing the current status of optical technology, Dr. McMillan noted that there was considerable skepticism about the possibility of fabricating mirrors in diameters greater than 60 inches. He therefore suggested that the initial MOL flights "be predicated on a mirror of approximately 60 inches of conservative design... to operate with or without a tracking mirror." He also recommended that General Greer initiate related development work, including advanced development of larger optical systems (with diameters {greater than 60} inches) which at a future date might be used in the MOL program. In addition, he provided guidance for the award of additional "black" contracts to Itek and Perkin-Elmer[‡‡].[39]

On 26 and 29 May, in advance of meetings with the MOL Policy Committee and with Drs. Brown and Hornig, the Air Force Under Secretary sat in on several more "dry run" presentations by AFSC and afterwards suggested some additional changes for the MOL Policy Committee briefing on 1 June. In attendance at this latter briefing were Zuckert, McMillan, Flax, Marks, McConnell, Blanchard, Schriever, and Ferguson. General Evans opened the presentation with a brief resume of the recent study activity and stated the principal conclusion—that "a large optical telescope could be built for manned orbital operations, that man could plan a useful role in the alignment and checkout of large structures in orbit, and that the program could be justified in terms of the high resolution obtainable {1 foot or less} through employment of man in orbit."[40]

General Evans was followed by Dr. B. P. Leonard of the Aerospace Corp., who reviewed overall MOL, vehicle characteristics and compared the results to be obtained in high resolution optical reconnaissance from the manned versus unmanned modes. The basic argument in favor of MOL, he said, was that the unmanned optics currently being flown were able to achieve ground resolution {better than one foot} and that, at best, an unmanned

system could approach {something less} as a limit[§§]. However, the latter would require a major advance in the state of the art, whereas in the manned mode {one foot} or better could be achieved "with existing technology, with growth improvement toward {better} resolution." Following Leonard's statement, General Maxwell briefed the Committee on the results of the MOL laboratory vehicle competition. The Source Selection Board had rated the four participating contractors in the following relative order of merit: Douglas Aircraft Company, General Electric Company, Boeing Aircraft Company, and Lockheed Missile and Space Company. General Maxwell stated that the first two companies showed a clear margin of superiority over the last two.[41]

The Committee consensus was that justification for the program should, as proposed, emphasize the higher resolutions that could be obtained from the manned system. The Committee approved submission of a proposed USAF MOL program to OSD but with certain changes to highlight the primary mission. A series of top level briefings followed. Dr. Brown was briefed on 2 June and the President's Science Advisory Committee on the 10th. Dr. Seamans and other NASA officials were briefed on the 23rd. Dr. McMillan, who was quite pleased by these presentations, congratulated General Schriever "on the high quality of the proposed MOL program recently submitted for approval." He said it was evident from the excellence of the final product that "much creative imagination, intelligent analysis, and plain hard work" had gone into it. The final briefings to the MOL Policy Committee and to Dr. Brown and PSAC, were, he said, of outstanding overall quality and "auger well for the future conduct of the MOL program."[42]

Several other factors at work during the first half of 1965 also tended to auger well for the program. One was the dramatic "spacewalk" on 18 March by Soviet Cosmonaut Pleksei A. Leonov, who maneuvered outside his space capsule for about eight minutes. No one was more impressed by Leonov's extravehicular activity—another Soviet "first"— than members of the House Committee on Government operations. In a report on U.S. space activities released to the public, the committee strongly recommended that Secretary McNamara "without further delay, commence full-scale development of a manned orbital laboratory (MOL) project."

---

‡‡ On 15 June 1965 General Martin advised McMillan that contractual actions had been initiated with Itek and Perkin-Elmer. (Msg. 8045, Martin to McMillan, 15 June 1965.)

§§ The figures given by Dr. Leonard for the manned system's capabilities were goals, not actual products. A later study (1967) of the various products obtained by the unmanned system showed that the best ground resolution ever obtained was 15.5 inches. Most flight produced results of 30 inches or more.

The House Committee said its recommendation was made "without prejudice to NASA's future requirements for manned space stations," fully recognizing that such vehicles would serve important civilian as well as military space purposes. But, concluded the Committee, the "compelling need of the moment is to overcome a military lag in space technology."[43]

ENDNOTES

1. Memo (S), Brown to McMillan, 4 Jan 65, subj: MOL.

2. Ibid.

3. Msg 7559 (S), SAFUS to SSD (For SSGS), info AFSC, CSAF, SSD, 8 Jan 65.

4. Msgs Whig 2240 and 2241 (TS-DORIAN), McMillan to Greer, 9 Jan 65.

5. Msg (TS-DORIAN), Martin to McMillan, 22 Jan 65; Whig 2334, SAFSS to Col Ford, 22 Jan 65.

6. Msg (TS-DORIAN), Greer to McMillan, 30 Jan 65.

7. Memo (TS-DORIAN), McMillan to Brown, 5 Feb 65, subj: MOL Security Guide.

8. Security Policies and Procedures for the DoD Manned Orbiting Laboratory, signed by Dr. McMillan, 19 Feb 65.

9. Ltr (C), Schriever to McMillan, 18 Aug 64.

10. Memo (U), Col Schultz to Gen Ferguson, 25 Sep 64, subj: MOL.

11. Ltr (C), Schriever to McMillan, 30 Oct 64.

12. Memo for Record (S-DORIAN), Brig Gen Evans, 6 Jan 64, subj: Mtg with Dr. McMillan on MOL.

13. Memo (C), Vance to McMillan, 7 Jan 65.

14. SAF Order No. 117.4, 18 Jan 65, subj: Spl Asst for MOL Program Mgt.

15. Memo (U), McMillan to Seamans, 5 Feb 65.

16. Memo (U), Zuckert to CSAF, 19 Jan 65.

17. Ibid.

18. Memo (C), Vance to McMillan, 25 Jan 65.

19. Minutes (S), Space Panel Mtg 65-3, 9 Feb 65.

20. Ltr (S), McMillan to Seamans, 8 Jan 65.

21. Ibid.

22. Ltrs (S), Brown to Seamans, 11 Jan 65; Seamans to Brown, 21 Jan 65; Brown to Seamans, 29 Jan 65.

23. Minutes (S), Mtg of NASA-DoD Study Team, 13 Jan 65.

24. Ltrs (S), Evans to Garbarini, 12 Feb, 26 Feb, 28 Feb, 1 Mar, 8 Mar, 8 Apr 65; Garbarini to Evans, 2 Mar, 10 Mar 65.

25. Ltr (S), Seamans to Brown, 17 Mar 65.

26. Minutes (C), E.N. Hilburn, MOL Study Cmte Review, 30 Apr 65.

27. Ltr (S-SAR), Seamans to Brown, 5 May 65; Interview, Berger with Yarymovych, 25 Jul 67.

28. Minutes (C), NASA MOL Study Cmte Review, 12 Apr 65.

29. Excerpts, SecDef Press Conference, 23 Jan 65; DoD News Release No. 52-35, 23 Jan 65.

30. Press Release, Decision on the MOL and Related Matters (U), 25 Jan 65.

31. Mr. Teague's letter was dated 25 Mar 65. He also is quoted in Lawrence J. Curran, "Committee Asks LBJ for MOL Ruling," *Missiles and Rockets*, 12 Apr 65, p 16.

32. Ltr (U), McNamara to the Vice President, 7 Apr 65.

33. Ltr (S), AFSC to Distributees, 24 Jan 65, subj: Request for Proposals No. RFP SSD-04-695-151, MOL Sys (Prelim Design Study); Minutes (C-SAR), AF MOL Policy Cmte Mtg 65-1, 25 Feb 65.

34. DoD News Release No. 113-65 (U), 1 Mar 65, subj: AF Selects Contractors for MOL Studies.

35. Msg 6942 (TS-DORIAN), Greer to McMillan, 30 Jan 65.

36. Memo (C), McMillan to DDR&E, 22 Mar 65, subj: MOL Funds for Sustaining Contractor Efforts; Memos (C), Brown to SAF, 5 May 65, subj: Approval of FY 65 RDT&E MOL Proj; Evans to SAF and Asst SAF (R&D), 8 Jun 65, subj: MOL Funds for Sustaining Contractor Efforts.

37. Memo for Record (S), Douglas R. Lord, NASA, 12 Apr 65, subj: Mid-Term Review of MOL Contracted Studies.

38. Msg Whig 3061 (TS-DORIAN), McMillan to Greer, 20 May 65.

39. Ibid.

40. Memo (S), Evans to Dep Exec Asst to SAF, 4 Jun 65, subj: Status Book for SAF (may Rpt); Memo for the Secretariat Record (S-DORIAN), 4 Jun 65, subj: Proceedings of AF MOL Policy Cmte Mtg 65-2; Minutes (U), AF MOL Policy Cmte Mtg 65-2, 1 Jun 65.

41. Ibid.

42. Ltr (U), McMillan to Schriever, 22 Jun 65.

43. See the 13th Report of the Cmte on Govt Operations, 89th Cong, 1st Sess., Government Operations in Space, 4 Jun 65.

MOL ASSEM...
INTEGRATION BUILD...

MATERIAL

PROPERTY OF THE
UNITED STATES GOVERNMENT

IF FOUND, DO NOT OPEN

BASELINE MOL MANNED MODE

USAF MOL/KH-...

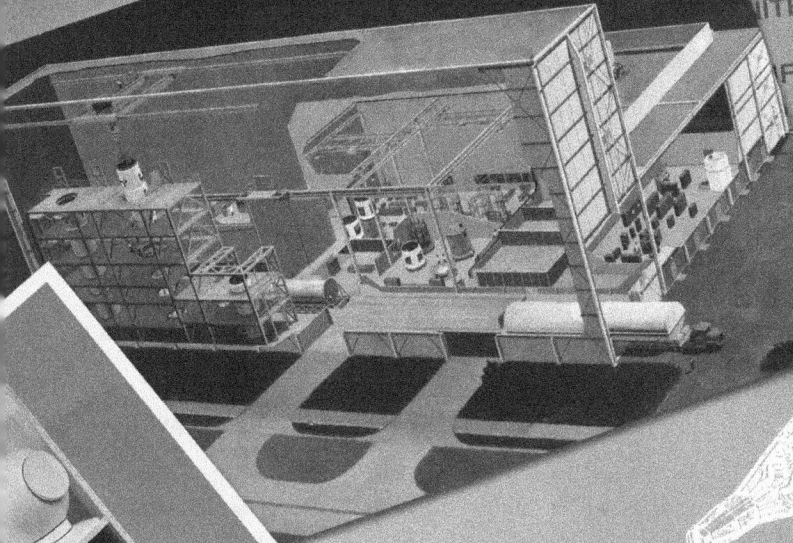

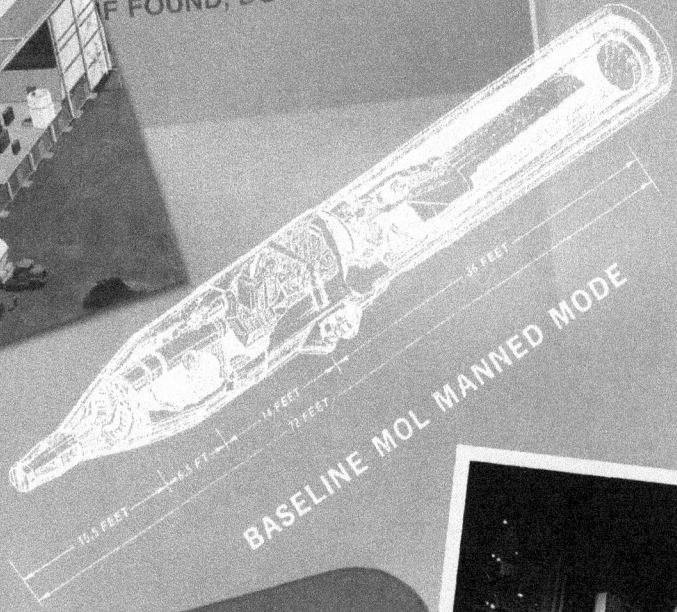

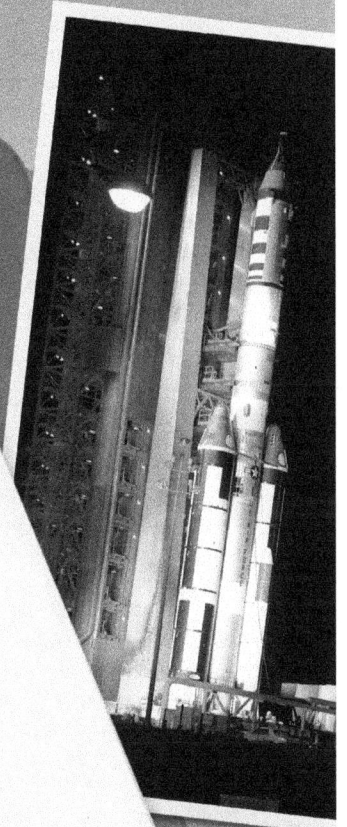

THE DORIAN FILES REVEALED:
A COMPENDIUM OF THE NRO'S MANNED ORBITING LABORATORY DOCUMENTS

CHAPTER VIII:
THE MOL PROGRAM DECISION
25 AUGUST 1965

THE MOL PROGRAM DECISION 25 August 1965

For several months prior to the MOL Policy Committee meeting of 1 June, General Evans and his staff had been collecting data and drafting papers to support an Air Force recommendation to OSD that they proceed with the program. This work was well along when the Committee authorized submission of a formal proposal to Secretary McNamara. Whereupon, during June 1965, General Evans, Col Lewis S. Norman, Jr., Lt Col Richard C. Randall, Maj Robert Spaulding, Dr. Yarymovych, and others intensified their writing efforts, completing a half-dozen drafts before a final docent was approved and forwarded on the 28th to OSD. This document, the culmination of 18 months of Air Force studies, analyses, and efforts going back to December 1963, consisted of a 14-page memorandum and eight lengthy appendices. One of the latter—a 9-page paper entitled "The Potential of Very High Resolution Photography"—was written by McMillan, who also reviewed and recast the covering memorandum to McNamara to give greater emphasis to the importance of acquiring a high resolution photographic capability.[1]

THE AIR FORCE PROPOSAL

In this memorandum, the Air Force recommended that DoD proceed with development of a manned orbiting laboratory using the Titan IIIC booster and the Gemini B spacecraft. It proposed a six-vehicle launch program— one unmanned and five manned—with the first manned flight taking place in late calendar year 1968 and the last one in early 1970. The cost of the program was estimated at $1,653 million. The Air Force advised the Defense Secretary that it would place primary emphasis on development, demonstration, and use of a manned optical reconnaissance system to provide resolutions of {one foot} or better on the ground. It expressed the belief that this order of resolution could be attained using an optical system of relatively conservative design having an aperture of 60 inches; such a system vas considered to be the primary payload for the early flights. It said parallel development also would be undertaken along a less conservative approach, leading to the possibility of a system of perhaps {larger than 60} inches aperture, capable of a resolution of {better than one foot} on the ground.[2]

The Air Force emphasized that the optics and optical technology to be developed for MOL would be directly applicable to unmanned systems. It planned to pursue development of elements such as image trackers, which were crucial to the performance of large unmanned systems. It said, however, that the development of MOL would produce a resolution of {better than one foot} much sooner" and with a higher probability of initial success" than a development based on an unmanned configuration. At {better than one foot} resolution, using current cost estimates, it predicted the manned system would be about as productive per dollar as an unmanned system "even setting aside the greater development difficulties and risks attaching to the unmanned system."[3]

Concerning the basic need for MOL, the Air Force argued that it was vital to have a high resolution photographic capability to acquire not only technical intelligence but also data on "tactical objectives" during times of international crisis. It noted that during the 1962 Cuban missile crisis, the United States undertook repeated photographic flights at very low altitude in order to identify details of military equipment "and in particular to determine the country of origin of some of this equipment. The credibility of findings based on such high resolution photography, the Air Force said, "can be crucial. Certainly it was essential, in the case of the Cuban crisis, for President Kennedy to have pictures whose credibility vas beyond his doubt, before he could make some of his crucial decisions."

The Air Force, consequently, concluded that there was a basic national need for satellite reconnaissance "at {one foot} resolution or better," that the manned program offered "the quickest and most assured way of reaching that goal," and that the MOL vas almost essential "if we are ever to develop systems manned or unmanned, having resolutions much better than {one foot}.[4]

In addition to the above formal program proposal: the Air Force submitted a second paper to OSD, describing the MOL management structure and plan to be adopted for the project definition phase, the rationale for selecting two contractors (Douglas and General Electric) instead of one to carry the development forward, NRO's relationship to MOL, and proposed security and public information policies.*

---

* See Chapter IX.

DR. HORNIG APPROVES, WITH QUALIFICATIONS

Several days after the Air Force proposal was received by OSD, the President's Science Advisor submitted his important evaluation of it to Secretary McNamara. As noted earlier, the Defense Secretary made it clear that before a MOL development would be authorized, Dr. Hornig, the Director of the Budget, ad he would have to agree that the project was worth the cost and would not duplicate any other approved space program.

To help evaluate the USAF proposal, Dr. Hornig earlier asked Dr. Purcell of the PSAC Reconnaissance Panel to submit a report to him and he also discussed MOL with Land[†]. After considering their comments, which he discussed informally with McNamara, on 30 June 1965 he forwarded his views to the Defense Secretary. To begin with, he said that since there was "very great value" in obtaining the highest possible photographic resolutions: he would be willing to pay a great deal to acquire a system that possessed such a capability. The Air Force, he said, had done "an exceedingly thorough analysis of both the manned and unmanned system alternatives." It had:

> ...documented a persuasive argument that: for equal total weights and total volumes, the manned system does have an advantage over the unmanned system and can be expected to provide a higher average resolution at an earlier time than the unmanned system. I therefore would support approval of the MOL program. I would point out that we should expect difficult technical problems in building the mirrors necessary for such a system. A capability is yet to be demonstrated. However, I believe that this risk is acceptable.[5]

But, Dr. Hornig noted, there were certain points to be noted about the USAF case. That is, he said, the Air Force's conclusions about the relative merits of manned versus unmanned systems were based on certain assumptions about existing technology which caused the latter to fall short by comparison. For the very sophisticated type of unmanned system being discussed, relatively little effort had been devoted to solving the problems inherent in automatic pattern recognition, image motion compensation, and precise pointing to

the accuracy required. Dr. Hornig said he believed that if sufficient competence, imagination and effort were devoted to the development of the necessary automatic subsystems, "the margin that now exists in favor of the manned system could in time be largely eliminated."

He also raised a number of related questions concerning manned versus unmanned systems. He said while available evidence "makes us reasonably confident that man is physiologically and psychologically capable of performing as required by MOL," this capability had not yet been demonstrated and it was possible that the flight tests would show that the manned system would not perform as well as predicted. Also, he thought it reasonable to anticipate the possibility "that either public reaction against MOL as an invasion of privacy or international opposition to manned overflights may prevent the use of a manned system." He said:

> Although both these risks are acceptable from the financial standpoint and should not therefore prevent initiation of the development of the MOL, they are serious enough politically to warrant our taking action to provide for the eventuality that an unmanned, rather than manned system will be required. In addition, it seems quite possible that' from an operational standpoint, an unmanned system will eventually be desired to complement the manned system by performing the more routine reconnaissance missions or be available in special circumstances, such as, for example, in the case of threats against the system by the other side.[6]

For these reasons, Dr. Hornig recommended that a major effort be made as an integral part of MOL, to develop subsystems which could be used for a high resolution unmanned system. There was no reason, he believed, why an immediate effort on the critical automatic subsystems "should perturb the progress of the MOL development program in its initial phases." In brief, he supported MOL program approval, provided the Air Force undertook "to concurrently develop an unmanned operational capability for the system."

In a separate report to the President[‡], Dr. Hornig advised that the MOL Program would provide a substantial increase in U.S. reconnaissance capability "by developing a system which could, for example, resolve so well that we could even discover {specified objects} in our overflight photography." Consequently, he informed the Chief Executive he was recommending that they initiate development of MOL and had so advised Mr. McNamara. However, he suggested that,

---

† Dr. Land submitted a lengthy paper on the USAF proposal to Hornig on 18 August. See Chapter X.

‡ Contained in a draft memorandum to the President, dated 30 June 1965, which he submitted to McNamara for review. He indicated to the Secretary he hoped to deliver the memo to the President the following day. (Ltr, Hornig to McNamara, 30 Jun 65.)

if they proceeded, they should be prepared to assume "serious political risks" when the flight tests began. However, he said:

> We should give consideration at the highest level to the contingencies which may occur so that one day we are not caught by surprise by the intensity of the reaction abroad as we were when the U-2 was shot down over the USSR. It is true that unmanned satellite reconnaissance has been used and accepted by both sides. However, it is possible that manned satellite surveillance could be considered 'overflight' with all its connotations. It is also possible that MOL will be construed by the USSR as a weapons system in space capable of launching bombs from orbit. We must certainly consider how likely it is that such an interpretation could be made, whether the leaders of the USSR could tolerate the existence of MOL if such an interpretation is made, and what their reaction might be...[7]

On the other hand, Dr. Hornig noted that manned activities in orbit had become somewhat routinely accepted over the past years and it was possible that MOL would also achieve acceptance if introduced to the public in a careful manner. If so, it might make a substantial contribution to the recognition of manned observation and surveillance as a normal mode of international behavior. He therefore recommended to the President that high level political oversight be given to: (1) the extent to which the public should be informed about MOL and the method by which the program was announced "so that we establish, right from the start, a picture of MOL which will give it the best chance of gaining acceptance by the international community," and (2) the contingencies that might arise if the flights were not accepted and the detailed plans for meeting those contingencies if they occurred.[8]

THE BUDGET BUREAU EXPRESSES DOUBT

After Sectary McNamara's staff received the Air Force's MOL program proposal and Dr. Hornig's comments, Colonel Clarence L. Battle, Dr. Hall's assistant in ODDR&E, began composing a memorandum on the subject for Secretary McNamara to send to the President. Col Battle's draft memorandum was reviewed by Dr. Brown, who made a number of changes in it and then forwarded copies to Mr. Charles L. Schultze[§], the Director of the Budget, Adm. William Raborn, Director of the CIA, and Vice President Humphrey, chairman of the National Aeronautics and Space Council.

On 8 July 1965 Schultze forwarded his comments to the Vice President and McNamara and questioned whether MOL's superiority as a reconnaissance system, as compared to a possible unmanned system, was worth the $1 billion of additional development costs and $200 million of additional annual operating costs. "I think," he said, "we must satisfy ourselves beyond a reasonable doubt that the probable superiority of the manned over the unmanned system is likely to be worth the additional cost before recommending to the President that the program proceed."[9]

The budget chief noted that the existing unmanned systems "have made and can continue to make essential, significant and spectacular contributions to intelligence and national security." He pointed out that the latest version, GAMBIT-3—which was under active development—was expected to provide {better than one foot} resolution at a development cost of same $200-300 million while an even better product {...} might be obtained with an improved unmanned system at a development cost of $600-800 million. On the other hand, he said, MOL would cost $1.6 billion more and it was not clear to him that {the promised} resolution photography had that much additional value for national security.

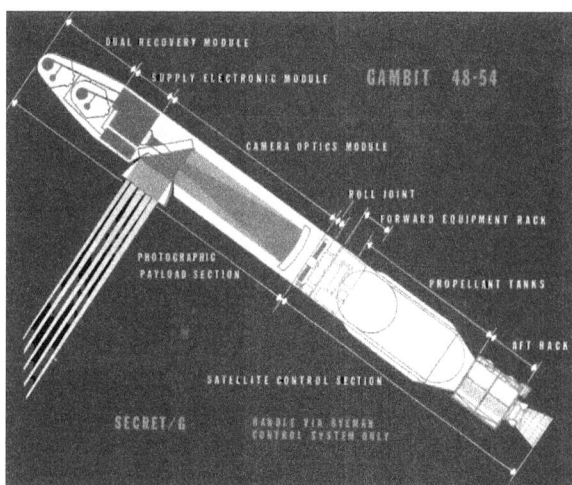

Figure 39. Gambit-3
Source: *CSNR Reference Collection*

---

§   Schultze succeeded Kermit Gordon on 1 June 1955

Schultze consequently concluded that—until the points he had raised were clarified—"there is no clear need to proceed with the manned system as now proposed." If there was a requirement to develop a system for obtaining higher resolution than GAMBIT-3, he thought they should proceed with development of an unmanned system. In this connection, he cited Dr. Hornig's comment that if sufficient competence, imagination, and effort were applied, unmanned systems could probably be developed with resolution capability approaching that expected from MOL.[10]

Later, however, after he was advised that further DoD studies indicated that the difference in cost between a manned and unmanned system would not be $300-1,000 million as originally thought but more nearly $300-400 million, he withdrew his objections. However, he requested, and Secretary McNamara agreed, that if studies during the next six months showed a cost difference substantially greater than $300-400 million, the MOL should be reappraised end a new decision made whether the additional benefits of the manned system were worth the costs.[11]

STATE DEPARTMENT AND CIA VIEWS

On 9 July 1965 the Space Council met to review the draft McNamara memorandum to the President, the problem of security and information handling of MOL, and a proposed public announcement—submitted by Mr. Webb—which the President might wish to consider. During this meeting the Space Council identified certain tasks for implementation prior to any public announcement on MOL, one being to coordinate with the State Department. Subsequently, Dr. Brown forwarded a copy of the McNamara memorandum to the State Department along with a proposed policy paper setting forth proposed information controls.

Figure 40. Dean Rusk
Source: CSNR Reference Collection

In response, on 16 August Secretary of State Dean Rusk advised McNamara that, while some international problems would likely arise, he did not consider these of sufficient negative importance to warrant advising against going ahead. He said, "if you are fully satisfied the project is justified in terms of potential contribution to national defense, I have no objection to your going forward with the recommendation to the President." Rusk said further that if a decision was made to proceed, it would be essential to maintain very tight control of the project and to carefully handle all publicity "if we are to succeed in safeguarding the sensitive aspects of MOL and deal effectively with whatever international problems arise." Commenting further on the information problem, the Secretary of State stated:

> I consider it most important that to the extent it can be controlled, everything said publicly about the MOL project emphasize its experimental and research nature, and that statements and implications that MOL constitutes a new military operational capability in space, or an intermediate step toward such a capability, be rigorously avoided. It would be useful to this end if fully knowledgeable people in this Department would work closely and continuously with your own people in devising detailed press and publicity handling guidelines, reviewing the text of key statements or releases, etc.[12]

The Director of the CIA also gave a general, if cautious, endorsement to the MOL Program. Admiral Raborn said, "It is in the interest of the United States to obtain the highest resolution of photographic coverage feasible over those areas of intelligence interest designated by the United States Intelligence Board, provided that such highest resolution will of course have to be weighed against the relative factors of cost, time, and relative importance of intelligence which could be obtained in an optimum balance of these considerations."[13]

Figure 41. William F. Raborn
Source: CSNR Reference Collection

MCNAMARA RECOMMENDS MOL PROGRAM APPROVAL

Having coordinated with all key individuals and agencies, Dr. Brown and Colonel Battle put the finishing touches to McNamara's memo to the President. The Defense Secretary reviewed the final draft on 24 August, made several minor language changes, and that same day carried it over to the White House where he recommended to the President that they proceed with MOL project definition beginning in fiscal year 1966.[14]

McNamara noted that Congress was currently in process of appropriating $150 million for the program (requested the previous January) and that he had previously indicated he would defer release of funds until such time as studies of the nature and value of the problem were satisfactorily completed. These studies, he told the President, had been completed and—based on his review of their conclusions—he now recommended release of the $150 million, initiation of a contract definition phase, and that the program proceed toward the following goals:

a. Semi-operational use beginning in late 1968 to secure photographs of {better than one foot} resolutions of significant targets. This is {significantly} better than the best satellite photography we are now obtaining, and {largely} better than the best U-2 photographs or the G3 satellite system, now under development, from which we expect photographs in about 15 months.

b. Development of high-resolution optical technology and systems for either manned or unmanned use. This technology will provide the {target} resolution and be aimed at ultimately even better resolution {than the initial target resolution}.

c. Provision of a facility for the development, test and use of other potential military applications such as SIGINT collection, radar observation and ocean surveillance, as the utility and feasibility of such applications became established.

d. Provision of an experimental program for determination of man's ability in assembling large structures, and in adjusting, maintaining and processing the output from complex military equipment in space.[15]

McNamara recommended that the MOL program be operated under the NRO security guidelines which already existed for military space projects. The idea, he said, was "to help avoid provocation in the international arena, and to forestall initiation of international action that might prevent the United States from using satellites for reconnaissance." He reported that DoD planned to pursue a modest and low key public information program and that the announced mission of MOL would continue to be expressed solely as "the investigation and development of orbital capabilities, manned and unmanned, associated with national defense."

The Defense Secretary advised the President that he had received the concurrence of Secretary Rusk, Admiral Raborn, Dr. Hornig, and Mr. Webb, and that Vice President Humphrey also endorsed program go-ahead. The Director of the Bureau of the Budget, he reported, had withdrawn his original objection, subject to a future program reappraisal of costs. McNamara said further that, in his view, there was a vital national need for reconnaissance photography with resolutions of {the planned resolution} or better. He noted that during the Cuban crisis the United States had made a special reconnaissance effort "to acquire pictures having the detail and the credibility that were necessary to verify and to convince others of the nature of the military activity in Cuba." In other future situations, he thought it might be important to accomplish these same ends. With {target} resolution, the nation also would be able to assess such military factors as the {...} nature of various Russian anti-missile deployments.[16]

The defense chief advised that he had incorporated several of Dr. Hornig's suggestions concerning an unmanned system¶ and that designs of the new devices needed for the unmanned operational mode would be pursued. He said:

It is my intention that the system will be designed so that it can operate without a man. It will operate somewhat differently, however, (and with improved overall effectiveness) with a man. Whether

---

¶   Dr. Hornig met with Dr. Brown on 23 August and the two men agreed that the Air Force would pursue development of the automatic system simultaneously with the manned MOL.

the system will produce poorer average resolution without a man depends on how well some of the ideas for such functions as automatic focusing and adjustment, automatic navigation and image motion compensation work out. But in any event, it is agreed that the man's ability to select targets, to override the automatic controls when they function less well than expected, to choose data for prompt transmission, will improve the overall utility of the data.. Furthermore, the presence of man in the development phase can be expected to shorten the development and improve the capability of the unmanned version of the system.[17]

THE PRESIDENT'S DECISION AND PUBLIC REACTION

McNamara's recommendation to the President, it should be noted, was made against the backdrop of six months of U.S. achievements which clearly proved that man would be able to function effectively in space. On 23 March, NASA launched its first two-man Gemini, successfully recovering the spacecraft and astronauts after three orbits of the earth. On 3-7 June, during its second Gemini flight, Air Force Maj Edward H. White became the first American to maneuver outside his space vehicle. White's 22-minute "space walk" exceeded that by Soviet Cosmonaut Leonov of the previous March. Finally, on 21 August 1965, NASA launched its third Gemini into a flight which shattered all existing orbital endurance records (astronauts L. Gordon Cooper, Jr., and Charles Conrad spent nearly eight days in a weightless state). On 24 August—the same day McNamara made his MOL recommendation to the President—Cooper and Conrad performed a number of military experiments which included sighting and photographing a Minuteman ICBM launched from Vandenberg AFB.

There was little doubt the President would accept the Secretary's recommendation. President Johnson decided, however, that he would personally make the announcement. The following day, 25 August, he opened a televised White House press conference with the following statement:

After discussion with Vice President Humphrey and members of the Space Council as well as Defense Secretary McNamara, I am today instructing the Department of Defense to immediately proceed with the development of a manned orbiting laboratory.

This program will bring us new knowledge about what man is able to do in space. It will enable us to relate that ability to the defense of America. It will develop technology and equipment which will help advance manned and unmanned space flight and it will make it possible to perform very new and rewarding experiments with that technology and equipment...

The Titan 3C booster will launch the laboratory into space and a modified version of the NASA Gemini capsule will be the vehicle in which the astronauts return to earth...

We believe the heavens belong to the people of every country. We are working and we will continue to work through the United States—our distinguished Ambassador, Mr. [Arthur] Goldberg is present with us this morning—to extend the rule of law into outer space.

We intend to live up to our agreement not to orbit weapons of mass destruction** and we will continue to hold out to all nations, including the Soviet Union, the hand of cooperation in the exciting years of space exploration which lie ahead for all of us.[18]

The initial press reaction to the President's announcement was critical. The New York Times, after commenting that the Presidential decision was "a fantastic, terrifying" measure of arms preparation, several days later editorialized that it had spread "disquiet across the world... Assuming that Russia has similar technical capacity to produce orbiting laboratories, outer space from 1968 onward could be full of manned spaceships with awesome potential." The Washington Post worried about assignment of the project to the Air Force which it said was committed to "total secrecy." Such secrecy,

---

** Both Moscow and Washington agreed to abide by a U.N. resolution, adopted 13 December 1963, which called upon all states to refrain from orbiting nuclear weapons or any other kinds of weapons of mass destruction.

the paper argued, "is bound to arouse international suspicions and alarms, particularly since the flights will be over Soviet territory."[19]

Some 45 private citizens expressed their opposition to the decision in letters they wrote to various administration officials, including the President, Secretaries McNamara and Rusk, and Administrator Webb. Their general theme was that the MOL would extend the arms race into space, in contradiction to U.S. policy favoring the use of space for peaceful purposes. A number of Congressmen also objected. Two feared that MOL might encourage a military space race, five argued that the project should be given to NASA, and another complained that it would lead to duplication of manned launch facilities on both coasts.[20]

Not all editorial comment (aside from the technical and professional journals: which generally approved the MOL announcement) was negative. For example, The New Republic, saw a positive aspect to the program:

```
It is possible that MOL will
demonstrate the feasibility of a
few American and Soviet space men
in their respective spacecraft
operating a continuous space watch.
If it does, and if both nations
exercise restraint, it could have a
stabilizing effect, as have our mutual
unmanned reconnaissance satellites.
If man can be an efficient observer in
orbit for extended periods, the time
may come when the U.S. should invite
the United. Nations to maintain a
continuous space control, with a
multinational crew to warn of any
impending or surprise attack.[21]
```

The Soviet reaction, as expected, was critical. Tass, the Russian news agency, commented pointedly that some of the orbiting laboratories would be launched from Vandenberg AFB, the firing site, it said, "for hush-hush spy satellites that fly over the territories of socialist countries several times a day." On 9 September Reuters reported the remarks of Col Gen Vladimir Tolubko, Deputy Commander-in-Chief of Soviet Strategic Rocket Forces. Echoing a West German news account which speculated that MOL would be able to bombard the earth with nuclear weapons, General Tolubko declared: "Now the Pentagon wants to use space laboratories not only for espionage but also to accomplish direct combat tasks."

Several weeks later, Izvestia published a lengthy article by Col M. Golyshev, not further identified, who attacked not only MOL but NASA's Gemini program. He reported that Astronauts Cooper and Conrad in Gemini 5 had carried out 17 military experiments, photographed missile launchings from Vandenberg, and performed "visual observations" of ground installations, in particular, the White Sands Proving Ground. He complained that Gemini 5 was used to check out "the possibilities of intercepting artificial earth satellites and carrying on reconnaissance from space." Colonel Golyshev concluded that MOL would be suitable "for creating command posts in space, intercepting foreign satellites and making reconnaissance. Such a wide range of combat capabilities gladdens the Pentagon strategists."[22]

To the distress of the U. S. Information Agency (USIA), the foreign press for the first time began to ask critical quest ions about the peaceful orientation of the American space program. Previously, the Mercury and Gemini flights had produced highly favorable publicity for the United States. The State Department, somewhat disturbed by the change in tone, dispatched an airgram in early September 1965 to all diplomatic posts. It included a copy of the President's MOL statement and emphasized that the new project had no "weapons in space" or "bombs in orbit" aspect whatever, and was neither illegal nor different in motivation and purpose from other defense research projects.[23]

On 7 September the Department also convened a meeting of an interagency committee (attended by State, DoD, CIA, and USIA representatives) to discuss the overseas reaction to Gemini 5 and the President's announcement. The Defense Department was represented by Lt Col Daniel C. Mahoney and Maj Robert Hermann[††]. The latter had been assigned as an information advisor to General Evans several months earlier.

The USIA official summarized for the committee the world press reaction to the military implications of the Gemini 5 flight and the MOL program, and he suggested a high policy statement was needed to counteract the unfavorable news coverage. In response, Major Hermann summarized DoD's public affairs policy for the military space program and noted that the National Space Act of 1958 had placed specific responsibility for military space activities on the Department of Defense. The MOL, he continued, did not represent a new policy by the U.S. government but was a logical step in providing for defense of the nation. As for countering unfavorable news coverage, he noted that Dr. Edward C. Welsh, executive secretary of the Space Council, had made a number of widely publicized speeches which possibly might satisfy the requirement for a high level statement of national policy.[24]

†† Colonel Mahoney was from the Office of the Assistant Secretary of Defense for Public Affairs. Major Hermann from the USAF Office of Information.

Sometime after this meeting[‡‡], Major Hermman met with members of the Space Council staff and discussed possible approaches to countering criticism of the program. Other officials also apparently contacted Dr. Welsh, who proved quite agreeable to restating U.S. national policy on MOL. On 28 October 1965—in an address to the American Ordnance Association—he cited the MOL as an example "of a highly valuable exchange of technology and experience by two operating agencies of the government." And he said further:

> Since I have mentioned the Manned Orbiting Laboratory, it is worth pausing right now to challenge forthrightly those who have asserted or intimated that it has something to do with a weapons race. We expect misrepresentations of that sort to came from unfriendly countries and sometimes from ignorant domestic critics. However, I was disappointed to find that a few otherwise well informed publications and individuals have asserted that MOL is a weapons carrier and a project contrary to our peaceful progress in space.

I assert as positively as I can that MOL is not a weapons system, is not a means by which aggressive actions can be perpetrated, and is in no way in conflict with the established policies, objectives, or methods of the United States. Rather, it is a program that will increase our knowledge of man's usefulness in space and will relate that ability to our national defense.[25]

[‡‡] The State Department later advised OSD that world press reaction to the MOL announcement, "while not laudatory, has not been as bad as it might have been."

ENDNOTES

1. Interview, Berger with Gen Evans, 29 Nov 67.

2. Memo (TS-DORIAN-GAMBIT), Zuckert to McNamara, 28 Jun 65, subj: MOL Program.

3. Ibid.

4. Ibid.

5. Memo (S-DORIAN), Hornig to SecDef, 30 Jun 65.

6. Ibid.

7. Memo (S-DORIAN), Hornig to the President, 30 Jun 65.

8. Ibid.

9. Memo (TS-TK DORIAN-GAMBIT), Schultze to Vice President Humphrey, 8 Jul 65, subj: MOL.

10. Ibid.

11. Cited in Memo (TS-DORIAN-GAMBIT), McNamara to the President, 24 Aug 65, subj: Manned Orbiting Laboratory.

12. Ltr (S), Rusk to McNamara, 16 Aug 65.

13. Cited in Memo (TS-DORIAN-GAMBIT), McNamara to the President, 24 Aug 65, subj: Manned Orbiting Laboratory.

14. Interview, Berger with Col Battle, 9 Nov 65.

15. Memo (TS-DORIAN-GAMBIT), McNamara to the President, 24 Aug 65, subj: Manned Orbiting Laboratory.

16. Ibid.

17. Ibid.

18. *N.Y. Times*, 26 Aug 65.

19. *N.Y. Times*, 29 Aug 65; *Wash Post*, 27 Aug 65.

20. Memo (U), Col J. E. Stay, Office of Information, To Asst Sec Def (PA), 1 Nov 65, subj: Resume of MOL Letters.

21. *The New Republic*, 11 Sep 65

22. *N.Y. Times*, 10 Sep 65; *Missiles & Rockets*, 22 Nov 65, p 17.

23. Memo for Record (S), Lt Col Daniel C. Mahoney, OASD/PA, 8 Sep 65, subj: MOL World Press Reaction.

24. Ibid; Interview, Berger and Maj Hermann, 20 Oct 65.

25. NASA SP-4006, *1965 Chronology on Science, Technology, and Policy*, p 494

MOL ASSEM...
INTEGRATION BUILD...

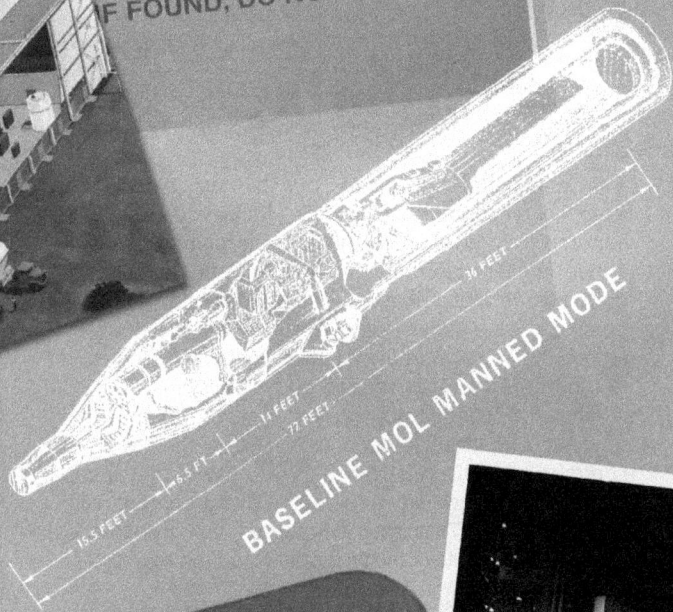

MATERIAL

PROPERTY OF THE
UNITED STATES GOVERNMENT

IF FOUND, DO NOT OPEN

BASELINE MOL MANNED MODE

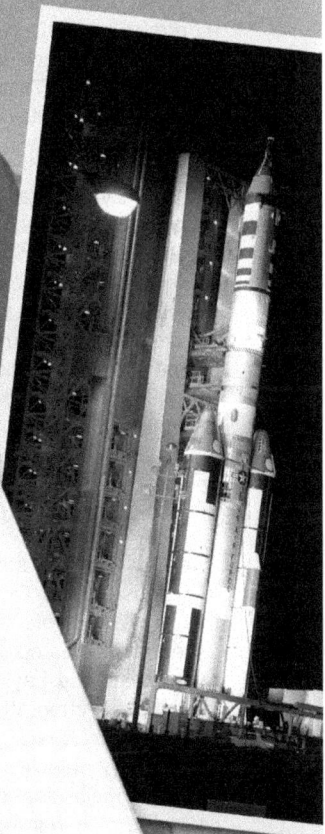

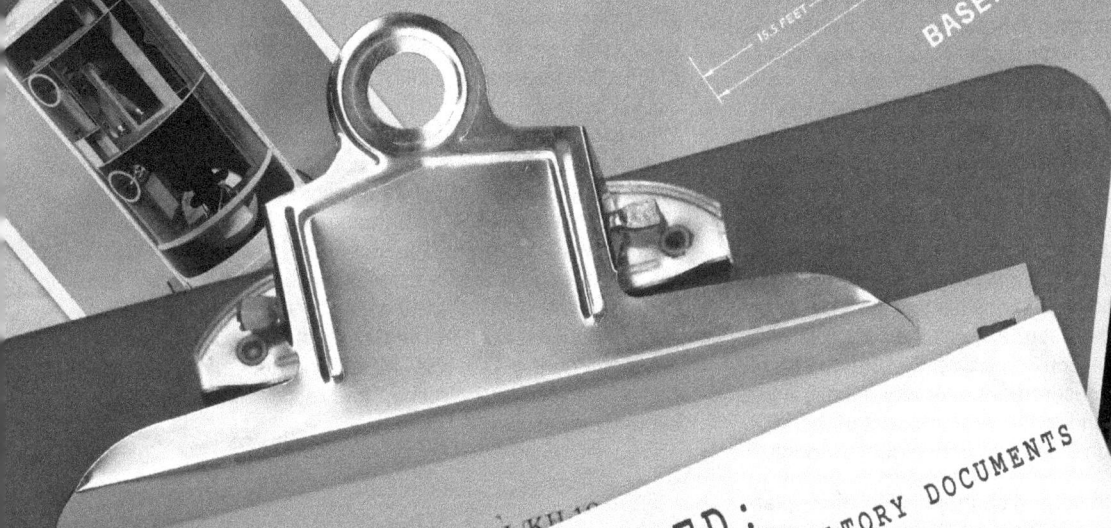

USAF MOL/KH-10

THE DORIAN FILES REVEALED:
A COMPENDIUM OF THE NRO'S MANNED ORBITING LABORATORY DOCUMENTS

CHAPTER IX:
ORGANIZING FOR
CONTRACT DEFINITION

ORGANIZING FOR CONTRACT DEFINITION

On 29 April 1965, some four months before the President's announcement, Dr. McMillan met with Generals Schriever and Martin to discuss the kind of management organization the Air Force should establish for the next MOL phase—contract definition*. At this meeting on the West Coast, the three men tentatively agreed that a strong autonomous system office should be organized there, supported by an appropriate AFSC structure. Subsequently, however, when Evans and Schriever undertook to put down on paper the details of a permanent MOL organization as they saw it, they found themselves embroiled in a major disagreement with General Martin. This issue concerned how the "black" and "white" aspects of the program should be managed.[1]

During the spring and early summer of 1965, General Evans' staff undertook to draft a paper on the proposed USAF management structure. It proposed the creation of a "strong, autonomous, integrated program implementation office" on the West Coast, headed by a general officer to be known as the Deputy Director, MOL. He would report to the MOL Program Director (General Schriever), who would be responsible to the Secretary of the Air Force, the Under Secretary, and the Director, NRC, for "total program direction." The Deputy Director, MOL, would be given "full procurement authority necessary to conduct both 'black' and 'white' procurement of the MOL program from funds provided him from higher authority."[2]

General Martin strongly disagreed with this plan. In a message to Dr. McMillan on 8 July, he declared that in view of his responsibilities to the NRO, it was essential that he control not only the development of all reconnaissance payloads, the reconnaissance payload section, and integration of all payloads into this section, but also all "black" contracts and "white" contracts affected by "black" contracts. He recommended that the responsibilities of the Deputy Director, MOL, be limited to "all non-reconnaissance and non-BYEMAN aspects" of the program, such as the laboratory section, Gemini capsule, boosters, launch facilities, etc. To insure essential coordination and "interface," he proposed

creating a "MOL payload office," which would be physically located adjacent to the office of the Deputy Director, MOL, to handle all black-related matters.[3]

General Schriever, however, felt that the above approach would fragment MOL management and was contrary to all the basic management principles AFSC had learned in the ballistic missile program. Writing to Dr. McMillan in early August, he argued that the management problem of dealing with the Directorate of Special Projects was "amenable to proven solutions from other programs that were no less complex than MOL." He urged that the final MOL management plan, which OSD had requested prior to program approval, provide for a "clear single channel of direction and responsibility" linking the Secretary of the Air Force or Director, NRO, and Director, MOL and the West Coast organization. There should be, he insisted, "unequivocal MOL program policy guidance from or through the Director, MOL, for all aspects of the program." (Endnote 4) The AFSC commander acknowledged that security was important but also noted that "The MOL is too big, and the image is too well established, to hide." In a second letter to Dr. McMillan, he reiterated his concern over "unnecessary fragmentation of the management authority of the Director, MOL, and the existence of multiple channels of direction and responsibility for MOL system acquisition."[5]

THE NEW MOL STRUCTURE: TWO MANAGEMENT CHANNELS

As the Director, NRO, however, McMillan's views were influenced by the "black" environment which had produced the highly successful U. S. unmanned satellite reconnaissance system. The conservative approach was to support General Martin's position, which he did†. Thus, the final MOL management plan sent to OSD on 24 August 1965—while it accepted the principle of "single, clear line of direction to a full" coordinated MOL program"—provided for two distinct management channels. The plan stated that since MOL would meet some of the NRO's proposed long-range objectives, the direction of it "should be responsive to policy, guidance, and approval of the Secretary of the Air Force with assistance and advice from DNRO."[6]

---

*   Originally known as project definition. Contract definition was defined as "that phase during which preliminary design and engineering are verified or accomplished, and firm contract and management planning are performed." The overall objective of contract definition was "to determine whether the conditional decision to proceed with Engineering Development should be ratified." (DoD Directive 3200.9, dated 1 July 1965.)

†   Dr. McMillan wished to maintain tight control over the "black" environment, from which the Air Staff and many other USAF agencies had been excluded.

Under this management plan, the Director, MOL—assisted by a full-time general officer serving as Vice Director‡—was designated "the principal operating agent" for the program. He was to organize a strong "integrated systems and program implementation office" at El Segundo, Calif, headed by the Deputy Director, MOL. The latter would be responsible for "system procurement, design, development, test and evaluation," overall mission operations, including man's safety during all phases of manned flight, etc., and would exercise "on-orbit control of the vehicle and reconnaissance payload in responsive to intelligence collection tasks established by the DNRO or his designee."

The basic MOL management structure would be completed with establishment of a "Sensor Payload Office" under General Martin. Its responsibilities were to manage all contracts for the high resolution photographic sensor payload, i.e., the primary optics, cameras and camera handling devices, etc. It would recommend the contractors to be selected, be responsible for detailed technical direction over the contracts, and provide "contracting services" to the Deputy Director, MOL, for all "black" contracts required by the latter.[7]

After McNamara approved this MOL organization (see chart, next page) and following the President's announcement, Secretary Zuckert formally designated General Schriever as MOL Director and sent him instructions which spelled out General Martin's responsibilities as follows:

> The Director, SAFSP, located at SSD, will be responsible to the DNRO for development, acquisition and test of the sensor payload in response to technical specifications and requirements provided by the Deputy Director, MOL. He will be responsible for all "black" contracting, and will establish a MOL sensor payload office, co-located with the MOL System Office, to carry out his responsibilities. He will review and approve implementation of BYEMAN security procedures. In addition, SAFSP will maintain surveillance over the utilization of the critical Air Force, Aerospace and industrial resources of the NRP including the MOL.[8]

Three days after receiving these instructions, Schriever convened a meeting at SSD of the key people who would be involved in managing the program. They included General Evans, who had been designated Vice Director, MOL; Brig Gen R. A. Berg, who was named Deputy Director, effective 1 October; and Generals Funk, Bleymaier§, and Martin. After discussing the essential supporting role that SSD would play during MOL's development, Schriever addressed Martin as follows: "I think it is important that any time anything goes on in the sensor area important enough to talk to your boss [DNRO], I should also be informed. We must not keep secrets from one another."[9]

Figure 42. Russell A. Berg
Source: CSNR Reference Collection

General Martin responded that he would try to keep Schriever informed, but he noted that in his discussions with McMillan, many subjects were covered with MOL frequently mentioned in that context. He said it would be improbable, therefore, that he would be able to inform General Schriever of details of each discussion, but would do his best to advise him of substantive issues. He understood that he was to be responsive to program guidance from the Deputy Director, MOL, and would

‡ The Vice Director was to organize and run the MOL Program Office in the Pentagon.

§ At this meeting Schriever announced Bleymaier's reassignment to become Commander of the Western Test Range. He had been serving as SSD Deputy Commander for Manned Space Systems.

Figure 43. John S. Foster
Source: CSNR Reference Collection

manage sensor activities to conform with such guidance. "Let me assure you," he said, "that we will work closely with the Deputy Director."[10]

During the next several months—while Evans, Berg, and Martin organized their respective offices¶—a new team of top-level civilian officials began moving into key USAF and OSD positions. On 30 September 1965, Dr. McMillan resigned to return to private life (Bell Telephone Laboratories) after serving four years with the Air Force and the NRO. He was succeeded as NRO Director by Dr. Flax, who continued in his post as Assistant Secretary of the Air Force (R&D). Mr. Norman S. Paul took over as Air Force Under Secretary. Another major change, effective 1 October 1965, was the selection of Dr. Brown to succeed Mr. Zuckert as Secretary of the Air Force. Brown was succeeded as DDR&E by Dr. John S. Foster, Jr.

At the first MOL Policy Committee meeting the new Air Force Secretary attended (on 14 October), General Schriever raised a question about the need for "streamlined management" above the USAF level to handle MOL "black" versus "white" program funds.[11] Subsequently, General Evans initiated a study of this requirement and on 19 October, after coordinating with Dr. Flax, he

¶ The MOL Program Office in the Pentagon, the MOL Systems Office on the West Coast, and the MOL Sensor Office.

forwarded to Dr. Brown a proposed draft memorandum to McNamara on "MOL management channels above the Secretary of the Air Force level." This memo noted that, while MOL was a part of the National Reconnaissance Program, because it was visible to the public and known to exist, its management was "not entirely amenable to procedures currently used for other parts of the NRP." One distinction related to the way funds were handled. Except for cleared personnel, the scope and existence of most NRP programs was not known and they were not defended in open sessions before Congress. In the case of MOL, all funding requirements were contained in white PCP's, only a portion of which were subject to special access, i.e., that pertaining to sensors.[12]

The memorandum further noted that DDR&E's staff had not normally been involved in the justification, review, and approval of the NRP. On the other hand, Dr. Brown or his Deputy Director of Space (Dr. Hall**) in recent months had been personally and intimately involved in the review, justification, and approval of the MOL program for both black and white portions. Since security dictated that there continue to be a visible MOL program, with certain aspects of its mission kept under wraps, it appeared necessary that DDR&E remain in an authoritative position to justify, review and approve various funding requests.[13]

It was therefore proposed that DDR&E, as an individual, be designated the MOL focal point for the Secretary of Defense and provided a Special Assistant who was cleared for all aspects of the program, had experience in the satellite reconnaissance field, and was known to people in that field. Dr. Brown subsequently discussed this proposal with Dr. Foster and Secretary McNamara and both agreed with the approach outlined. Later, Dr. Foster designated Mr. Daniel J. Fink, Deputy Director (Strategic and Space Systems), as his principal staff advisor and assistant to assure that MOL received proper support.[14]

PROCEDURES FOR GUIDANCE AND DIRECTION OF NRO ASPECTS OF MOL

On 22 October Dr. Flax as the new NRO Director met with General Martin to discuss the latter's responsibilities relative to MOL. He agreed that MOL funds would be made available to the Directorate of Special Projects in the same manner "as presently employed" in other NRO programs, but that General Schriever would be permitted "to follow" overall funding for both "black" and "white" activities. He further agreed that Martin might receive directives directly from General Schriever,

** Dr. Hall resigned from this post in October 1965 and returned to private life.

after he (Dr. Flax) had approved in each case, "with an information copy simultaneously going to General Berg," and that the Director of Special Projects "must comply with overall system integration and overall system engineering instructions from General Berg concerning the integration of the payload into the complete MOL system," etc. However, Dr. Flax also determined that, since the DORIAN payload was "an NRO payload," General Martin would be held responsible to him "for development, acquisition, and test, including decisions as to the configuration of the payload.¹⁵

The fact that he had little control over the payload portion of the program remained troubling to General Schriever. In October, at his direction, Evans and his staff drafted a paper on the MOL organization to "clarify" the management principles outlined in the 24 August plan. Their view was that, unlike previous "black" projects, where the reconnaissance sensor itself was "the major element around which overall system integration is postured," MOL was different because of the introduction of man into the system and because of "the currently expressed national policy of overt and unclassified admission of the existence of MOL." Consequently, the suggestion to conduct MOL as a covert program was denied, although "conduct of covert activities within the program itself" was not.

Because of this special nature of the program, they argued that MOL was different from other NRO activities and that the Director, MOL, was and should be responsible for "all" its aspects. According to this interpretation of the 24 August management plan, "guidance and direction issued by Director, MOL is clearly competent and authoritative, not only for the 'white' aspects of the MOL program, but also for 'black' aspects—those which interface with the NRP." This was believed consistent "with the spirit and intent of MOL management since it would preserve a single, clear line of direction" and permit integration of "black and white" guidance and direction at the most effective management level.¹⁶

The above viewpoints were submitted to Dr. Flax in a paper titled "Procedural Considerations for MOL Program Management." Dr. Flax in turn asked General Stewart, the Director of the NRO Staff, for his comments. After reviewing the paper, General Stewart suggested a number of changes and in a memorandum to General Evans on 18 October 1965, he suggested that Generals Berg and Martin also be asked to comment before a proposed new management directive was submitted to Flax and Schriever. In early November, after the comments of Flax, Stewart and Berg were incorporated, the "procedural considerations" paper was submitted to General Martin for his comments and/or concurrence.¹⁷

On 12 November Martin responded with a lengthy critique sent to Dr. Flax and General Schriever, in which he challenged and disagreed with the basic thrust and intent of the paper. He argued that, in view of guidance he had received from Dr. Flax, he retained complete responsibility for development, acquisition, and test of the MOL reconnaissance payloads in the same manner as the other NRO payloads. He rejected the view that MOL was different from other NRO programs because of man's presence or the policy of publicly admitting the existence of the project. He said:

> The MOL reconnaissance sensor dominates the entire configuration of this project. Its influence is far greater than the presence – of man, even in the "manned-only" configuration. The requirement of manned/unmanned capability further extends the influence of the reconnaissance sensor on the configuration of the project...

My present MOL responsibilities are assigned to me by the Director, NRO, who has informed me that he holds me responsible for carrying them out. I, therefore, cannot agree with the "procedural principles" as written in the draft. I submit that it is axiomatic that my direction should come to me from the person to whom I am responsible for carrying it out, with no intervening modification of interpretation, and with no constraint on my direct access to such person for questions, clarification, response or discussion of such responsibilities and direction.¹⁸

To ease some of the MOL Director's complaints, however, General Martin proposed adopting some alternate procedure: which would be consistent "with the present assignment of responsibilities. "Thus, for example, he suggested that when Dr. Flax sent him written guidance or direction, an information copy be simultaneously provided to General Schriever. Copies of messages and letters sent from his office to Flax similarly would be dispatched to the MOL Program Office.¹⁹

Martin's stand killed the plan to change the existing dual management structure. General Stewart, however, argued informally with Evans that they would follow the Martin-Berg arrangement. It was his view that all MOL program direction should be issued by the MOL Director, requiring only that instructions sent to Mart in first have Dr. Flax's (or his authorized representative's) concurrence, and that the flow of information upward should go through the same channels. Subsequently, General Stewart assigned Colonel Battle to General

Evans' staff to provide the necessary NRO-MOL Program Office coordination. By late 1965 the entire management question went into limbo and not until many months later, when the program entered Phase II engineering development, was the organization revised.[20]

MOL FINANCIAL PROCEDURES

Discussions of "black" MOL financial procedures began in the summer of 1965 when the NRO Comptroller {... } proposed that the Air Force include a single MOL program element in the defense budget which would incorporate both "black" and "white" funds, with the MOL Program Office exercising a substantial level of control over the former. {The NRO Comptroller's} plan became the basis for a series of discussions and reviews which led, on 4 November, to the signing by Dr. Flax and General Schriever of a formal agreement governing "MOL Black Financial Procedures."[21]

Under terms of this agreement the NRO Comptroller and the MOL Program Office would work together to prepare current and future year cost estimates of MOL black requirements. These would be reviewed and approved by both the Director, NRO, and Director, MOL, before issuance. The responsibilities of the Director of Special Projects would include providing "black" cost estimates, coordinating with the Deputy Director, MOL, and forwarding them to the NRO Comptroller and the MOL Program Office. Authority to obligate the "black" funds would be issued by the NRO Comptroller directly to the Director of Special Projects, who would be held accountable for them.[22]

A companion agreement on "White Financial Procedures," approved by Dr. Flax and {the NRO Comptroller} in December 1965, also was promulgated. Signed by Assistant Secretary of the Air Force Leonard Marks, Jr., it provided that MOL white funds would go through normal AFSC channels to SSD for the MOL Systems Office. The "white" financial agreement also outlined procedures for making budget estimates, preparing program change proposals, and submitting other financial papers required by OSD, the Bureau of the Budget, and Congress.[23]

THE MOL SYSTEMS OFFICE

The day President Johnson announced the United States would build MOL, General Schriever dispatched Program Directive No. 65-1 to the Space Systems Division. This document, which provided authority "to establish and commence management functional activity for the Manned Orbiting Laboratory System (Program 632A)," required the Deputy Director, MOL, to submit by 15 September 1965 a MOL Systems Office management plan, organizational charts, job descriptions of key positions, etc., plus an Aerospace Corporation management plan in support of the program.[24]

The program directive was received by the MOL project office, which had been created in March 1964 under General Bleymaier, the SSD Deputy Commander for Manned Space Systems, and Colonel Brady, the System Program Director. By August 1965 this SPO had a staff of 42 military and 23 civilian personnel. In accordance with Schriever's directive, Bleymaier and Brady immediately initiated an office reorganization. The realigned SPO included separate divisions or offices for Program Control, Configuration Management, Engineering, Test Operations, Procurement and Production, Requirements, Bioastronautics, Facilities, and Navy liaison. On 23 September, after General Berg, the new Deputy Director, briefed him on the proposed new MOL Systems Office structure, General Schriever approved the changes, which became effective 1 October.[25] (See chart, next page.)

On 28-29 September, General Schriever held the first MOL Program Review Conference at SSD, attended by General Evans, Funk, Berg, Martin, and Bleymaier and Dr. Yarymovych, Colonel Brady and Aerospace's Drs. Ivan Getting, Allen F. Donovan, Byron P. Leonard, and Walt Williams. They discussed the planned approach to MOL field management and agreed that the basic principles enunciated in the AFR 375 series of regulations would be applied. The conferees recognized, however, that judgement would have to be exercised "in that the MOL was not going into the operational inventory in the typical sense; and, hence, the series of regulations could not be totally adapted to the MOL program."

Later, General Schriever formally authorized Berg to deviate from standard system acquisition policies and procedures in the 375 series of regulations.[26]

During this meeting Aerospace Corporation officials briefed the USAF officers on their planned organizational structure for general systems engineering and technical development (GSE/TD) support of MOL. They indicated they would establish a MOL Technical Director within the Corporation's Manned Systems Division. Gen Schriever objected to this arrangement, since it would place the Technical Director at the fourth organization level. He also noted that Aerospace had not adequately defined the Director's responsibilities. Emphasizing the need for "vertical organization to totally support the MOL program," Schriever said that all elements of the various corporate divisions "must be responsive to the MOL Director commensurate with the unique Air Force management structure."[27]

He asked General Berg to work with Aerospace to provide a more acceptable organizational structure. Subsequently the corporation took steps to reorganize itself so that it paralleled the MOL Systems Office and was "in line with the overall MOL management concept." On 14 October it established a MOL Systems Engineering Office headed by Dr. Leonard, Aerospace vice president, who would report directly to Dr. Donovan, senior technical vice president of the corporation.[28]

THE CONTRACTOR TEAM

On 25 August 1965, the President in his announcement named the two successful MOL contractors—Douglas Aircraft and General Electric—contrary to the original DDR&E plan, which had called for selecting one contractor for Phase I definition. This decision to go ahead with two industrial concerns had origins in Air Force-DDR&E discussions which followed the award of the 60-day laboratory vehicle study contracts on 1 March to Boeing, Douglas, General Electric, and Lockheed.

Thus, on 18 March General Evans reported to McMillan that he had found "there are arguments within the Air Force and certainly within DDR&E against proceeding with a single contractor." These parties suggested that by continuing with more than one firm during the definition phase, a competitive atmosphere would be preserved "from which, hopefully, will emerge better cost and schedule information and perhaps new ideas." General Evans was opposed to this suggestion. He said that "we cannot afford the luxury of another competitive study period" and expressed the view that there were more advantages to proceeding with a single contractor. It would be cheaper, he said, the security problem would be less with one contractor and the administration of a single contract would be easier.[29]

By late May and early June 1965, after the MOL Source Selection Board had submitted its evaluations of the four contractor proposals, a new rationale was offered for proceeding with two of the firms which gained general approval. That is, it was argued that the program could be strengthened by integrating the two strongest contractors into a single team. On 12 June General Schriever formally proposed to General McConnell, the Chief of Staff, that they adopt the two-contractor approach. He explained that while "contractor A clearly offers the best overall technical program and management approach, the proposal of contractor D is superior in a few important respects that bear on mission capability. This suggests it may be particularly advantageous to the government to include contractor D in the program in those areas where his capabilities will strengthen the development team."[30]

The Chief of Staff thought this proposal worth pursuing, whereupon a study was undertaken to identify "the allocation of specific tasks" in the program between the two contractors. It was quickly found that the tasks to be done divided readily into three major categories involving the laboratory vehicle, the sensor module, and the payload of sensor package itself. In a memorandum to General McConnell on 25 July, the AFSC Commander described these as follows:

1. The laboratory vehicle contractor would be the system integration contractor, responsible for structural analysis of the entire system through the launch phase and the successful operation through the 30-day mission of all elements except those actually contained in the sensor module.

2. The sensor module contractor would be responsible for receiving requirements from the sensor designer and Systems Program Office, and would "define, assign, and engineer a discrete structural carrier for the sensor package and associated crew displays." He also would assemble and test the sensor module elements and prepare them for launch, and provide interface requirements to the laboratory vehicle contractor for his own equipment as well as those from the sensor contractor.

3. The sensor contractor would be responsible for the design, fabrication and test of the sensor elements and package. The sensor contractor would interface principally with the sensor module contractor.[31]

The two-contractor approach was discussed further on 30 June by Secretary Zuckert, Dr. Flax, and other USAF officials. On 1 July additional information was sent to the Chief of Staff and gained his endorsement. Subsequently, in the management plan sent to OSD on 24 August, Secretary Zuckert reviewed the Source Selection Board results and reported that it had rated the four competing firms in the following order of merit:

Douglas, General Electric, Boeing and Lockheed. He noted that "the total MOL development program confronting us is a very complex and important one which will demand the very best talent and experience in the industry." Because of the need for comprehensive knowledge and appreciation of systems integration and first hand experience in fabricating large structures, he suggested that:

> ...it may be particularly
> advantageous to draw on the
> capabilities of two outstanding
> contractors to accomplish the task
> originally envisioned for a single
> laboratory vehicle contractor. I
> have carefully reviewed the report
> and findings of the selection board
> and have completed an additional
> examination of contractor past
> experience and performance, and of
> security factors pertinent to the
> MOL Program. The Douglas Company
> offered the best overall technical
> and management approach. Its past
> experience and performance as a
> system integrator on weapons such as
> Thor, Genie and Nike Hercules/Zeus
> is good and considerably broader
> than that of the General Electric
> Company. General Electric, on the
> other hand, showed superiority in
> important aspects that bear on
> mission capability. They have current
> experience in space vehicle operation
> as well as expertise in handling the
> complex interface with large optical
> systems. They have over 1000 people
> immediately clearable for DORIAN
> work. The Douglas Company have very
> few cleared people.[32]

In addition, Zuckert noted, these contractors possessed in aggregate "a most imposing array of existing test facilities" available for support of the program. He therefore concluded that it was in the government's interest to include both these contractors on the industrial team. Secretary McNamara, who several days earlier had approved the designation of Eastman Kodak as the primary DORIAN optical contractor,[33] agreed with the two-contractor approach as did the President.

INITIATING CONTRACT DEFINITION

In late August 1965 the Titan III booster was the only segment of MOL hardware for which a Phase I definition contract had been let, although negotiations with McDonnell Aircraft for Gemini B definition had been completed and a contract was expected to be issued momentarily. Following the President's announcement, SSD procurement officials moved to place Douglas under contract by submitting a request for proposal and work statement on the laboratory vehicle and asking the firm to reply by 20 September. General Electric also was sent a request for proposal and work statement on the mission module.[34]

The MOL SPO, in addition, distributed a "contract compliance document" which was to serve as a vehicle to direct the contractors. Designated the "Government Plan for Program Management for the Manned Orbiting Laboratory System Program," its aim was to provide "an early overview of the key features" of the entire project to top management officials in government and industry who might become directly associated with MOL. The plan included a projected MOL flight schedule, a breakdown of the primary hardware and functional elements of the system, and also identified the agencies responsible for each program segment, explained the interrelationship of the several organizations involved, and described their roles and missions.[35]

General Evans, to support contract definition activities, on 31 August forwarded to Dr. Flax current requirements and requested the Assistant Secretary to seek DDR&E's approval. The following breakdown of costs included funds for continuation of certain pre-MOL activities:

	MILLIONS
Titan IIIC	$7.0
Gemini B	7.0
Laboratory Vehicle	12.0
Mission Module & Capsule	12.0
Flight Crew & Equipment	1.0
Facilities	0.5
Systems Analysis	2.0
Mission Control Equipment	0.5
Test Support	0.5
Navy	2.0
Pre-MOL	10.5
Total	$62.0

On 10 September, Dr. Flax asked DDR&E to release $12 million if fiscal year 1965 and $50 million in fiscal year 1966 funds so that the Air Force could begin contract definition. He reported that the Air Force had initiated actions to award the necessary contracts and he estimated Phase I definition would be completed in approximately six months. On 30 September, after he was briefed on the above costs, Dr. Brown authorized release of the requested $62 million[tt]. Flax, in the meantime, signed a "Determination and Findings" authorization for procurement associated with continuance of pre-MOL activities and the start of contract definition.[36]

Meanwhile, Douglas submitted its proposal and fixed price estimate for the MOL definition phase. Formal negotiations, begun on 30 September, were completed on 17 October, when agreement was reached on a fixed price "minimum level of effort" contract totaling $10.55 million. The contract's effective date was 18 October and Douglas agreed to submit definition phase data and Phase II proposals by 28 February 1966. The tentative date for Phase II go ahead was set for 1 May 1966.[37]

In mid-October contract negotiations also were completed with General Electric. The cost of its definition phase activities was set at $4.922 million, most of it in the "black." The "white" portion of the G. E. effort, extracted and prepared as a separate contract, came to $0.975 million of the $4.922 million total.[38]

Even before its Air Force contract was signed, Douglas moved to invite subcontractors to bid on five major MOL subsystems. These were in the areas of environmental control, communications, attitude control, fuel cell electrical power, and data management. On 28 October Douglas made its first selection with MOL Systems Office concurrence[‡‡] naming Hamilton-Standard to work on the environmental control and life support system. Of the three competitors for the communication system contract, on 9 November it selected Collins Radio. Douglas also awarded the attitude control and translation system subcontract to Honeywell on 16 November, and the electrical power system subcontract to Pratt & Whitney on 22 December.[39]

The one subcontract the Air Force rejected during this period was Douglas' award to IBM of the data management system contract. An Air Force/Aerospace evaluation team challenged that firm's superiority over a competing proposal submitted by Univac, which Douglas had rated only several points below the former's proposal.

The team noted that while IBM had a better technical proposal, its estimated cost was $32 million compared to Univac's $16.8 million. After a further review of the two proposals, Douglas decided, with USAF approval, to let study contracts to both firms to obtain more specific technical data and cost information.[40]

Concerning this subcontracting process, General Schriever prepared and on 5 November briefed General Berg on policy to be followed to avoid congressional criticism. He said the Air Force should avoid becoming involved in the contracting action between the prime contractor and the subcontractor, to avoid charges that it had influenced in any way the award to a particular firm to the detriment of another. He said the Air Force's basic interest was to assure that its requirements were fully stated and its interests protected in the specifications and provisions of the subcontract, the bids of subcontractors were responsive, etc.[41]

[tt] Total approved fiscal year 1965 MOL expenditures came to $36,500,000; for fiscal year 1966, $50,000,000.

[‡‡] Douglas accepted the Air Force's source selection procedures and agreed to work closely with the MOL Systems Office, giving the latter the opportunity to review the paperwork leading up to the company's choice of a subcontractor.

ENDNOTES

1. Memo (U), Col J.S. Chandler, Office of Asst Dep Cmdr for Space (MOL), AFSC, to MSF-1, SAFSL, 3 May 65, subj: Telecon w/Gen Evans; Memo (TS-DORIAN), Evans to McMillan, no date, subj: Mgmt of the MOL Program.

2. Memo (S-DORIAN), Evans to McMillan and Schriver, 5 Jul 65, subj: Mol Mgt.

3. Msg 8226 (TS-DORAIN), Martin to McMillan, 8 Jul 65.

4. Ltr (TS-DORAIN), Schriver to McMillan, 5 Aug 65, subj: MOL Prog Mgt.

5. Ltr (TS-DORAIN), Schriever to McMillan, 9 Aug 65, no subj.

6. Memo (TS-DORIAN), Zuckert to McNamara, 24 Aug 65, Tab A, Mgt of the MOL Prog.

7. Ibid.

8. Memo (TS-DORIAN), Zuckert to Schriever, 25 Aug 65.

9. Memo for Record (TS-DORIAN), Col L.S. Norman, Jr., 28 Aug 65, subj: Conference With Gen Schriever, 28 Aug 65.

10. Min (TS-DORIAN), AF MOL Policy Cmte Mtg 65-3, 14 Oct 65; Secretariat 14 Oct 65.

11. Ibid.

12. Memo (TS-DORIAN), Evans to Brown, 19 Oct 65, subj: MOL Mgt.

13. Ibid.

14. Memos (TS-DORIAN), Col Brian Gunderson, Dep Exec Asst, OSAF, to Gen Evans, 2 Nov 65; Evans to Flax, 3 Nox 65, subj: Status Report on OSD Relations.

15. Memo for Record (S-Special Handling), prep by Gen Martin, 25 Oct 65.

16. Draft (TS-DORIAN), Oct 65, subj: Procedural Considerations for MOL Program Mgt (SAFSL BYE 37596-65).

17. Memos (TS-DORIAN), Stewart to Dr. Flax, 14 Oct 65; Stewart to Evans, 18 Oct 65, Evans to Schriever, 2 Nov 65.

18. Ltr (TS-DORIAN), Martin to Flax & Schriever, 12 Nov 65, subj: Procedures for MOL Program Mgt.

19. Ibid.

20. Memo (TS-DORIAN), Stewart to Flax, 24 Nov 65, subj: MOL Management; Interview, Berger with Stewart, 1 March 1968.

21. Msg Whig 3404 (TS-DORIAN) to Col Gerry Smith, 22 Jul 65; Talking Paper for Gen Evans (TS-DORIAN), 16 Sep 65, subj: MOL Black Financial Procedures; Agreement, MOL Black Financial Procedures (TS-DORIAN), 4 Nov 65, signed by Dr. Flax and Gen Stewart.

22. Agreement cited above.

23. Memo (TS-DORIAN), Evans to Asst SAF (FM), 17 Dec 65, subj: MOL Financial Procedures; Memo (U), Leonard Marks, Jr., Asst SAF to Vice CSAF, 18 Jan 66, subj: MOL Financial Procedures.

24. MOL Program Directive No. 65-1 (TS-DORIAN), 25 Aug 65, subj: Management of the MOL Program.

25. Ltr (U), Col Brady to All MOL Personnel, 1 Oct 65, subj: MOL Organization and Structure; Memo (S), Col Brady, Asst Deputy Director, MOL, to Program Control Office, 28 Sep 65, subj: MOL Mgt Reporting System.

26. Ltr (TS-DORIAN), Evans to Berg, 15 Oct 65, w/ Atch Memo for Record (TS-DORIAN), 12 Oct 65, subj: MOL Program Review #1, 28-29 Sep 65; Memo (U), Schriever to Berg, 21 Nov 65, subj: Application of 375 Series Mgt Procedures to MOL.

27. Memo for Record (TS-DORIAN), 12 Oct 65, subj: MOL Program Review #1, 28-29 Sep 65.

28. Msg (S-DORIAN), 9008, Berg to Schriever, 7 Oct 65; *Missiles and Rockets*, 18 Oct 65, p 16.

29. Memo (C), Evans to McMillan, 19 Mar 65.

30. Ltr (S-DORIAN), Schriever to McConnell, 12 Jun 65.

31. Ltr (S-DORIAN), Schriever to McConnell, 25 Jun 65.

32. Memo (S-DORIAN), Evans to Blanchard, 1 Jul 65, subj: MOL Contractor Selection; Memo (TS-DORIAN), Zuckert to McNamara, 24 Aug 65, subj: MOL Management.

33. Msg Whig 0001 (TS-Spl Handling), McMillan to Martin, 22 Aug 65.

34. Memo (C), Evans to Flax, 1 Sep 65, subj: Initiation of MOL Definition Phase.

35. Space Systems Division Government Plan for Program Management for the MOL System (S-SAR), Aug 1965.

36. Memos (S), Flax to DDR&E, 10 Sep 65, subj: Initiation of MOL Definition Phase; Brown to Flax, 20 Sep 65, subj: Approval of (1) USAF FY 1965 RDT&E MOL Program; (2) USAF FY 1966 RDT&E MOL Program.

37. Msg 9096 (S-DORIAN), Berg to Schriever, 19 Oct 65; Memo (TS-DORIAN), Schriever to SAF, 8 Nov 65, subj: MOL Monthly Status Report.

38. Memo (TS-DORIAN), Schriever to SAF, 8 Nov 65, subj: MOL Monthly Status Report.

39. Memo (U), Evans to Schriever, 19 Nov 65, subj: Status of MOL Subcontractor Support; Msg (U) SSML 00006, MOL SPO to OSAF, 22 Dec 65.

40. Memo (TS-DORIAN), Schriever to SAF, 12 Feb 66, subj: MOL Monthly Status Report, p 4.

41. Memo (U), Schriever to Berg, 4 Nov 65, subj: Policy Guidance Regarding Selection of Subcontractors to Avoid Congressional Criticism.

MOL ASSEMBLY
INTEGRATION BUILDING

MATERIAL

PROPERTY OF THE
UNITED STATES GOVERNMENT

IF FOUND, DO NOT OPEN

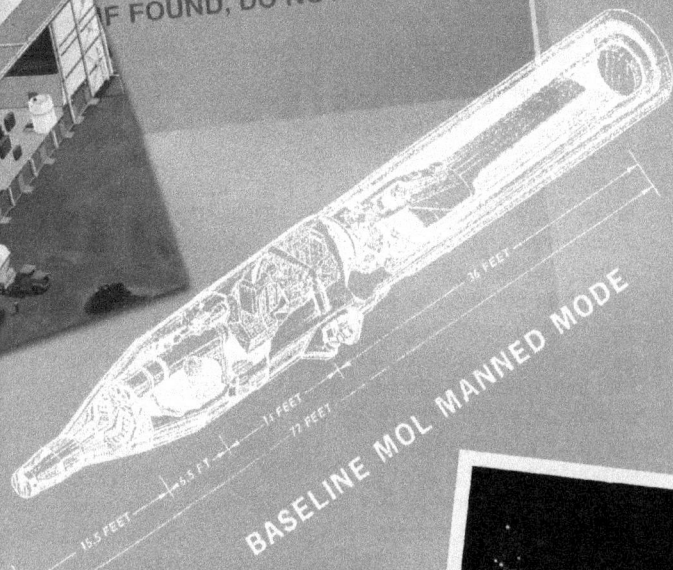

16.5 FEET   6.5 FT   11 FEET   36 FEET
72 FEET

BASELINE MOL MANNED MODE

USAF MOL/KH-10

THE DORIAN FILES REVEALED:
A COMPENDIUM OF THE NRO'S MANNED ORBITING LABORATORY DOCUMENTS

CHAPTER X:
THE MANNED/UNMANNED
SYSTEM STUDIES 1965 – 1966

THE MANNED/UNMANNED SYSTEM STUDIES 1965 - 1966

Contract definition activities had scarcely gotten under way when USAF and NRO officials found themselves involved in several new studies, one of a "wholly unmanned" system which some feared might lead to termination of the "manned MOL" even before engineering work began. It will be recalled that after the President's Science Advisor and the Director of the Budget expressed interest— for different reasons—in an unmanned system, Dr. Brown and Secretary McNamara agreed the MOL also would be designed to operate without a man. The question of an unmanned system also was raised by Dr. Land of the PSAC Reconnaissance Panel. On 18 August 1965, in a lengthy memorandum reviewing the original USAF arguments for proceeding with MOL, he informed Dr. Hornig that not enough time had been devoted to exploring "alternatives to the use of man." Specifically, Land challenged certain Air Force statements about the unmanned version. He said it seemed to have assumed that a family of inventions was required to make the unmanned navigation system work and to eliminate photographic smear.* The PSAC Panel was puzzled by these assumptions since it saw "quite clearly" the feasibility of "adapting what is already known in both of these domains for use within a vehicle." He said further :

> A solution to these problems would permit the unmanned system, operating with essentially the same camera, to achieve the same ground resolution on prescribed targets as the manned system. It would also contribute significantly to the manned operation by relieving the observer of much of the routine tracking and identification task, and making the pointing and selection of area of interest less critical.

Figure 44. Edwin H. Land
Source: CSNR Reference Collection

> The conclusion that an unmanned vehicle would result in a lower resolving power seems to us, therefore, unwarranted†; the further implied conclusion that the solution of the problems involved when a man is not employed to direct the telescope, would seriously delay the program also seems to us unwarranted. Indeed, it appears that the limiting factor in the schedule will probably be learning how to design the mechanics of very large mirrors so that they will retain their shape in their mounts in space. We, therefore, recommend... the MOL system... Camera payload be designed as a completely automatic system. This device could then be flown with or without a man depending upon a national judgment on each occasion about the need or desirability of

---

* The USAF statement read: "...to get in an unmanned system the kind of performance, in toto, that we expect of a manned system will take some new inventions and will call for a photographic system of much greater complexity than that needed when the man is present." See Memo (TS-DORIAN/GAMBIT), Zuckert to McNamara, 28 June. 1965, subj: Proposed MOL Program.

† The USAF memorandum stated, in part: "... From our knowledge of man's ability to point and track, and from our estimates of the better level of adjustment that he can maintain, we conclude that the manned system would statistically show a medium resolution of {better than 1 foot} against one of {better than 1 foot} for the unmanned...We feel that a manned system will get us an operational resolution of {better than 1 foot} more quickly and more reliably than an unmanned." See Memo, (TS-DORIAN-GAMBIT), Zuckert to McNamara, 28 June 1965, subj: Proposed MOL Program.

adding the special human capabilities for target selection, selection of data to be transmitted to ground station and verbal reporting.[1]

After reviewing the above memorandum, Dr. Hornig asked DDR&E to meet with him on 23 August to discuss the issues PSAC had raised. During this meeting Drs. Hornig and Brown agreed that the Department of Defense would undertake to develop MOL with a {better than 1 foot} capability, either manned or unmanned. They also agreed that a flight demonstration of the unmanned system would be conducted nine months after the first manned flight.[2]

The importance of the unmanned system also was emphasized by Under Secretary McMillan just prior to his departure from the government. In instructions he sent to General Martin on 29 September 1965, he said a decision was needed early in the program "as to how the alternative unmanned capability will be developed." To help with this decision, McMillan directed Martin to initiate a two- to-three month analysis of both the manned and unmanned versions to "identify the critical aspects of the two approaches, including the impact on spacecraft and system design." He also requested an immediate study be initiated to determine the critical automatic subsystems which would be needed for the unmanned system and asked that a report be forwarded to the Director, NRO in mid-December.[3]

The manned/unmanned question was reviewed by MOL officials at the program review meeting convened by General Schriever on the West Coast on 28-29 September. During the discussion several officers voiced concern that man's potential would not be sufficiently exploited if the MOL design was optimized for unmanned operations. Following this meeting, the MOL Program Office prepared a talking paper for General Schriever aimed at convincing top officials to alter direction "to permit optimization of the telescope for manned operation, perhaps accepting as a consequence degraded performance in an unmanned mode." The talking paper, however, did not progress beyond the draft stage since, during the first MOL Policy Committee meeting he attended as Air Force Secretary, Brown pointed out that he was committed to provide a development plan using the same general optical system for the manned and unmanned versions and was committed to PSAC to provide an unmanned launch within nine months of the first manned flight.[4]

At this meeting on 14 October, Dr. Flax and General Schriever assured Secretary Brown that the program did provide for the unmanned requirement. They said the approach being taken was to optimize the sensor "with man in-the-loop" and then automate the functions necessary to provide for unmanned operation of the system. General Schriever reported that the MOL Systems Office and Directorate of Special Projects were studying the problems involved and stated that the results of their investigations were due in mid-December‡.[5]

PSAC REEMPHASIZES THE UNMANNED SYSTEM

To make sure its views were clearly understood, the PSAC Reconnaissance Panel convened a meeting on 8 November 1965 of key MOL personnel. Among the Air Force representatives in attendance were Dr. Flax, Generals Berg and Martin, and Col Lew Allen and Lt Col Frank Knolle of the Directorate of Special Projects. Others present were Dr. Leonard of the Aerospace Corporation; Mr. Fink and Samuel Koslov of ODDR&E; Mr. Thomas of the Bureau of the Budget; and Mr. Hermann Waggershauer, Arthur Simmons, John Sewell, and Dr. F. C. E. Oder of Eastman Kodak. PSAC members at this meeting included Drs. Hornig, Land, Purcell, Joseph F. Shea, Richard L. Garwin, Allen E. Puckett, James Baker, Marvin L. Goldberger, D. P. Ling, and D. H. Steininger.[6]

Figure 45. Lew Allen, Jr.
Source: CSNR Reference Collection

---

‡ A small integrated task group was set up to do this work and to guide the prime contractors—General Electric, Douglas, and Eastman Kodak—in studies of a baseline MOL configuration.

In his opening remarks to the group, Dr. Land emphasized that both the manned and unmanned system could make use of a considerable amount of interchangeable components, "provided that the original design was carried out, from the beginning, on this basis." The Panel's view was made clear by Dr. Shea, who was Deputy Director for NASA Manned Space Flight. He cited the space agency's experience with Project Apollo, which included a good example and a bad one of how to approach the manned/unmanned capability problem. He said the good example had involved the Lunar Excursion Module (FEM) development. From its earliest inception, agency officials levied a requirement on the designers to fly the same basic equipment in unmanned as well as manned modes. With everyone accepting this requirement from the beginning, the solution proved relatively simple. On the other hand, the example of how not to do the job occurred in the Command Service Module (CSM) Project. The planners "barreled along" for a while, designing everything for the manned-only mode of operation. Later, when they tried to convert to add the unmanned capability, the result proved very complex. These remarks of Dr. Shea's were clearly endorsed by the panel.[7]

Following the PSAC meeting, General Martin informed Dr. Flax that the current Phase I MOL project definition effort was oriented "entirely on a manned-only minimally automated design" and that consideration of an unmanned capability had been limited to study tasks based on converting the manned-only design to unmanned operations. No efforts to date, he reported were based on the single design with dual operating modes, as urged by the PSAC panel. Quite the contrary "all our efforts so far have been in the direction which they clearly don't want."[8]

The PSAC panel views on the MOL were restated formally by Dr. Hornig on 22 November in a memorandum to Dr. Flax. The panel believed that a {better than 1 foot} resolution could be obtained by a properly designed unmanned as well as manned system. It thought that MOL officials should pursue an operational program which could use both elements of the system." It interpreted the Hornig-Brown agreement of 23 August to mean the initial system definition would produce a design consistent with the above guidance and that the separate modules and conversion equipment necessary for automatic operation would be developed and built "concurrently with the manned MOL."[9]

As a consequence of this· PSAC guidance, MOL Program officials initiated studies of the system design to determine which manned functions would have to be automated and how to do it. On 30 November Colonel Allen reported to a meeting of the MOL Policy

Committee that analyses and investigations were under way which addressed the points raised by the panel. He said he believed these studies would be sufficiently comprehensive to· enable decisions to be made by mid-December on the basic MOL configuration. At this meeting General Schriever, who wished to emphasize the manned system, reported he had talked with Astronauts L. Gordon Cooper and Charles Conrad, Jr.[§], and that both were skeptical of a design approach for a manned vehicle which stressed the use of automatic modes.[10]

The unmanned/manned studies and investigations— conducted by West Coast agencies and the contractors— were completed on schedule by mid-December. They indicated that automatic alignment, tracking, focus adjustment, and image motion compensation for the sensor were feasible and that, as Eastman Kodak expressed it, "full automation of the system was not a major problem." The investigators concluded that the unmanned mode resolution would approach that of the manned version; however, the latter would offer important advantages in targeting, readout, and visual reconnaissance. They also agreed that the manned/ unmanned configuration should be established "as baseline" but noted that incorporation of automatic features into the manned vehicle would create a weight problem, and create other difficulties.[11]

The MOL Systems Office moved at once to revise its general performance and design requirements specification to incorporate the dual approach. On 23 December General Berg formally notified the MOL contractors of the change. He directed them to include, within the basic MOL design, "such automaticity and redundancy that with removal of the Gemini and selected laboratory components, and the addition of appropriate kits to the laboratory, the system can be flown unmanned." He also informed them that the contract definition phase was extended to 1 May 1966; the system acquisition phase would begin 1 September 1966; and the first manned flight would take place in September 1969.[12] However, concerning this schedule, Eastman Kodak advised that the first set of optics would not be available before late 1969 and a January 1970 flight date was the earliest that optics availability could support[¶].[13]

Meanwhile, the MOL Program Office arranged to brief the Reconnaissance Panel on the steps taken by the Air Force to automate the MOL mission module. At an informal meeting on 7 January 1966, Dr. Flax and Colonel Allen reported to Drs. Hornig, Land, Purcell,

§ Cooper and Conrad, pilots of Gemini 5, completed 120 revolutions of the earth during their eight days in space, 21-29 August 1965.
¶ This was but the first of a number of program slippages attributable to difficulties associated with development of the unique optics package.

Garwin, and Steininger on the manned/unmanned baseline configuration. A more detailed presentation was made to the panel on 9 February 1966. MOL officials reported that one of the key conclusions of their recent studies was that the ground resolutions obtainable with the unmanned system "is the same as that of the manned, provided the unmanned system is operated with a perigee of 70 N. M." They stated further that the unmanned system also could perform for 41 days at the lower altitude with 1,400 pounds of added fuel for orbital "sustenance." Both the January and February briefings were well received by the panel members, whose sole recommendation was that a specific flight be selected for the first unmanned mission. This suggestion was acted upon and Flight No. 6 was so designated.[14]

THE ROLE OF MAN IN THE MOL

As indicated, General Schriever was worried about the effect the unmanned system might have on MOL development planning, being strongly opposed to any possible decision to eliminate the manned version. On 29 December, during a conversation with Gen Evans, he proposed they undertake an operational analysis of "manned and unmanned capabilities for reconnaissance." He was particularly interested in the "quantitative difference " in the reconnaissance "take" of the two modes and also in a qualitative comparison of the resolutions on specific targets and the reliability of the two configuration on a 30-day mission.[15]

Figure 46. MOL Astronaut Flight Suit
Source: CSNR Reference Collection

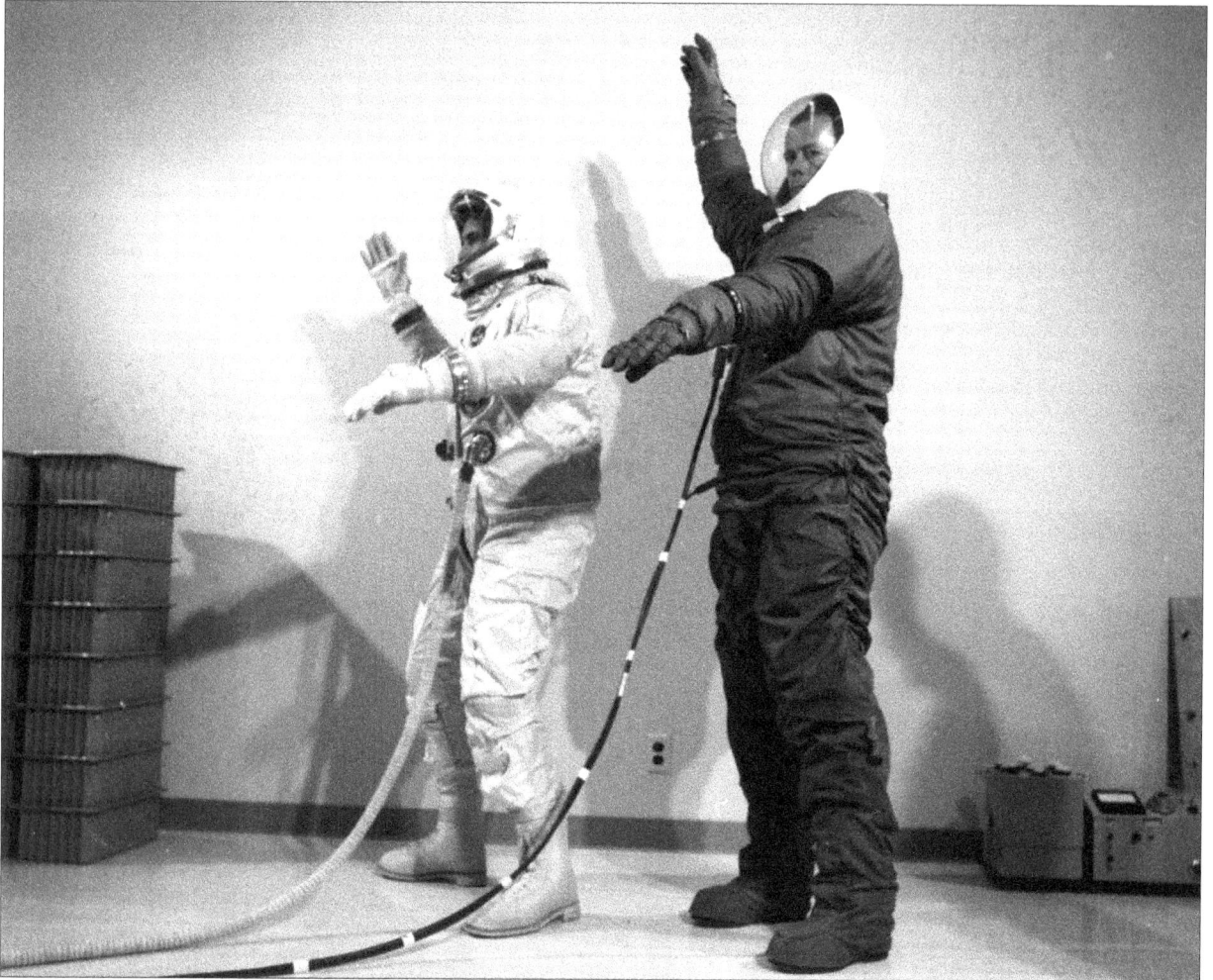

Figure 47. MOL Astronaut Flight Suit
Source: CSNR Reference Collection

This suggestion was formally embodied in a Schriever directive to General Evans on 17 January 1966, instructing him to initiate a study which would bring into sharper focus man's role in MOL. In particular, Evans was to consider NASA's experience with manned space flight and the Air Force's extensive accomplishments "in the effective utilization of man in the performance of unique and highly complex functions under conditions of extreme stress"— as typified by the F-12, X-15, XB-70 and other flight test programs. The AFSC Commander thought a fresh look at this problem might suggest actions "that we should take to exploit more completely man's contributions in the conduct of MOL missions, and in particular the high resolution optical reconnaissance mission."[16]

General Evans shortly thereafter organized a study group under the chairmanship of Col Lewis S. Norman, Chief of the MOL Program Office's Mission Planning Division. Members of the group were Lt Cols Stanley C. White, Benjamin J. Loret, and Arthur D. Haas, and Maj Kenneth W. Weir. Beginning in February, these officers began to compile basic data on various aspects of the subject and during the next several months they interviewed more than 60 individuals, including astronauts, scientists, engineers, contractor personnel, etc. Their preliminary findings were presented to General Schriever on 25 April and a final report, the bulk of it written by Colonel Loret, was completed on 25 May 1966.[17]

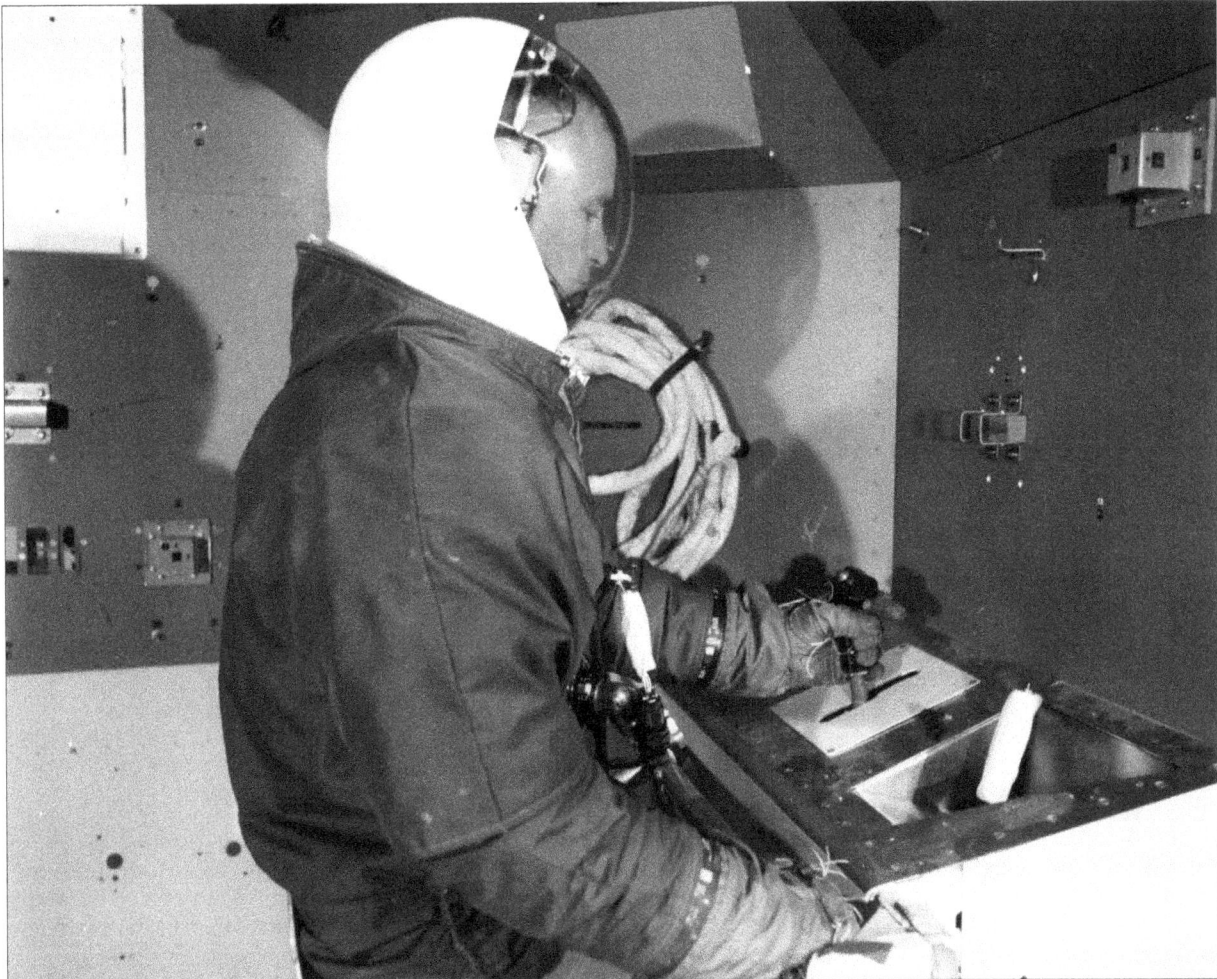

Figure 48. MOL Astronaut Training
Source: CSNR Reference Collection

In this report the group stated that at the beginning of the study, the members were concerned that the rationale for including man in the MOL had deteriorated since program approval in August 1965. They cited developments in automatic equipment, which appeared to undercut the original USAF argument "that a manned system appeared capable of achieving {better than 1 foot} ground resolution whereas an unmanned system probably could not, or at least not as soon." But contrary to the group's expectation, it found as it completed its work "that the argument for man is as strong now or even stronger than it was when the program was first approved."[18]

Thus, the study group maintained that the original rationale in McNamara's 24 August 1965 memorandum to the President—his point that conducting the development program with a manned spacecraft would improve the prospect of achieving resolution in-the {better than 1 foot} class**—remained valid, even taking into account that technological progress in development of automatic devices would provide greater assurance that the unmanned configuration would produce {better than 1 foot} resolutions. The group argued that, even if it were postulated that a completely unmanned system

** McNamara's statement was: "Beyond the initial objective of producing {better than 1 foot} ground resolution photography, successful automation will be increasingly difficult. Conducting the development program with a manned spacecraft will improve the capability of achieving resolutions in the {better than 1 foot} class."

would be more cost effective in the long run than the current MOL manned/unmanned configuration in achieving {better than 1 foot} resolutions, "the need for early achievement of this capability and for ultimate growth to higher resolutions make it mandatory that the program proceed in accordance with the current plan, i.e., to retain man in the system." The group continued:

> We believe the essence of today's argument is that, from a current program viewpoint, inclusion of man will virtually guarantee an earlier {better than 1 foot} resolution capability—and earlier useful "take"—even for the unmanned MOL configuration than would be possible in a wholly unmanned system. Further, we believe that a system capable of {better than 1 foot} resolution will be more cost-effective in a manned configuration if, "in fact, {better than one foot} resolution is possible at all with an unmanned system.[19]

THE BUDGET BUREAU ASKS A REEXAMINATION OF MOL

Even before the group's preliminary finding on man in the MOL were submitted to General Schriever, the importance of a favorable outcome was reemphasized by a 21 March 1966 memorandum from Mr. Schultze, the Budget Director, to McNamara. In this memorandum, Schultze reminded McNamara about the Budget Bureau's prior reservations about the cost of the manned system as compared to the unmanned version[††]. He referred to the recent DoD studies which indicated that the unmanned system could achieve "substantially the same resolution as a manned system." Consequently, he suggested that the MOL program should be reexamined "to determine whether the benefits other than resolution justify the cost of a manned system." He requested DoD to undertake to develop cost estimates for the unmanned system to provide a basis for a joint review "to determine whether or not reconsideration of the original decision is justified."[20]

In his reply on 25 March, Secretary McNamara advised Schultze that the Air Force was pursuing "a dual development approach" in the MOL program which had the advantage of ultimately permitting it to fly the system either manned or unmanned. He confirmed that recent studies indicated the unmanned version would ultimately approach the {better than 1 foot} ground resolution range

of the manned system. However, he emphasized that man served "a dominant role in the on-orbit development process to achieve the high quality capability" and that by proceeding along this road, he felt a system could be built that would be "close to optimum in each mode" while retaining the benefits of both.

The defense chief further stated that while the Department ultimately expected high performance in the unmanned version, to proceed unmanned from the start "would certainly lower our confidence in the time and reliability with which we hope to achieve the desired performance." He said:

> These considerations are still valid, as they were last summer. We now better understand the techniques and inventions that must be developed to achieve an unmanned capability. As a result, we can make more meaningful estimates of the costs required to go the wholly unmanned route. This effort[‡‡] is in process and should be complete in about two months. At that time we will be able to evaluate the wholly unmanned approach and compare it with the present option of retaining man in the development process and subsequent operations.[21]

As a follow-up to the above exchange of correspondence, DDP&E on 6 April requested Dr. Flax to have the following considerations included in the Air Force's studies of the manned versus unmanned approach:

1. A wholly unmanned system configured to provide the same quality and quantity of reconnaissance-intelligence information as the MOL.

2. The difference and risks of obtaining equal intelligence content with the wholly unmanned system and the development and operating costs to achieve it.

3. The expected performance of the unmanned and manned versions of MOL.

4. The operation effectiveness of the two approaches with regard to numbers of missions required to insure equal target coverage (i.e., objectives seen per day, per week, per month), and ability to select and/or discriminate between

---

†† See pp 127-129

‡‡ For a discussion of costs and funding, see Chapter XI.

target systems. The Air Force also was to assess the manned and unmanned systems' ability to circumvent weather phenomena.[22]

On 8 April Flax instructed Martin to "let us quickly as possible" two conceptual system study contracts for an unmanned reconnaissance satellite system employing the DORIAN optical subsystem. The unmanned system was to have a lifetime goal of at least 30 days on orbit and make use of existing and projected technology, components, and subsystems to the maximum extent possible. The minimum product desired from each contractor, he said, was "a conceptual system design with appropriate analyses," which identified and analyzed critical technology, components and/or subsystems. The contractors also were to provide an estimated system schedule and detailed costs of a follow on operational program at a rate of about 5-6 launches per year.[23]

Flax also requested the Directorate of Special Projects to undertake a separate in-house conceptual study of a wholly unmanned system, to be submitted with the results of the contractor studies by 6 June 1966. The three studies—except for technical information inputs—were to be managed "apart from the MOL Program Office and of your DORIAN project office." Five days later the Directorate of Special Projects awarded two 60-day study contracts ($220,000 each) to Lockheed and General Electric to develop a conceptual system design and schedules and costs for a wholly unmanned system.[24]

These investigations were well under way when General Schriever—two days after being briefed on 25 April on the preliminary conclusions of the Man in the MOL study group—directed Gen Berg to initiate three new studies with results to be made available to him about the same time the wholly unmanned system investigations were completed. He asked, first of all, for a cost comparison study of an optimized manned and unmanned system to include a projection of probability of mission success of each version. Secondly, he requested that another investigation be undertaken similar to the above which concentrated on "an advanced DORIAN system capable of {better than one foot} ground resolution, comparing again optimum manned and optimum unmanned configurations." His third requirement was for a "broadly based parametric study of all relevant factors of experience in past space flight manned and unmanned."

Schriever emphasized the importance of these studies "in the support and justification of the MOL program." In the final analysis, he considered that all of the studies should contribute "and must be woven into a logical rationale" and theme which clearly showed "a current,

defensible contribution of man to the current MOL system, and an expanding capability for follow-on systems which can be exploited in an effective manner."[25]

While work on these new studies began, the MOL Policy Committee on 29 April reviewed the situation. Concerning the wholly unmanned DORIAN system investigations resulting from Mr. Schultze's request, Dr. Brown commented that the Budget Bureau had not received much sympathy "from either DoD or Dr. Hornig." He reported that he and Dr. Flax had discussed Schultze's request with Secretary McNamara and had reviewed the cost effectiveness of MOL as compared to GAMBIT 3. He commented that while an unmanned photographic system to operate for 30 days might be feasible, it also would be costly. Dr. Flax remarked that as the unmanned system studies progressed, it was becoming more evident that "man is a neatly packaged system to do many tasks."[26] When Schriever commented that the desired capability could be attained sooner with the manned approach, Dr. Brown reiterated that neither McNamara nor Hornig were against the manned system. The Secretary of Defense, he said, wanted the intelligence data and only questioned the best way to get it. However, Brown added that cost increases or schedule slips could change the current OSD bias favoring the manned MOL.[27]

Because of a delay in completing the various studies, it was not until mid-June 1966 that the results of the two contracted investigations and the in-house Special Projects analysis were available. In Washington, the NRO staff also had performed an in-house study of the manned and unmanned versions of MOL from the standpoint of the total number of intelligence targets which might be photographed during a typical 30-day mission. In forwarding a copy of the NRO study to OSD, General Evans noted that:

> The results of the study show that, with the astronauts performing a weather avoidance role, the manned system will successfully photograph significantly more intelligence targets than will the unmanned system on a comparable mission. Various cases were examined and the improved factor of the manned system over the unmanned ranged from 15 to 45 percent. The general conclusion reached by the study is that on identical missions against average Sino-Soviet weather, the manned system with the astronaut providing

a weather avoidance function and
having the option of photographing
pre-designated alternate targets,
can be expected to successfully
photograph 18 to 20 percent more
targets than the unmanned system.[28]

In late August an overall final report on "Manned/Unmanned Comparisons in the MOL"—which included data submitted by General Berg's staff and a summary of the wholly unmanned DORIAN system investigations—was forwarded to Dr. Foster as the formal Air Force response to his memorandum of 6 April 1966. Forty-seven pages in length, this consisted of a covering memorandum from Dr. Brown and four major sections, which answered in detail the major points which Foster had requested the Air Force consider.

In his memorandum, Brown repeated the earlier conclusion that either the automatic version of MOL or a completely unmanned configuration potentially could give the same resolution as a manned system. On the other hand, he noted that many of the automatic devices had never before been used in an orbital reconnaissance system, and while it was believed they ultimately could be made to perform reliably, there was uncertainty how long it might take. For this reason, the Air Force was convinced that the risk against early achievement of {better than one foot} resolution was "considerably greater with an unmanned vehicle"; that is, to the extent that man's participation in the development proved effective, "the {better than one foot} resolution unmanned capability should be achieved earlier in the automatic mode of MOL."[29]

The recent studies, he said further, had considered an unmanned DORIAN system flown on both a 30-day and 60-day mission. He agreed that the latter would be more economical—provided that the component reliability could be developed to acceptable levels. In either case, he said that the absence of man increased the development risk. Citing the various new features which were planned for automatic operation during both manned and unmanned flights, he noted that man would be able to override or compensate for most of the failure modes envisioned for this equipment. However, any "out-of-specification performance" in automatic functions could defeat the {better than one foot} resolution objective, whereas retaining man during the orbital development period "not only will enable us to increase the output and quality of reconnaissance data acquired in this period but will assist in identification and correction of equipment deficiencies.

Figure 49. MOL Vehicle
Source: CSNR Reference Collection

Brown reported that the latest estimated development cost of the 7-launch MOL program was $1.818 billion. On completion of the development, the system could be operated manned or unmanned (automatic mode), with the latter able to perform in orbit for 60 days, subject to the same qualifications on reliability stated above for the wholly unmanned system. The development cost of the unmanned (automatic) MOL was estimated at $1.50 billion. When compared in this manner, the difference in development cost for including a manned operating mode and a manned development program was estimated at $318 million. This difference, the Secretary said, would be "almost entirely offset" if the manned system's weather compensation potential proved to be only 20 percent, since it would result in an increased photographic "take" during the development cycle.

Brown also referred to the potential of the manned system to provide a superior intelligence content per day on orbit. The results of operator-reaction tests conducted on a laboratory- simulator showed, he said, that "crew participation in target selection could yield almost three times as many photographs of high-intelligence- value targets as could be taken by an unmanned system on the same mission." The type of operations that the crew could perform included locating significant military vehicles, inspecting special radar equipment, detecting a silo with an open door, detecting a missile being moved, etc.

The Secretary said further than there appeared to be distinct advantages in having a man select cloud-free targets. He pointed out that, in a typical unmanned mission, the photographic loss was 50 percent from cloud cover alone. On a manned mission, the operators could employ their spotting scopes—which would have been pre-programmed against targets along the path—to determine which targets were in the clear "and then orient the main optics for photographing the clear targets." Other advantages of having the man in the system was that he could decide the best viewing angle from which a target should be photographed. If the MOL, for example, approached a parked aircraft from the rear and needed intelligence of its front end, the man could wait until he had passed over and then snap a backward looking picture. He also could, on command from the ground, insert aerial color film, infrared and other special film in the secondary camera so that their special characteristics could be brought into play. Such films might prove of value in detecting camouflaged targets or in acquiring information on the nature and level of enemy industrial plant activity.

Brown cited a number of additional advantages of having man aboard. During times of crisis the MOL could be transferred from its nominal 80-mile orbit to one of approximately 200-300 miles. In this higher orbit the system would have access to all targets in the Soviet Bloc approximately once every three days and be able to take photographs at resolutions of about one foot. The crew could employ the acquisition and tracking scopes, which would provide a resolution of about nine feet, for intelligence by direct viewing. They could detect the absence or presence of aircraft, ships in port, cargo accumulations, parked vehicle build-up, railroad activity, etc. The MOL could enter orbits of about 200 miles after one to 21 days and still remain in orbit 30 days, permitting daily reports of activities of significant value in determining the posture and state of readiness of Soviet forces.

{...}The Secretary further stated that the MOL laboratory module possessed sufficient flexibility to support other missions besides high resolution reconnaissance, such as communication intelligence or ocean surveillance, should they be approved. The manned system in addition had the potential of providing a unique laboratory environment for conducting scientific experiments, having 1,000 cubic feet of pressurized volume and up to 3,000 cubic feet (8,000 pounds) of unpressurized experiment space.[30]

THE RECONNAISSANCE PANEL BRIEFING, AUGUST 1966

Much of the material contained in the above report Dr. Brown sent to Foster also was presented in a day-long briefing given the PSAC Reconnaissance Panel. At this meeting, held on 13 August, were Drs. Land, Baker, Puckett, Shea, Garwin, Steininger, and D. P. Ling of PSAC. The Air Force representatives included Dr. Flax, General Evans, Stewart and Berg, Dr. Yarymovych, and Colonels Battle and David Carter. Mr. John Kirk and Samuel Koslov represented DDR&E and Messrs. Thomas and Fisher the Bureau of the Budget. The main presentations were made by Mr. Michael Weeks, Samuel Tennant, and Dr. Leonard of the Aerospace Corporation.[31]

Mr. Weeks reported to the PSAC members on the studies of the design of the baseline MOL, the steps taken to provide automaticity, even in the manned mode, and plans to provide for reliability through redundancy rather than extensive on-board manual maintenance. His report not unexpectedly was well received by the panel since it reflected previous PSAC guidance. Mr. Tennant then reviewed the "wholly unmanned DORIAN system" and the problems such an approach entailed. He was followed by Dr. Leonard, whose presentation covered the relative effectiveness factors of manned and automatic versions of MOL and the wholly unmanned DORIAN system. His statement that man could perform a better function in weather avoidance was not challenged by the panel; his argument that an added benefit of man's presence was target photography verification was not accepted. Concerning this point, Dr. Garwin suggested that the Itek image motion sensor mechanism could lend itself very well to the verification task by means of recording the output of the device.

Dr. Leonard's major thesis for using a man in the system was that he possessed the ability to detect active indicators and enhance the intelligence "take" by increasing the number of special photographs shot on a mission. While the panel was interested in this concept, doubts were expressed about the validity of the laboratory simulations, which the PSAC members thought were not sufficiently representative of "real viewing conditions." The panel suggested an extensive simulation program would provide more valid data. It also expressed reservations about whether the design of the acquisition and tracking scope (5-inches) was sufficient to enable man to spot active indicators, suggesting a 15-inch aperture might be closer to what was required. Doubts about whether man could actually stay in space for 30 days were voiced by Mr. Koslov.

At the conclusion of the presentations, Dr. Land summarized for the panel. He said it wanted assurances that, as a matter of national need, an unmanned reconnaissance capability would be provided because man might not be able to go on certain missions for political reasons. He did not object if the Air Force put a man in the system for some mission enhancement, which he was quite prepared to accept, but this should not be done at the expense of compromising the stated requirement to build an unmanned reconnaissance capability.

The panel wanted the Air Force to proceed with the various studies needed to answer the several questions raised during the meeting§§. However, he concluded (most importantly from the Air Force viewpoint) that the panel also was adamant "that we should not hold up any contractual proceedings while these questions were being settled."[32]

Several days following this meeting, one of the PSAC members, Dr. Steininger, remarked during a visit with the NRO staff that the DoD was "killing itself in attempts to justify the man." He said the man did not need to be justified to the panel, which accepted his presence. "MOL is an experiment in which man is the experimenter," he said. "We should keep it that way." Further, he stated that the panel insisted on automating all MOL functions so that the man "could stay loose and be an experimenter." The panel, and Dr. Land in particular, thought that a sensor to do man's weather avoidance task also might be built, but felt that DoD was so busy justifying man in the system, "it won't really want to work on the sensor." In summary, Dr. Steininger said, the panel wanted "to release man to do his job."[33]

GENERAL SCHRIEVER'S FAREWELL REMARKS

At the end of August 1966 General Schriever retired from active duty as head of the Air Force Systems Command and as MOL Program Director¶¶. From the earliest days of the nation's missile and space programs in which he had played a prominent role, he had been convinced that man would utilize space for a variety of military purposes, reconnaissance being only one of them, and that it was essential that the Air Force move vigorously into this new realm. On the eve of his retirement, he wrote to Secretary Brown and restated his conviction "that the conduct of manned military missions in space will become indispensable to the defense of the nation in the future." Citing the tremendously expanding and accelerating technology, and a restless international environment, he expressed concern that in the manned military aspects of space, "our pace has been conservative." He said he thought NASA's manned space flight experiences, which had "brought to the forefront the values of man as an integral and essential element in the conduct of space missions of great national significance," were not without implications for the military. He said:

> The inception of the Manned Orbiting Laboratory Program has given us the opportunity to bring into sharper focus a broader appreciation of the potentials of military space by now encompassing the uniqueness, flexibility, and responsiveness of man. Our experience in recent years in military conflict has shown the wisdom of configuring our military materiel to permit its flexible employment in a spectrum of uses. Thus, we are enabled to respond effectively to the new and unpredictable military and political circumstances which inevitably arise. Our experience, likewise, shows that realization of this flexible responsiveness is largely dependent upon man. I see a close parallel between our experience with utilization of conventional military material and that which we will and must employ in space.

§§ On 22 August General Evans directed the MOL Systems Office to undertake a series of new studies to provide the information requested by the panel. See Msg Whig 5623 (S-DORIAN), Evans to Berg, 22 Aug 66, subj: Study Requirements Resulting from PSAC Mtg of 13 Aug 66.

¶¶ He was succeeded as Director by Gen James Ferguson, Deputy Chief of Staff, R&D, Headquarters USAF, who took over as AFSC commander.

It is my firm conviction that conduct
of a vigorous manned military space
program is essential in preparing to
respond to hostile activity in the
space environment. As operational
space functions become more complex
and more sophisticated with time,
the need for the development of truly
effective manned systems emerges
with increasing urgency. There is
no true alternative for a manned
system...[34]

Some 33 days after Schriever's retirement, McNamara decided the unmanned/manned question. After considering the data provided him he advised Mr. Schultze he intended to proceed with "the present MOL Program at the optimum engineering development pace dictated by the development cycle for the optical payload." His reason was that he was more confident that the manned system could achieve {better than one foot} resolution than the wholly unmanned system "because of the engineering development problems in precision subsystems."[35]

ENDNOTES

1. Memo (S-DORIAN), Edwin H. Land to Donald F. Hornig, 18 Aug 65.

2. Ltr (S-DORIAN), Hornig to Brown, 24 Aug 65.

3. Msg Whig 0501 (TS-DORIAN), McMillan to Martin, 29 Sep 65.

4. Talking Paper for Gen Schriever, Oct 65, subj: Manned vs Unmanned (in a Chronological Listing of an Annotated Bibliography, Apr 66); Memo for Secretariat Record (TS-DORIAN), by Gen Evans, subj: Proceedings of MOL Policy Cmte Mtg 65-3, 14 Oct 65.

5. Memo for Secretariat Record, cited above; Memo (TS-DORIAN), Schriever to SAF, 8 Nov 65, subj: MOL Monthly Status Report; Min (TS-DORIAN), AF MOL Policy Cmte Mtg 65-3, 14 Oct 65.

6. Memo for Record (TS-DORIAN), Brig Gen Martin, 10 Nov 65, subj: 8 Nov 65 PSAC Recon Panel Discussion on DORIAN; Memo (TS-DORIAN, Schriever to SAF, 9 Dec 65, subj: MOL Monthly Status Rpt.

7. Ibid.

8. Msg 9312 (S-DORIAN), Martin to Flax, info Schriever, 12 Nov 65.

9. Memo (TS-DORIAN/TALENT-KEYHOLE), Hornig to Flax, with Atch, 22 Nov 65.

10. Min (TS-DORIAN), AF MOL Policy Cmte Mtg 65-4, 30 Nov 65.

11. Memo (TS-DORIAN), Schriever to SAF, 4 Jan 65, subj: MOL Monthly Status Rpt.

12. Ltr (S-SAR), Berg to Evans, 28 Dec 65, subj: Information to Contractors on MOL; Msg 9709 (S-DORIAN), Berg to Schriever, 29 Dec 65.

13. Msg Whig 4585 (S-DORIAN), Schriever to Berg and Martin, 30 Dec 65; Memo (TS-DORIAN), Schriever to SAF, 4 Jan 65, subj: MOL Montly Status Report.

14. Memo (TS-DORIAN), Evans to SAF, 12 Feb 66, subj: MOL Monthly Status Rpt; Min (TS-DORIAN), AF MOL Policy Cmte Mtg 66-1, 8 Feb 66; Memo (TS-DORIAN), Evans to SAF, 4 Mar 66, subj: MOL Montly Status Rpt.

15. Memo for the Record (S-DORIAN), by Gen Evans, 4 Jan 66, subj: Mtg with Gen Schriever on 29 Dec 66.

16. Memo (S-DORIAN), Schriever to Evans, 17 Jan 66.

17. Memos (S-DORIAN), Col Norman to Vice Dir/MOL, 25 May 66, subj: Final Rpt on the Role of Man in the MOL ; Evans to Director, MOL, 25 May 66, same subj.

18. Rpt on Man in the MOL (S-DORIAN), 25 May 66.

19. Ibid.

20. Ltr (TS-DORIAN), BOB (Schultze) to McNamara, 21Mar 66.

21. Ltr (TS-DORIAN), McNamara to Schultze, 25 Mar 66.

22. Memo (TS-DORIAN), Foster to Flax, 6 Apr 66, subj: IDL vs A Wholly Unmanned Sys Dev & Mission Comparison Study.

23. Msg Whig 5063 (TS-DORIAN), Flax to Martin, 8 Apr 66.

24. Ibid.; Memo (S-DORIAN), Evans to SAF, 7 May 66, subj: MOL Monthly statis Rpt.

25. Msg SAFSL (S-DORIAN), Schriever to Berg, 27 Apr 66.

26. Min (TS-DORIAN-GAMBIT), AF MOL Policy Cmte Mtg 66-2, 29 Apr 66.

27. Ibid.

28. Memo (TS-DORIAN-GAMBIT), Evans to Dep Dir, Def ense Res & Engr (Strategic and Space Sys), 9 Aug 66, subj: Comparison of Manned and Unmanned Versions of the DORIAN Sys.

29. Memo (TS-DORIAN-GAMBIT), Brown to DDR&E, 26 Aug 66, subj: Manned/ Unmanned Comparison Study.

30. Ibid.

31. Memo f or Record (S-DORIAN-GAMBIT), by Dr. Yarymovych, 17 Aug 66, subj: Brief ing to PSAC on MOL.

32. Ibid.

33. Memo (TS-DORIAN), Werthman to Flax, 19 Aug 66, subj: PSAC Panel Comments on the MOL.

34. Memo (U), Schriever to SAF, 31Aug 66, subj: Manned Mil Missions in Space.

35. Ltr (TS-DORIAN), McNamara to Schultze, 3 Oct 66.

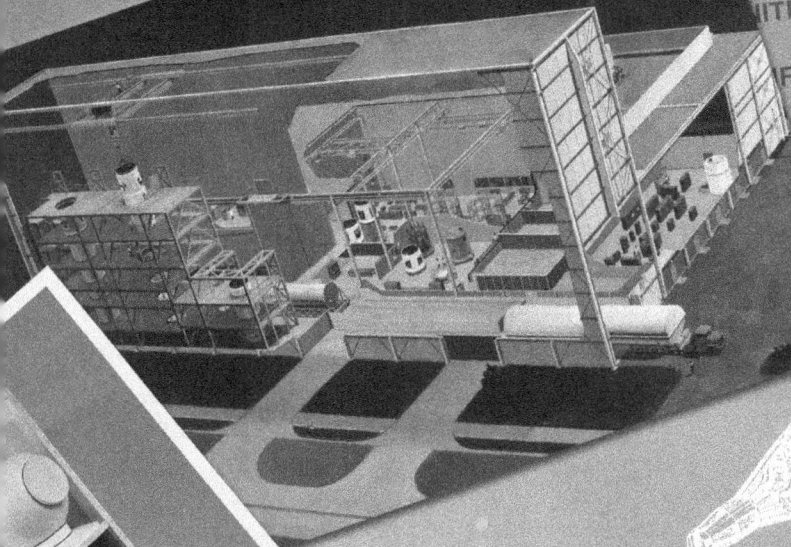

MOL ASSEM
INTEGRATION BUILD

COU
MATERIAL

PROPERTY OF THE
UNITED STATES GOVERNMENT

IF FOUND, DO NOT OPEN

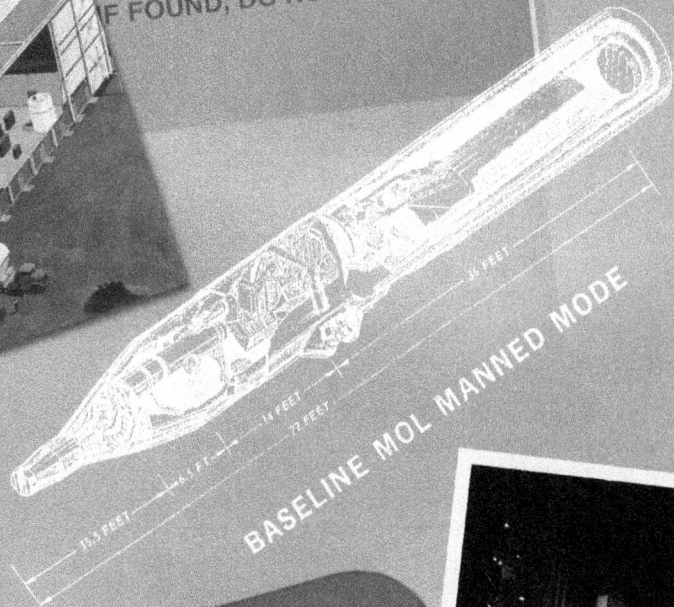

BASELINE MOL MANNED MODE

USAF MOL/KH-10

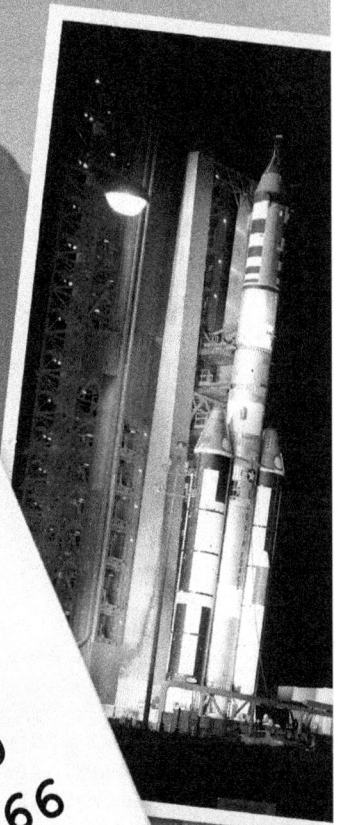

THE DORIAN FILES REVEALED:
A COMPENDIUM OF THE NRO'S MANNED ORBITING LABORATORY DOCUMENTS

CHAPTER XI:
BUDGET, DEVELOPMENTAL, AND
SCHEDULE PROBLEMS—1965–1966

BUDGET, DEVELOPMENTAL, AND SCHEDULE PROBLEMS—1965-1966

Even as Air Force officials were reacting to PSAC's insistence that they incorporate an unmanned configuration into the MOL program*, a severe financial problem arose that threatened and finally delayed early system acquisition. Some 60 days after the President authorized the Air Force to go ahead with the program, Daniel J. Fink, Deputy Director (Strategic and Space Systems), ODDR&E, asked the MOL Program Office to review and substantiate its fiscal 1967 and 1968 budget requirements for Phase II engineering development. This review was the beginning of a critical OSD evaluation of the Air Force's budget requests. Apparently anticipating an OSD rejection, Dr. Flax in early November 1965 cut $20 million from the Air Force 1967 MOL request, bringing the total down from $395 million to about $374 million. Schriever later remarked to Evans that, while he could not quarrel with the Assistant Secretary's cut, "because it is arbitrary," he disagreed with the procedure.[1]

The MOL Program Director, however, soon had much more to worry about than a $20 million reduction. From the Secretary of Defense's office came word that McNamara intended to limit the MOL program in 1967 to $150 million—the same sum provided in 1966. Secretary Brown immediately wrote to the defense chief to voice concern. He pointed out that the original MOL plan had projected the first manned flight in late calendar year 1968, based on a schedule requiring the start of engineering development in January 1966. Dr. Brown reviewed several alternate schedules which might be adopted to reduce fiscal year 1967 funding requirements. But these, he advised Mr. McNamara, would have the effect of slipping the first manned flight three to 12 months. He said a development schedule with a goal of a first manned launch in April 1969 would require about $294 million during the year. He further stated that a $230 million budget would be "the lowest fiscal year 1967 funding compatible with maintaining continuity of contractor efforts already under way" and an early manned flight.[2]

On 18 November, analyzing the MOL funding problem for DDR&E, Deputy Director Fink expressed agreement with the above arguments. That is, he said that if $294 million were provided, the first manned MOL could be flown in early calendar year 1969. If an additional slip of three to six months was considered acceptable, funding could be reduced to approximately $230 million. Fink

recommended to Dr. Foster that the MOL budget not be reduced below the $230 million level. (Endnote 3) These arguments failed to convince McNamara, who, in the final thrashing out of the fiscal year 1967 DoD budget request, concluded that $150 million was sufficient for MOL[†].

This severe cut in 1967 spending plans was a main topic of discussion between Brown, McConnell, Schriever, Paul, Flax, Marks, and Ferguson on 30 November 1965 at a meeting of the MOL Policy Committee. Dr. Brown reported that Foster and Flax were working on alternate plans to slip the program either six or nine months, with fiscal year 1967 funding needs to be calculated in each case. Flax said that DDR&E was in agreement that $150 million would not be enough to support the program during the year. However, Assistant Secretary Marks advised the Committee that there was little hope for reinstatement of the bulk of the requested MOL funds.[4]

Marks' view was soon borne out as OSD rejected an Air Force reclaim and the $150 million total was incorporated into the President's defense budget for submission to Congress in January 1966. On 9 December Evans informed Berg of MOL's unhappy financial prospects. He said there was a possibility that the final budget might provide a slightly higher level but, in any event, the Systems Office should "cost out alternative MOL programs" based on various funding levels, which in each case would insure a balanced program.[5]

Subsequently, on 29 December 1965, General Berg reported to the MOL Program Office on possible actions that might be taken to minimize MOL expenditures in fiscal year 1967. His plan called for completing Phase IB studies by 1 May 1966 and initiating Phase II engineering development by 1 September. This could be done, he said, within a total 1967 budget of $237 million "plus the $100 million carryover of fiscal year 1966 funds." He indicated, if only $150 million were provided, there would be a further stretch-out of the program.[6]

---

* See pp 164 ff.

[†] The fiscal year 1967 budget, prepared in late calendar year 1965, was the first to feel the impact of the accelerating war in Southeast Asia. In the years that followed, the Vietnam War came to require enormous sums to the detriment of many defense projects including MOL.

SENSOR DEVELOPMENT SLIPPAGE

The Systems Office's costing exercises had scarcely gotten under way when MOL officials received more bad news. On 9 December, the Eastman Kodak Company, the DORIAN sensor contractor, dispatched a letter to Gen. Martin advising that the firm would be unable to fulfill its original commitment to deliver the first optical sensor in January 1969 for a planned April 1969 first manned launch‡. Company officials stated they would require a 10-month extension, with delivery of the first flight optics taking place about 15 October 1969 and the first manned launch slipping to mid-January 1970.

This unexpected development was discussed at a West Coast management review meeting on 20 December 1965, attended by Foster, Flax, Schriever, Evans, Martin, and other officials. They decided that General Martin should immediately initiate an investigation and review of the Eastman Kodak schedule, while Dr. Foster made arrangements to travel to Rochester, N.Y. (on 22 December) to discuss the problem with company executives.[7] General Schriever's reaction was that the Air Force should not accept the new Eastman Kodak schedule "at this time." On 30 December he suggested to Berg and Martin that, in drawing up their plans, they continue to aim for a late 1969 launch.[8]

They agreed, pending a detailed review of Eastman Kodak's schedule and costs, not to accept as final the proposed schedule slippage, although Berg noted that a lengthy schedule slip would have at least one beneficial effect of nearly fitting "the constraint of FY 67 expenditures of approximately $230 million."[9] Meanwhile, General Martin organized a committee of Special Projects officers who proceeded to Rochester (5-8 January 1966) to review in some detail the company's schedule information, the reasoning and philosophy behind it, and its physical and personnel resources. The information collected was subsequently compared with the Special Project Directorate's several year's experience in acquiring several unmanned reconnaissance systems. Among other things, they noted that the development time for three unmanned systems—with sensors substantially smaller than the planned DORIAN optics—ranged from 19 to 33 months (from time of program go-ahead to the first fight), whereas the proposed new Eastman Kodak schedule would require 51 months (from the October 1965 contract signing to the first flight in January 1970).[10]

On 20 January, after being briefed on the above review, Dr. Flax also travelled to Rochester to discuss the MOL sensor schedule with company officials and

examine ways of compressing it to achieve an earlier launch.[11] During his conversations with these officials, Flax suggested that they consider a less conservative approach. On his return to Washington, the NRO Director wired General Martin requesting he prepare at least two DORIAN schedules—one for the "baseline" Eastman Kodak proposal and the other a compressed schedule which would provide a launch "at least six months earlier."[12]

On 28 January, in response to Flax's suggestion, Eastman Kodak submitted a new DORIAN development schedule to the Directorate of Special Projects, designated Plan B. It would eliminate the prototype compatibility model flight article and accelerate delivery of the qualification model even before the latter had been completely qualified.§ By taking this approach, the company stated it could deliver the first Flight Model (FM-1) in April 1969 as opposed to October 1969. However, Special Projects considered this plan undesirable from the viewpoint of "quality assurance" and it asked the firm to take another look at the schedule problem.[13]

Whereupon, Eastman Kodak prepared and submitted a third alternate proposal (Plan C) on 9 February. It called for delivery of FM-1 in July 1969 while still retaining the compatibility model and completing qualification model testing prior to the launch of the first flight sensor. Applying the normal delivery-to-launch time span of three months, this meant that the first "all-up" DORIAN sensor could be launched in October 1969. Company officials warned, however, that they required a prompt go-ahead on construction of essential new facilities to maintain this schedule. On 15 March, after further meetings and discussions with Special Projects personnel, the firm submitted a revised schedule which called for delivery of a Camera Optical Assembly (COA) at Rochester 37 months from the day the Air Force authorized new facility construction¶.[14]

Several weeks later, however—at the MOL monthly management review meeting on 2 April 1966—Mr. John Sewell of Eastman Kodak advised there would be a new two-month slip in the delivery of the FM-1 optical system, from 15 July to 15 September 1969. He attributed this to Air Force delay in authorizing construction of facilities, the problem of acquisition of long lead items of equipment,

---

‡ These dates were agreed upon in July 1965 during discussions between company officials and Dr. McMillan.

§ The compatibility model, integrated with the Mission Module forward section and the Laboratory Vehicle, was to be used to check out interfaces and system operation. The purpose of the qualification model was to demonstrate the system's ability to meet performance requirements in a simulated space environment.

¶ These facilities included a new steel frame building and a masonry building about 141,200 square feet to house several test chambers, plus various items of equipment. Total estimated costs of facilities and equipment was $32,500,000. [See Memo for Record (TS-DORIAN), prep by Col R. C. Randall, 25 Mar 66, subj: Status on DORIAN Facilities and ASE Requirements]

and the firm's "underestimate " of the time needed to debug the planned optical test chamber.[15] To eliminate the first obstacle, Dr. Flax on 4 April asked DDR&E for authority to proceed with the purchase of the unique facilities and support equipment needed to develop the primary optical sensor. Foster quickly approved, whereupon Flax authorized Martin to sign a contact with Eastman Kodak and proceed with the necessary facilities construction.[16]

Meanwhile, following the 2 April 1966 management meeting, the MOL Program Office took another critical look at the firm's proposed 15 September 1969 first manned launch. Working back from the first date, it became clear to MOL officials that "only 31 months is available to EKC," not the 37 months the firm stated it would need to deliver the camera optical assembly. The resulting six months gap would therefore slip the proposed first all-up launch to April 1970.[17]

MOL Program officials were thus faced with the fact that less than eight months after the President had announced the first manned launch would take place in late calendar year 1968, it had slipped into calendar year 1970. This situation was particularly embarrassing to those OSD and Air Force officials who had recently testified before Congress. On 23 February 1966, for example, Secretary Brown told the House Appropriations Subcommittee: "Our best estimate at this time is that the first manned flight will not occur prior to mid-1969, which is a slip of about nine months from what we said last year." On 8 March Dr. Foster, also advised the Senate Armed Services Committee that the first manned launch would take place "about mid-calendar year 1969."

Citing the repeated schedule slips in the program, General Evans on 7 April expressed apprehension to the MOL Director that there might be "a very adverse effect on the program as a whole and the Air Force's management image." He said that, if the schedule slipped further—due to the still unresolved funding problem—the program might not survive "as a manned reconnaissance system." He urged steps be taken to reduce expenses and that all contractors be advised "that we have a major cost problem, that we need their assistance, and that they should be creative in exploring ways of reducing program costs."[18]

THE FY 1967 BUDGET REVIEW

The MOL budget was already under intensive study by Air Force officials. In late December 1965, acting on news that McNamara intended to limit 1967 program funding to $150 million, General Schriever ordered a thorough budget review of financial requirements. He directed Evans to establish a budget review committee to meet with top contractor officials to analyze the latter's cost proposals for Phase II engineering development.[19] Subsequently, Evans proposed, and Flax approved, establishment of several task forces and senior cost review boards.

One task force, headed by Col Robert Walling of the MOL Program Office, consisted of eight officials who beginning in April 1966 embarked on an intensive review and evaluation of contractor white financial requirements. Their results were subsequently submitted to a MOL Senior Cost Review Board, chaired by Schriever and including Evans, Stewart, Yarymovych, Major General G. F. Keeling and Brig Gen W. E. Carter of Headquarters AFSC and Major General David M. Jones of NASA. The second task force, chaired by {the} NRO Comptroller, performed the same kind of review in the black area. Its membership included three representatives from the MOL Program Office and two from the Office of Space Systems.[20] In this area General Martin had major responsibility for compiling the DORIAN payload cost estimates. However, because Eastman Kodak and General Electric could not complete their final Phase II cost estimates by 1 May, Martin's DORIAN report (submitted on 22 April 1966) contained only the "best cost data" and were subject to change when firm contractor proposals were received.[21]

On 29 April, at a meeting of the MOL Policy Committee, General Evans presented the results of the above cost reviews. He reported there was a substantial difference between the contractors' preliminary or interim estimates ($2.6 billion) and the Air Force's estimate ($1.978 billion) of program costs. Evans reviewed six different program options for the committee to consider: two of them would reduce the number of flights by one or two, with program costs dropping to either $1.817 billion or $1.714 billion.[22]

After the briefing, the Committee directed that an additional seven-shot schedule—two unmanned flights in 1969 followed by a first manned flight in December 1969 with fully-qualified DORIAN optics—also be "costed out." The Committee recognized the difficulties in getting early delivery of the first qualified optics package, but still wished to see the first manned flight take place in calendar year 1969. Concerning the optics problem, Flax and Schriever agreed they would visit Eastman Kodak prior to the next Committee meeting to discuss the matter with company officials. As for the Air Force's "public posture" on possible further schedule slippage, the Committee directed that no unclassified announcement be made "at this time."[23]

To assist Flax and Schriever during their visit to Rochester, the Directorate of Special Projects prepared a detailed background paper on Eastman Kodak's situation. The paper noted that during many discussions and reviews with company officials, Special Projects personnel had found it extremely difficult to single out critical hardware items that could be given special attention. For example, the contractor maintained that in his judgment he needed all his allocated time spans to do the various jobs, such as mirror polishing, fabrication, etc. The Directorate said there were three significant reasons behind the contractor's stance:

> First, he has undoubtedly factored into his planning the bitter experience he is presently having in attempting to meet G3 schedules**. Secondly, he is undoubtedly concerned about the availability on schedule of the large new facility and the unknowns facing him in the area of simulated zero gravity testing of 72" light weight mirrors. Thirdly, he must produce specification performance {better than one foot} resolution) on the first flight. Considering the cost per flight, he is not disagreeing with this rationale; however, past programs have started initially with lower specifications and worked up to specified performance [only] after a number of flights.[24]

Eastman Kodak's conservative personnel policies also were an important factor affecting the schedule. The company believed in "a well groomed organization" and felt it could only be achieved by increasing personnel strength at a modest rate. New personnel, after being sent through a short indoctrination course, were assigned to a job with well-defined responsibilities and inter-relationships, thus preserving a solid "teamwork" attitude and approach to a project. The firm believed this approach was more economical and necessary to insure a satisfactory end product.[25]

With this background information in mind, on 17 May Flax and Schriever met with Eastman Kodak officials at Rochester to again review the entire problem. On the basis of this meeting, during which the Air Force officials emphasized the importance of an early manned launch, an agreement was reached on a MOL schedule which provided for the first manned flight with DORIAN optics in December 1969 using the prototype compatibility model and the launch of FM-1 in April 1970. Eastman Kodak agreed to do what it could to insure the compatibility model was provided high quality optics. On his return to Washington, Flax directed Martin and Berg to review their plans in light of this decision.[26]

On 20 May 1966 the MOL Policy Committee met again to review the schedule problem and the latest program costs for Phase II engineering development. In attendance were Brown, Paul, Flax, Schriever, Ferguson, and Gen William H. Blanchard, sitting in for the Chief of Staff. General Berg briefed the Committee on the program's estimated costs, which he said had risen substantially above the original August 1965 program costs of $1.5 billion. For the currently-approved nine-shot baseline program[††], the major contractor estimated their overall costs at $2.805 billion. This compared to the System Office's estimate of $2.058 billion.[27] In this regard, Schriever and Flax both remarked that they had stressed to the contractors the importance of cutting costs and had warned of the danger of project termination if costs went too high.

The Deputy Director, MOL, also reported on "Option 6," the proposed seven-shot program (a first manned launch in December 1969) which the Committee had requested be costed out. He said this option would reduce the funding requirement to $1.75 billion. Dr. Brown and General Schriever agreed Option 6 constituted the best schedule, although the Secretary noted it would still require extra funds in fiscal year 1967. The Committee formally determined that Option 6 should be adopted and it so directed.

Under this schedule there would be seven flights— three in calendar year 1969, three in 1970 and the last in 1971. Flight No.1 would be the Gemini B qualification flight. Flight No.3 would be the first manned flight and would carry the compatibility model camera-optical sensor fully operational, Flight Nos. 4 and 5 would be manned-automatic, and Flights 6 and 7 would be automatic. The Committee directed that a firm program cost baseline be established for this seven-flight program after Phase II contractor negotiations were completed.[28]

** An advanced unmanned reconnaissance system {better than one foot resolution}, G3 was initiated, in February 1964 with a first flight scheduled for July 1966.

†† The original MOL schedule included an initial booster development flight, and one unmanned and five manned flights. In December 1965 this seven flight program was increased to nine to provide two vehicles to fly the unmanned system recommended by PSAC.

INITIATING MOL ENGINEERING DEVELOPMENT

When Option 6 was adopted by the Committee, the contract definition studies (Phase IB) were not yet completed. The MOL Program Office found it necessary to provide additional funds to extend the studies into Phase IC, described as preparation for engineering development. On 31 March Dr. Flax approve an extension of the Eastman Kodak contract for another 120 days, and a similar four-month extension was authorized to cover General Electric and Douglas activities into the first quarter of fiscal year 1967. On 24 June, at the request of the MOL Program Office, DDR&E released $60 million in fiscal year 1966 funds, bringing the total made available to the authorized $150 million. Of that amount, approximately $108 million was eventually spent on contract definition activities (Phases IB and IC).[29]

While all these activities were under way, MOL officials had continued to search for additional program funds. As noted, the Air Force had advised OSD that it required a minimum of $230 million in 1967 to maintain a balanced program. When Secretary McNamara refused to increase the budget beyond $150 million, General Evans and his staff sought the assistance of the Congress requesting that the fiscal year 1967 DoD appropriation for MOL be increased an additional $80 million. Beginning in early calendar year 1966, they briefed key Senators and Congressmen and selected members of the House and Senate Appropriations and Armed Services Committees. In May 1966 this effort bore some fruit when the House Armed Services Committee agreed to provide the requested additional $80 million. The Senate committee however, failed to act and it seemed the entire effort had been lost. In the meantime, on 30 June 1966 a sympathetic Dr. Foster advised MOL Program Office officials that he would, through internal OSD action, increase 1967 program funds another $28 million. This brought the total to $178 million, leaving a shortage of about $52 million.[30]

Subsequently, the MOL Systems Office prepared a "MOL Program Plan and Funding Requirements" document which identified a still higher 1967 funding requirement—$253.9 million. In forwarding this document to Secretary Brown on 20 July‡‡, General Schriever remarked prophetically that it was "difficult to be sure that development costs will not exceed the $1.75

billion estimate." That total, he pointed out, was based on contractor studies and in-house investigations and did not include any contingency funds.[31]

In a separate paper submitted to Dr. Brown the same day, the MOL Director listed several alternate approaches for proceeding with MOL development. He noted that if the program was limited to the OSD apportionment of $178.4 million in 1967, it would suffer a major schedule slip and delay the first manned reconnaissance flight to approximately June 1970. This schedule also would generate very high funding requirements (estimated at $550 million each) in 1968 and 1969. A second alternative—if 1967 total funding were raised to $208.4 million—would provide a first manned reconnaissance flight in April 1970. The Director, MOL also listed a third alternative, which he advocated. That is, he proposed that the Air Force:

```
Proceed     initially     with...the
recommended program schedule [the
first manned flight in December 1969]
with the proviso to reschedule the
MOL Program no later than-January
1967 based on the realities of
negotiated contract prices and FY
67 fund availability. Also this
approach would allow the subsequent
reprogramming action to take into
account the level of FY 69 funds
provided in the [impending] FY 68
budget. The merit of this approach is
that it affords the least disruption
to the program until contract
negotiations have been completed and
proceeds with the program development
build up to a point in time where
contractor effort could be held to
proceed at a level based on a program
schedule dictated by the end FY 67
and FY 69 funding availability.[32]
```

Several weeks elapsed without a response from the Secretary's office whereupon General Evans on 18 August wrote to Drs. Brown and Flax and asked for a decision on proceeding with full-scale engineering development. He brought to their attention certain OSD policies governing the start of engineering development, pointing out that two important program elements—the Titan IIIM and the DORIAN sensor payload— had with OSD approval already entered into Phase II development

‡‡ Originally submitted to Dr. Brown on 22 June 1966, the document was returned to the MOL Program Office with a request for certain changes. It was resubmitted on 20 July.

because they required long lead times.§§ Based on the approval already granted, Evans suggested there was clear "intent and willingness" on the part of OSD "for the Air Force not only to proceed with the Engineering Development Phase but also to protect development lead-time where necessary." The only limitation or hindrance to going into full engineering development on all MOL segments was the funding deficit in 1967 and subsequent years. In view of the above, Evans urged Drs. Brown and Flax to authorize the MOL Program Office to proceed with engineering development.[33]

On 20 August Secretary Brown accepted the above recommendation. He authorized the MOL Program Office to obligate fiscal year 1966 and 1967 funds at the necessary rate to protect development lead time with requirements for 1967 funds being limited to $208 million. He said this authorization would apply "only until program approval for full-scale development and, in any event, will not apply beyond January 1, 1967." He asked that every effort be made to hold 1967 funding to a minimum, consistent with the primary objective for the first manned flight.[34]

Whereupon General Evans on 30 August directed Berg to continue his negotiations with all major contractors in accordance with the flight objectives and schedules defined in the Program Plan and Funding Requirements document. He was requested to prepare a briefing on total program costs resulting from these negotiations, and was authorized to obligate 1966 and 1967 funds as needed to protect schedules and development lead times, up to 1 January 1967. He also advised that, pending a review by higher authority of the final negotiated program costs, authority to proceed with full-scale MOL development would be withheld.[35]

General Berg took immediate steps to implement this directive. His office issued "pre-contract" cost letters to Douglas, General Electric, and McDonnell for the month of September 1966, limiting them to expenditures of $4.0 million, $2.0 million, and $1.789 million respectively. To provide contractual coverage for the above, the MOL Systems Office planned to negotiate amendments to Phase IC contracts to cover this interim effort until engineering development contracts were approved.[36] In a report to Dr. Brown on the above actions, General Evans advised that he hoped to be able to provide him "by late

October with the firm cost data you require to support a decision on full-scale development of the MOL.¶¶ Prior to the decision, I will continue to protect the flight schedule, within the funding constraints you have stipulated."[37]

In the meantime, on 25 August 1966, the MOL Program Office received good news from Capitol Hill, where a Joint House-Senate Conference approved a compromise appropriation of $50 million to be added to the $150 million requested by the President in his January budget. The gift came on the first anniversary of the President's announcement that the United States would proceed with the Manned Orbiting Laboratory Program and brought the total fiscal year 1967 appropriation to $228.4 million.[38]

---

§§ As noted, in early April 1966 the Eastman Kodak facilities/equipment package was approved by DDR&E. Subsequently, on 1 August, a $258,471,000 negotiated contract for the sensor engineering development phase was awarded the contractor. In the case of Titan IIIM, in early 1966 the four contractors involved were provided $20 million in 1966 funds to begin engineering design and some hardware development. They were: Martin Marietta Corp., United Technology Center, Aerojet General Corp., and AC Electronics Division of General Motors.

¶¶ Despite the fact that "full-scale" MOL development was not authorized, and only segments of the program were fully funded, the date of 1 September 1966- which Berg had suggested the previous December as the date for initiating Phase II—was adopted by the MOL Program Office as the official start of engineering development.

ENDNOTES

1. Memo (C), Evans to Schriever, 3 Nov 65, subj: MOL FY 66-67 Funding. Schriever's written remarks are scrawled on this memorandum.

2. Memo (TS-DORIAN), Brown to McNamara, 4 Nov 65, subj: MOL Funds for FY 67.

3. Memo (C), Fink to Foster, no date. Handwritten comment states this memo was "Given to John Kirk, 18 Nov 65."

4. Min (TS-DORIAN), AF MOL Policy Cmte Mtg 65-4, 30 Nov 64.

5. Msg (S-DORIAN), Whig 4467, Evans to Berg, 9 Dec 65, subj: MOL Reprogramming Alternatives.

6. Msg (S-DORIAN), 9709, Berg to Schriever, 29 Dec 65.

7. Memo (TS-DORIAN-GAMBIT), Evans to Flax, 5 Jan 66, Subj: Talking Paper for the MOL Prog, Tab 7; Atch to Memo (S-DORIAN), Evans to Schriever, 13 May 66, subj: EK Visit on 17 May, with atch, "Background Information on EKC Scheduel."

8. Msg (S-DORIAN), Whig 4585, Schriever to Berg & Martin, 30 Dec 65.

9. Msg (S-DORIAN) 9783, Berg to Schriever, 6 Jan 66.

10. "Background Information on EKC Schedule," cited in note 7 above; Msg (TS-DORIAN/GAMBIT), Martin to MOL Program Office, attached to Memo (S-DORIAN), Evans to Schriever, 13 May 66, subj: EK Visit on 17 May.

11. Msg (S-DORIAN) 9901, Col Smith to Flax, 19 Jan 66.

12. Msg (TS-DORIAN), Whig 4681, Flax to Col Smith, info Schriever & Berg, 24 Jan 66.

13. "Background Information on EKC Schedule," cited in note 7.

14. Ibid.

15. Memo (S-DORIAN), prep by Col Walling, 13 Apr 66, subj: Program Review, MOL Systems Office, 2 Apr 66.

16. Msg (S-DORIAN), 0650, Martin to Flax, 4 Apr 66; Memo (TS-DORIAN), Foster to D/NRO, 66, subj: Approval of Spl MOL Facilities at EKC.

17. Memo (S-DORIAN), Norman to Evans, 5 Apr 66, subj: The Schedule Problem.

18. Msg (S-DORIAN), Whig 5055, Evans to Schriever, 7 Apr 66.

19. Memo for Record (S-DORIAN), by Gen Evans, 4 Jan 66, subj: Mtg with Gen Schriever on 29 Dec 65.

20. Memo (U), Berg to Evans, 8 Mar 66, subj: Cost Review Team; Msg (TS-DORIAN), Whig 4897, Flax to Martin, 10 Mar 66; Memos (S-DORIAN), Evans to Flax, 29 Mar 66, subj: Establishment of a MOL Cost Review Team; (S-DORIAN), Flax to Evans, 15 Apr 66, same subject.

21. Ltr (TS-DORIAN), Martin to Flax, Schriever, 22 Apr 66, subj: DORIAN Cost Estimates.

22. Mins (TS-DORIAN), AF MOL Policy Cmte Mtg 66-2, 29 Apr 66; Memo (TS-DORIAN), Evans to SAF, 6 May 66, subj: MOL Monthly Status Report.

23. Msg (S-DORIAN), Whig 5230, Flax/Schriever to Martin and Berg, 29 Apr 66; Mins (TS-DORIAN), AF MOL Policy Cmte Mtg 66-2, 29 Apr 66.

24. Memo (S-DORIAN), Evans to Schriever, 13 May 66, subj: EK Visit on 17 May with Atch prepared by Dir/Spl Projects, "Discussion Points for EK Visit," and "Background Information on EKC Schedule."

25. Ibid.

26. Msg (TS-DORIAN), Whig 5296, Flax to Martin, Berg, Evans, 20 May 66.

27. Mins (TS-DORIAN-GAMBIT), MOL Policy Cmte Mtg 66-3, 20 May 66.

28. Ibid.; Msg (TS-DORIAN), Whig 5301, Flax/Schriver to Berg and Martin, 23 May 66, subj: MOL Policy Cmte Mtg 66-3, 20 May 66.

29. Msg (S-DORIAN), Whig 5033, Flax to Martin, 31 Mar 66; (TS-DORIAN), Whig 5159, to Martin, 14 Apr 66; (S-DORIAN), 1023, Lt Col Buettner to Randall, 4 May 66; (S-DORIAN), Evans to Flax, 11 May 66; Memo (C), Foster to SAF, 24 Jun 66, subj: Approval of FY 66 RDT&E MOL Program.

30. Memos (U), Evans to Maj Gen D. L. Crow, Dir/Budget, 3 Jun 66, subj: MOL Funding; (C), Foster to Flax, 30 Jun 66, subj: FY 67 MOL Funding; (U), Evans to Schriever, 8 Jul 66, same subject.

31. Memos (S-DORIAN), Schriever to SAF, 22 Jun 66, subj: MOL Program Plan and Funding Requirements; (S-DORIAN), Schriever to SAF, 20 Jul 66, same subject.

32. Memo (S-DORIAN), Schriever to SAF, 22 Jul 66, subj: MOL Program Plan and Funding Requirements.

33. Memo (S-DORIAN), Evans to Flax, 18 Aug 66, subj: Authorization for MOL Full-Scale Development.

34. Memo (S-DORIAN), Brown to Dir/MOL, 20 Aug 66, subj: Authorization to Proceed with the Engineering Development Phase of the MOL Program.

35. Msg (S), SAFSL 9087l, Evans to Berg, 30 Aug 66.

36. Memo (S-DORIAN), Evans to Brown, 7 Sep 66, subj: MOL Monthly Status Report.

37. Memo (S-DORIAN), Evans to SAF, 7 Sep 66, subj: Engineering Dev Phase of the MOL Program.

38. Memo (S-DORIAN-GAMBIT), Evans to SAF, 7 Sep 66, subj: MOL Monthly Status Report.

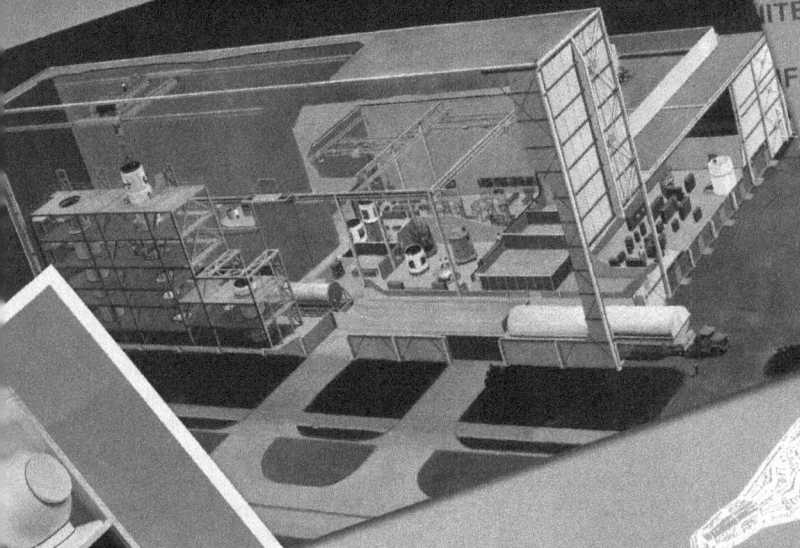

MOL ASSEMBLY INTEGRATION BUILDING

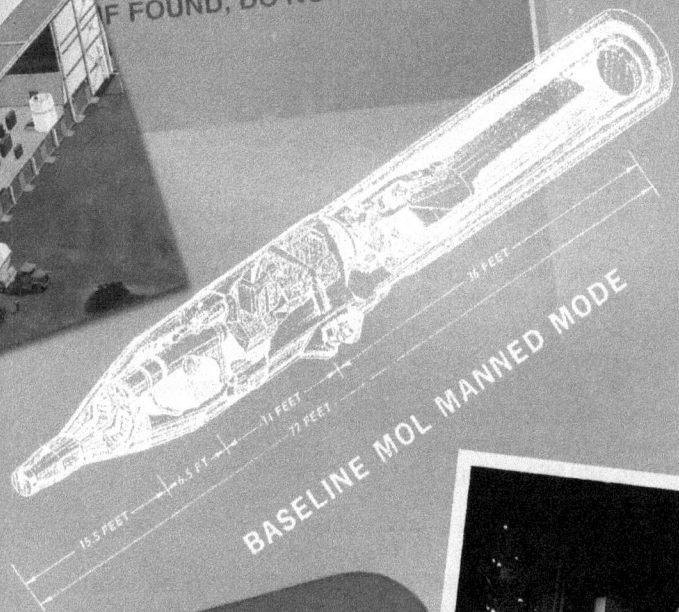

MATERIAL

PROPERTY OF THE
UNITED STATES GOVERNMENT

IF FOUND, DO NOT OPEN

BASELINE MOL MANNED MODE

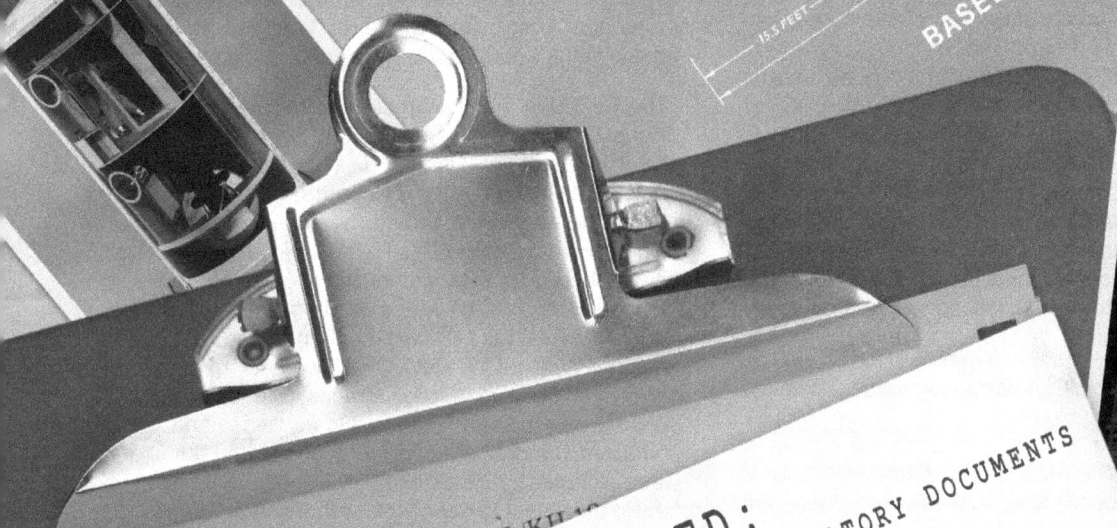

USAF MOL/KH-10

THE DORIAN FILES REVEALED:
A COMPENDIUM OF THE NRO'S MANNED ORBITING LABORATORY DOCUMENTS

CHAPTER XII:
CONGRESS, MOL SECURITY AND
THE RANGE CONTROVERSY

CONGRESS, MOL SECURITY AND THE RANGE CONTROVERSY

When the MOL project received Presidential approval in August 1965, Air Force officials recognized they would soon be called to testify before Congress. To smooth their path on Capitol Hill, they worked closely with representatives of the Office of Legislative Liaison, OSAF, in particular with Col William B. Arnold, who was extraordinarily helpful to the program. One concrete example of his aid was his persistent and persuasive work with key members of the appropriations committees, which led to the Joint House-Senate Conference decision to add $50 million to the fiscal year 1967 budget.[1]

In looking ahead to appearances before Congress, Program officials were troubled by the problem of how to preserve MOL/DORIAN security in the face of expected committee inquiries*. One of the first things they determined to do in this instance was to give DORIAN briefings to key staff members of the House and Senate space committees. Thus, on 10 September, James J. Gehrig, Staff Director of the Senate Committee on Aeronautical and Space Sciences, and on 21 September W. H. Boone, Chief Technical Consultant to the House Committee on Sciences and Astronautics, were DORIAN-briefed.[2]

The first request for MOL information came, however, from Chairman Chet Holifield of the House Military Operations Subcommittee. In a letter to the Air Force in early November, he requested a briefing for two of his staff, Herbert Roback and Daniel Fulmer. Since the Holifield's subcommittee was primarily interested in missile and space ground support equipment, General Evans was able to avoid entirely the sensitive mission area. On 17 November he and his staff presented a three-and-one-half hour "Secret" briefing to the two Congressional aids which apparently satisfied the Chairman's requirements.[3]

However, the next Congressional query—received on 14 December from the House Committee on Science and Astronautics—immediately posed a security problem. On that date the committee informed the Air Force that it would hold hearings in January 1966 on "the operational aspects" of MOL and "how the MOL program complements and/or duplicates the NASA Apollo Applications Program."[4] Apparently, the committee's

plan was not coordinated with its Chief Technical Consultant, Mr. Boone, who had been DORIAN-briefed several months before. On 16 December, accompanied by Colonel Arnold, Evans met with a different member of the committee staff, Mr. Peter Girardi, who requested that the MOL briefing in January be conducted at the "Confidential" level. The Vice Director replied it could not be done, that "with some tolerance on the part of the Committee," the Air Force might be able to give a Secret briefing. To Girardi's questions concerning MOL experiments and payloads, Evans stated that DoD security regulations required special access to those areas and he said they would be "troublesome ones to handle in the hearings from a security standpoint."[5]

The next day the Congressional security matter was discussed at a meeting attended by Flax, Evans, Arnold, and Brig Gen L. S. Lightner, Deputy Director, Office of Legislative Liaison. After reviewing possible courses of action, the conferees agreed the committee chairman, Congressman George P. Miller of California, should be approached. Dr. Flax authorized Colonel Arnold and General Berg to visit Miller at his home at Alameda, Calif., to explain the difficulties of an open hearing and to suggest it not be held. They were to offer, as an alternative, to give a special access MOL briefing to selected committee members. On 20 December Arnold contacted Congressman Miller by phone and a meeting was arranged for 29 December. Several days later he flew to the Coast to coordinate with General Berg.[6]

On 29 December the two officers met with the Committee Chairman at his Alameda office and discussed the MOL security problem, in general terms, for several hours. During their conversations—described by General Berg as "affable and interesting"—Miller stated that he under- stood the need for security and did not wish to disturb it. He also said that he, personally, did not want to know MOL's mission and did not believe it necessary to receive highly classified information. It was his policy, he explained, not to accept invitations to visit classified projects in his district. However, he rejected their suggestion that a classified briefing be given to selected members of his committee, saying that he believed he should hold a regular hearing. But he said his mind was open on the matter and he asked Arnold to contact him in Washington in January to arrange a meeting between himself, Secretary Brown, and General Schriever.[7]

---

* The problem was unique in that unlike other completely black programs, MOL was both black and white, had been publicized by the President, and information on it was expected to be demanded by Congress.

Several weeks later Dr. Brown directed Arnold to contact the chairman and advise that the Air Force Secretary would like to call at his Capitol Hill office to discuss MOL. Miller proposed, instead, to visit Brown's office in the Pentagon. At this meeting on 14 January, the Congressman and the Secretary agreed hearings would be held but that they would be confined to seven areas: the MOL booster, life support systems, tracking stations, ships, recovery areas, schedules, and rendezvous. They also agreed that Flax and Schriever would appear before the committee, the exact date to be determined by the chairman[†].[8]

The House committee hearings, initially set for 31 January, were rescheduled on 7 February. To prepare for them, the MOL Program Office—assisted by General Berg's staff—undertook an intensive effort to prepare non-DORIAN classified and unclassified statements for General Schriever's expected appearance. Unfortunately, several days before the Committee was to meet, the White House released a report on the nation's space program which drew unexpected and unwanted public and Congressional attention to the program.

THE FLORIDA UPRISING

On 31 January President Johnson submitted his annual report to Congress on the U.S. aeronautics and space program (for calendar year 1965). Simultaneously, copies were released to the press and within days—based on its content—the news media of Florida was angrily denouncing the Air Force and its MOL plans. The outcry became so great that Senator Anderson agreed that his committee, which had been planning general hearings on the military space program, would hold a special session devoted solely to the MOL project.

The President's report to Congress noted that a number of unmanned MOL launches would be made from the Eastern Test Range, using Titan IIIC vehicles, and that at least five manned launches would be "flown out of the Western Test Range." This information was not new. Dr. Flax, in response to a query from Florida Congressman Edward J. Gurney (who represented Brevard County including the Cape Kennedy area), had reported on the above plans in general terms in a letter dated 2 September 1965. Various trade publications, such as Aviation Space Technology also had noted that the Air Force would launch MOL from the West Coast.[9] Unfortunately, the President's report went on to state that during 1965, the

Air Force had completed its Titan III.Integrated-Transfer-Launch (ITL) facilities at the Cape—"a dual launch pad facility" which possessed "a high launch rate capability" (and cost $154 million) and also had begun work on "an initial launch capability at the Western Test Range."[‡] The latter, it said, would provide support "for polar or near polar orbit mission requirements that would be degraded if flown from Cape Kennedy."[10]

It was this information that the Florida media seized on to raise a cry of "costly duplication" of facilities. In a lengthy front page story, the Orlando Sentinel on 4 February castigated Air Force planners for "cutting loose completely" from the Eastern Test Range and saddling the U.S. taxpayer with unnecessary costs. It derided the Air Force for claiming it was necessary to launch MOL into polar orbit from Vandenberg, and cited unnamed "veterans of the space program" as declaring such a requirement was "nonsense." Fulminating against "certain Air Force space empire builders," the Sentinel urged Florida's Congressmen and Senators, state and local government officials and the citizens of the state "to stop this threatened waste of national resources."[11]

Consequently, when Chairman Miller called the House Space Committee into executive session on 7 February to take testimony on MOL, Congressman Gurney was primed for attack. It was immediately evident that the Air Force's informal contacts with the Chairman had paid off. In an opening statement, he declared that "the Committee had no interest in [MOL's] mission or characteristics as there could not possibly be any duplication in these areas." He advised the members to concentrate their attention on those things common to the Air Force and NASA programs to check duplication and he listed the seven areas previously coordinated with Dr. Brown.[12]

The first witness, Dr. Seamans, Associate Administrator of NASA, began with a strong endorsement of the MOL program. The space agency, he said; agreed the program met national requirements and had assisted the Air Force in a variety of ways, including conducting studies, dealing with vehicle design. When Congressman Gurney's turn to ask questions came, he asked Dr. Seamans whether NASA had ever successfully launched vehicles into polar orbit from the Cape. Yes, it had, Seamans replied, but in each case the space agency had required State Department and DoD approval because of the safety/overflight problem. Gurney pursued the matter, declaring that, in his view, the total danger for an ETR polar launch was no greater than a WTR launch and, "if NASA could perform polar orbits from the Eastern Range, why couldn't DoD?"[13]

[†]   The chairman earlier agreed that Mr. Girardi might be briefed on the program and this was done on 5 January. Senator Anderson, Chairman of the Senate Space Committee was given a DORIAN-level briefing on 21 January and, on 4 February, Sen. Margaret Chase Smith, a member of his committee, was also indoctrinated into the program.

[‡]   This data was provided by the Office of the Deputy Chief of Staff, Research and Development, Headquarters USAF,

Seamans responded that a "national decision" had been made to launch certain operational programs from Vandenberg into polar orbits and to use the ETR for equatorial orbits. He noted that NASA planned to launch operational weather satellites from the West Coast because they required polar orbits. Gurney asked whether NASA and DoD had coordinated on studies comparing MOL launches from both ranges. Seamans said he was aware such work was under way but NASA was not involved and General Schriever would be a better person to ask. He remarked further that to launch into polar orbit from Cape Kennedy required a "dog leg" in the initial boost phase. Such a maneuver, he concluded, was "scarcely within the [weight] limitations of the Titan IIIC/MOL.[14]

The next day, 8 February, the committee reconvened with General Schriever in the witness chair. Once again Chairman Miller cautioned the members "their interest did not lie in the mission of MOL" and that they should concentrate on NASA/DoD possible duplication of efforts. After opening the meeting, Miller left the room and Schriever began reading a lengthy paper to the committee. He gave the history of the MOL program, described the system, and reported on planned schedules, the MOL booster, life support system, tracking stations, etc. He ended his statement with a review of Defense Department policy requiring mutual exchange of information and cooperation with NASA on their individual space projects.[15]

After he had answered various questions dealing with the program, Congressman J. Edward Roush of Indiana finally asked the "forbidden" one: "What is the ultimate purpose of MOL and why is it that everything the Air Force is doing cannot be done by NASA?" Schriever replied that the mission was military in nature, was not of interest to NASA, and did not fall within the space agency's area of responsibility. At this point Chairman Miller returned to the hearing room and remarked: "It is not necessary to ask this type of question if you have confidence in the U.S. military."[16]

When Congressman Gurney was recognized by the Chairman, he began his interrogation by proclaiming himself as a strong advocate of military man in space. However, he reminded the MOL Program Director that the Air Force had invested "$150 million" in its Cape Kennedy launch facility, which he claimed it was abandoning. He noted that polar launches had already been made from the Cape, that NASA was planning a polar orbit manned

mission from that site,[§] and he challenged Schriever about the Air Force's "exaggerated" safety requirements for ETR polar flights.

General Schriever replied by reminding the committee that several years before a Thor missile launched from the Cape had impacted on Cuban soil. He admitted that a polar orbit was technically feasible from the Eastern Test Range but said there was a weight penalty which made it impractical for MOL. The Air Force, he said, had initiated a study on possible MOL launches from the ETR, but he said that "if you attempt the launching in the necessary 180 to 185 degree direction, it will fly over Miami and Palm Beach. Neither the Saturn IB or Titan IIIC can make the turn necessary for a safe polar launching and still boost the full MOL payload into orbit." He said he would submit to the committee information on the exact loss of payload weight during such a maneuver.[17]

Following this statement and other questions and answers on possible duplication between the Air Force and NASA space programs, another Congressman—Representative William F. Ryan of New York—insisted Schriever explain the mission of the MOL. Once again, Chairman Miller interjected with a reminder that the committee would not inquire into the mission. Whereupon, Ryan asked why NASA couldn't accomplish all that the Air Force planned to do? Schriever answered that the 1958 Space Act had definitely stated that the Department of Defense would be responsible for military applications in space and "the MOL program is definitely a military application.[18]

Schriever's testimony concluded the executive hearings of the House committee. With the important help of its chairman, the problem of a breach in MOL security was overcome and the question of duplication apparently answered to the satisfaction at least of Congressman Miller. Thus, he stated to a press representative that he felt there was no major duplication of effort between MOL and NASA's Apollo Applications program. He also declared he supported the Air Force's decision to launch the MOL from Vandenberg.[¶] Cape Kennedy, he said, was the best site for near equatorial launchings, but the Western Test Range was best for polar orbit launches.[19]

---

§   At this time NASA was considering possible polar orbit launches from Cape Kennedy, whenever the reliability of the Saturn IB launch vehicle was established.

¶   The fact that Chairman Miller was a Californian certainly did not, of course, hinder the program.

SENATOR HOLLAND REQUESTS A MEETING

Nevertheless, the issue continued to roil the Floridians. On 10 February, Congressman Gurney requested General Schriever to answer 11 questions concerning Air Force launch plans, the requirement for polar orbits, the cost of Vandenberg facilities, etc. The same day, Mr. Francis S. Hewitt, a staff member of the Senate Appropriations Committee, informed the Air Force that Sen. Spessard L. Holland of Florida—a member of both the Senate Appropriations and Space Committees—wished MOL officials to attend a meeting in his office on 15 February to review the program. Hewitt advised that the Air Force representatives should be prepared to answer the specific question: "Why can't ETR be used to launch the MOL."[20]

Also, on 10 February, Senator Holland wrote to Chairman Anderson of the Senate Space Committee and requested a "thorough hearing" into the Air Force's plans to launch MOL from Vandenberg. In his letter, Senator Holland said that:

> The people, officials and news media of central Florida are complaining vigorously about this proposal which they tell me will cost our country unnecessarily many millions of dollars of added expense and will deteriorate the fine joint effort of NASA and the Air Force which has been conducted so effectively at Cape Kennedy. They also feel that such a move would cause unnecessary hardship to many families not settled in the Cape Kennedy area. They feel, and strongly assert, that there is no sound reason whatever for making this proposed move.[21]

Senator Anderson approved Holland's request and Gehrig, the Staff Director, advised the Air Force that the full committee would meet on 24 February to take testimony "on the MOL as it relates to facilities at Patrick versus Vandenberg." Gehrig remarked that this was "a continuation of the pressure tactics of the Florida delegation to attempt to keep as much of the Air Force Space Program as possible at Patrick."[22]

On 14 February, Gehrig met with a member of Schriever's staff, Col James M. McGarry, Jr., to discuss the proposed Senate committee hearings. During their meeting, they worked up a series of questions which they

agreed General Schriever and other Air Force officials should be prepared to answer. These covered such topics as the ETR "dog leg," the cost of West Coast facilities, whether MOL was ever considered for launching from Cape Kennedy, and related matters. Gehrig advised that if Senator Holland could be told that it had never been contemplated that MOL would be launched from Florida, it would help turn aside much of the criticism.[23]

Also, on 14 February, acting on a request from Dr. Flax, General Berg forwarded his evaluation of problems involved in launching MOL from ETR. He said the existing facilities at Cape Kennedy could be used if it was acceptable to (1) run a black reconnaissance program at ETR in conjunction with unclassified NASA programs; (2) risk disclosure of the program by discovery of payload elements in Cuba or some Central or South American country should the booster fail; and (3) risk human life to launch due south from Cape Kennedy. Since he considered these unacceptable, Berg submitted several pages of detailed information on current MOL planning for WTR polar launchings.[24]

The next day the Florida delegation gathered in Holland's office with the DoD delegation—headed by Dr. Flax—in attendance. Besides Holland, the Florida contingent included Sen. George A. Smathers and Congressmen Gurney, Charles E. Bennett, James A. Haley, and A. Sidney Herlong, Jr. Others present were Gehrig and Dr. Glen P. Wilson of the Senate Space Committee staff and Mr. Charles Kirgow of the Senate Armed Services Committee staff. Lt Col James C. Fitzpatrick, of the Directorate of Development, Headquarters USAF, opened the meeting with a presentation on the Titan III family of boosters, costs of the Cape Kennedy facility and Air Force plans for both ranges. He was followed by {a representative} of the MOL Program Office, who described the basic MOL flight equipment and program schedule.**

The briefings of Fitzpatrick and {the MOL program representative} were punctuated by many questions from the Florida delegation, especially on the costs of facilities on both coasts. The Air Force representatives also spent a considerable period trying to explain the value of polar orbit, and this matter was not made clear until a "hastily acquired globe" had been brought into the room. At this point, the Floridians demanded to be told MOL's mission. Dr. Flax replied that it was highly classified and known to only a few people. Gurney and Holland persisted, but Dr. Flax stated that divulging of this sensitive information was subject to control by the

---

** The DoD representatives present included Dr. Yarymovych, Colonel Arnold, Capt Howard Silberstein, DDR&E and Lt Col William R. Baxter, Director of Range -Safety, ETR.

Executive Branch and the chairmen of the committees involved. When Holland threatened to take the subject to the Senate floor and have it aired, Dr. Flax said he thought that was something the Senator wouldn't want to do. "That's for us to decide," the Senator retorted.[25]

At the conclusion of the meeting, the Florida contingent remained dissatisfied. "This is not the end of this little tete-a-tete," Holland said. "We are going after this and we are not going to stop here. We are not going to lose this like we let Houston get away!" His reaction reflected the great pressures he and other Florida representatives were being subjected to from home. A flood of letters and telegrams—from real estate dealers, developers, citizens, and other local interests—had poured into their offices. Some were artificially stimulated by the local press. For example, the Melbourne Daily Times published for its readers a clip-out form letter protesting the proposed move of MOL to Vandenberg; more than 2,500 of these made their way to Washington.[26]

Among letters sent directly to the White House was one from the editor of the Orlando Sentinel, Martin Anderson, one of the most vociferous critics of the Air Force.[††] In the President's response, prepared by the MOL Program Office, Mr. Johnson explained that the MOL was a military space program and that polar launches were required to accomplish its principal missions. He said:

> While it is true that some polar launches have been conducted from the Eastern Test Range using the "dog leg" maneuver, this does result in a reduction of physical capacity. In the case of the Manned Orbiting Laboratory, the 10-15 percent loss in payload required by performing this maneuver is sufficient to jeopardize seriously the success of the program. Furthermore, there is a risk in the case of failures of impacting classified military payloads in areas where classified information might be compromised.
>
> The facilities which will be built at Vandenberg to launch the MOL will be considerably simpler than those available at the Integrated Transfer and Launch Titan III facility at Cape

Kennedy. We have every intention of using the Cape Kennedy facilities to the maximum advantage in our space program. In particular, there are 10 remaining launches in the Titan III R&D program which will be used to orbit such important programs as the Defense communication satellites and nuclear test detection (Vela) satellites. Current Air Force plans beyond those R&D launches involve approximately four launches per year from Cape Kennedy..."[27]

Meanwhile, with the MOL Program Office facing further interrogations by the Anderson committee, Schriever directed Berg to organize an ad hoc task group to study all aspects of the controversy and to report to him on 18 February. Col Walter R. Hedrick, Jr., was later named chairman of this group, which convened the afternoon of the 16th to begin its work. By the 19th it had completed and gave to General Schriever and Dr. Flax a preliminary "secret" briefing, which emphasized the necessity of launching into "80-100° orbital inclinations" to meet program objectives. Concerning facility costs on both coasts, the ad hoc group noted that while the Cape facilities were cheaper by some $60 to $70 million, the decreased payload resulting from yaw steering— and reduced number of days on orbit—made WTR launchings more economical for long-term operations.[28]

THE SENATE SPACE COMMITTEE HEARINGS

On the morning of 24 February Chairman Anderson opened "Secret level" hearings into the MOL Program. The main witnesses were Drs. Foster and Flax representing DoD and Dr. Seamans of NASA. To place the matter before the committee into proper perspective, Sen. Margaret Chase Smith introduced into the record excerpts from hearings held in January 1965, in which Secretary Vance had reported on DoD plans to begin Titan III facility construction at Vandenberg. Senator Holland then made an opening statement and introduced into the record five editorials and stories from Florida newspapers, all of which were highly critical of the Air Force's plan to "move" MOL to California.[29]

Dr. Foster, the first witness, began by reading a lengthy statement which emphasized that the MOL program was aimed at fulfilling military requirements. He said:

---

†† Not everyone protested. Orlando TV Station WFTV editorialized on 15 February 1966: "How could the Cape be losing something it never had...Local citizens and businessmen are becoming unduly upset...The MOL is a military project that may involve maximum security. It is possible that the whole Cape area could be as closed as a tight security area as is the case of Vandenberg AFB."

To satisfy these requirements, there is no question but that we must place the MOL payload in near polar orbits. Orbital inclinations from 80⁰ to 100°are considered mandatory. To assure overall success of the program and minimum system costs, we have given careful attention to maximum use of NASA developed subsystems, the minimum weight payload which can meet the requirements, and the most effective launch vehicle approach. We have also considered the range constraints which might limit our ability to launch payloads as planned during early flight of the MOL Program and any follow on which may develop as a result of the MOL program.[30]

Dr. Foster reviewed the problems of possible land impact of the MOL and security of the payload in such a circumstance. He pointed out that ETR launch trajectories would involve land overflight, with the vehicle passing directly over southern Florida and Miami, and that this was "'totally unacceptable in my opinion due to the hazards involved." He said that, in the case of MOL, a decision had been made a year earlier to avoid unnecessary land overflight, "particularly since this program involves repeated launches of classified military payloads." He assured the committee that the Defense Department planned to continue various Titan III launch operations from Cape Kennedy.[31]

Senator Holland began his lengthy interrogation of Dr. Foster by remarking that other scientists (who were not named) had challenged the basic premises of the MOL program, particularly the need for polar orbits. He continually pressed Foster to answer "why polar orbit" was needed. Part of the colloquy went as follows:

Senator Holland: What I am trying to ask is, if you will, state why the polar orbit is the sole and exclusively chosen one, under the thinking of the Air Force.

Dr. Foster : I am sorry, I can only say that it is a requirement of the program.

Senator Holland: Yes, but, in other words, you are not going to state to this committee why you choose the polar orbit rather than the other courses that can be fired to greater advantage and more cheaply out of Cape Kennedy.

Dr. Foster: No, other than to say that in order to fulfill the purpose of the program, these inclinations are required.

Senator Holland: Doctor, this committee is composed of Senators of the United States who are entitled to know something about this program, and so far as the Senator from Florida is concerned, he thinks he is just as safe to trust with knowledge of this program...as yourself or anybody else, and I want to know why the polar flight is the only one that will fulfill the requirements of the Air Force...[32]

At this point Sen. Howard W. Cannon of Nevada came to the rescue of the besieged witness. He posed a series of questions to the Defense Research Director which elicited the general information that MOL flight objectives required "that areas be overflown in a polar orbit that cannot be overflown in an equatorial orbit."‡‡[33]

Holland also raised questions about several Air Force reports he had learned of, which he said confirmed the view that the Cape could be used to launch MOL. Dr. Flax replied that the reports mentioned had been prepared by Air Force officials at Cape Kennedy who were "not fully aware of the MOL requirements, have no responsibility for the MOL program," and were "merely speculating across the board" on all possible applications of the Titan III family flown out of the Cape. He agreed to provide the Committee copies of these reports.[34]

During the afternoon session, the main witness was Dr. Seamans of NASA. He had earlier assured the Air Force that he would take "a real hard position" on MOL launchings from the West Coast and "remain firm,"[35] and he did so. He reported on NASA's future plans for launchings from both coasts, emphasizing that Cape Kennedy would be the primary base. He described the

‡‡ Two documents, "Titan IIID Comparison, WTR vs ETR," and "Briefing on Titan III Capability at Cape Kennedy," were forwarded to the Committee by Dr. Flax on 25 February. [Ltr, Flax to Gehrig, Cmte on Aeronautical and Space Sciences, 25 Feb 66

type of vehicle the space agency would be launching from Vandenberg, and said it was desirable to conduct regular operational launches from the West Coast base. He reiterated that NASA officials not only had supported the MOL program "but we have also supported the necessity for the MOL's launching from the Western Test Range."[36]

The testimony taken on 24 February—the united front of DoD and NASA—had the effect of taking some of the pressure off all parties concerned. At day's end, Senator Holland, like his colleagues, sought to emphasize the positive aspect of the Cape Kennedy situation. He told a reporter for the Miami Herald that, "despite the MOL move," there would be a substantial number of Air Force launchings from the Cape in the future, that it would not cut down on its personnel there, and that the Vandenberg investment would be "relatively small" compared to ETR.[37]

Subsequently, a number of other Congressional committees also sought information about ETR-WTR facilities "duplication." Replies were made to all inquiries[38] and by early spring 1966 the Floridians had all but dropped the issue. Thus, during a floor debate in the House of Representatives on 3 May on NASA's authorization bill, Colonel Arnold observed that "not one of the nine Members of the Florida Delegation present rose to protest the planned use of the Western Test Range for MOL."[39]

ENDNOTES

1.   Intrvw, Berger with Maj Robert Hermann, SAFSL, 12 Sep 68. Memo (U), Evans to Maj Gen T.G.Corbin, 2 Sep 66, subj: Appreciation to Colonel William B.Arnold.

2.   Memo (S-DORIAN) , Schriever to SAF, 7 Oct 65 , subj: MOL Monthly Status Report.

3.   Memos (U), Evans to Brown, Flax, Schriever, Ferguson, 12 Nov 65, subj: Request for MOL Briefing; Evans to Flax, 19 Nov 65, subj: Briefing for Staff Members of House Mil Ops Subcmte on MOL Program.

4.   Memo (U), Brig Gen L. S. Lightner, Dep Dir, SAFLL, to SAF, et.al., I Dec 65, subj: Hearings on MOL .

5.   Memo for the Record (U), by Gen Evans, 16 Dec 65, subj: Intrvw with Mr. Peter Girardi.

6.   Memo for Record (U) by Col W. B. Arnold, 20 Dec 65, subj: MOL Hearings, House Cmte on Science & Astronautics.

7.   Memos for the Record (S-DORIAN), by Berg, 5 Jan 66, subj: Visit to Congressman Miller; (U), by Arnold, 10 Jan 66, subj: Mtg with Chairman Miller.

8.   Memo for the Record (U), by Arnold, 13 Jan 66 subj: MOL Hearings Before House Cmte on Science & Astronautics; Memo (C , Col B.s.Gunderson to Flax, 14 Jan 66, no subj.

9.   Ltr (U) Flax to Repr Edward J. Gurney, 2 Sep 65; Aviation Week & Spae Technology, 6 Sep 65.

10. Rprt to the Congress from the President of the United States, 31 Jan cc, on U. S. Aeronautics and Space Activities, 1965, pp 50-51.

11.   *Orlando Sentinel*, 4 Feb 66.

12. Memo for the Record (S), by Col Arnold, 7 Feb 66, subj: MOL Hearings.

13.   Ibid.

14.   Ibid.

15.   Memo for the Record (S), by Col Arnold, 8 Feb 66, subj: MOL Hearings.

16.   Ibid.

17.   Ibid.

18.   Ibid.

19.   *Space Daily*, 15 Feb 66.

20.   Memo for Record (U), by Hyman Fine, Msl & Space Systems Div, Dir/Budget, 10 Feb 66, subj: Congressional Inquiry Re MOL.

21.   Ltr (U), Sen. Holland to Sen. Anderson, 10 Feb 66, in Hearings before Senate Cmte on Aeronautics and Space Sciences, 89th Cong, 2nd Sess, *Manned Orbiting Laboratory*, p. 9.

22. Memo (U), Col Walter T. Galligan, Dep Chief, SAFLL, to Col Arnold, 11 Feb 66.

23. Ltr (U), Col James M. McGarry, Jr., Dir, Off of External Activities, Hq AFSC, to Gen Schriever, 14 Feb 66, subj: Location of MOL at VAFB.

24.   Msg (S-DORIAN), 0145, Berg to Schriever, 14 Feb 66.

25.   Memo for the Record (S), by Col Arnold, 16 Feb 66, subj: MOL Briefing for the Florida Delegation.

26.   Ibid.; Memo (U), Lt Col Edgar L. Secrest, Jr. to Gen Corbin, 30 Mar 66, subj: Weekly Rprt on MOL.

27.   Ltr (U), President Johnson to Publisher Martin Anderson, *Orlando Sentinel*, 18 Feb 66. Anderson ran the letter in his paper on 25 February.

28.   Msg (S) SAFSL 90483, SAF to SSD (For Berg), 15 Feb 66; Rprt (S), Ad Hoc Talk Gp, 16-18 Feb 66; Memo for the Record (S), by Col Richard Dennen, 18 Feb 66, subj: ETR vs WTR Launch from MOL.

29.   Hearings before Senate Cmte on Aeronautics and Space Sciences, 89th Cong. 2nd Sess, *Manned Orbiting Laboratory*, pp 2-9.

30.   Ibid. pp 10-13. For the unexpurgated version of Dr. Foster's statement, see MOL Program Office Read File, Statement for the Record (S), 24 Feb 66.

31.   Ibid., pp 11-12.

32.   Ibid., pp 35-36.

33.   Ibid., p. 39.

34.   Memo (U) , Col B.R. Daughtery, Exec Asst, USAF, to Flax, 11 Feb 66.

35.   Hearings before Senate Cmte on Aeronautics and Space Sciences, p 39.

36. Ltr (S), Schriever to Gurney, 24 Feb 66; Ltr (U), Flax to Herlong, 2 Mar 66; Memo (S), Evans to Dir of Dev, 11 May 66, subj: Questions for Mil Const Subcmte (Sikes), House Appns Cmte; Ltr (U), L. Mendel Rivers, House Cmte on Armed Services to Secy Brown, 1 Mar 66, Ltr (U), Brown to Rivers, 14 Mar 66.

37. Memo (C) , Col Arnold to Gen Corbin, 4 May 66, subj: Weekly Report on MOL.

MOL ASSEM...
INTEGRATION BUILD...

MATERIAL

PROPERTY OF THE
UNITED STATES GOVERNMENT

F FOUND, DO NOT OPEN

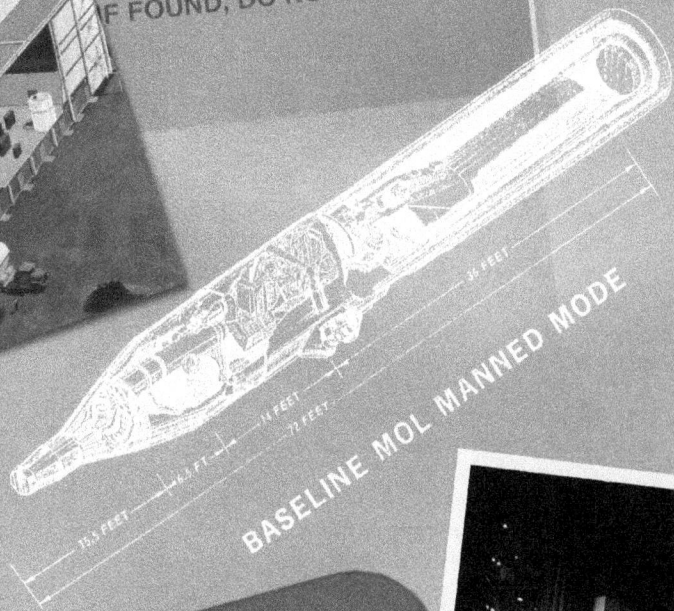

BASELINE MOL MANNED MODE

155 FEET

USAF MOL/KH-10

THE DORIAN FILES REVEALED:
A COMPENDIUM OF THE NRO'S MANNED ORBITING LABORATORY DOCUMENTS

CHAPTER XIII:
AIR FORCE/NASA COORDINATION

AIR FORCE/NASA COORDINATION

By 1965-1966 Air Force and NASA manned space programs had evolved to the point where the competition between the two agencies had manifestly declined. Deeply involved in its Gemini program, NASA at this time was also laying the ground work for its multi-billion dollar Apollo moon-landing project.* The Air Force, meanwhile, was working energetically to get going with the MOL, which it believed would provide the vehicle that would conclusively demonstrate the value of putting a man into space to perform various military missions, beginning with reconnaissance. This period saw increasing coordination of the efforts of both agencies. Thus, in 1965 the Evans/Garbarini group had worked closely together on the Apollo/MOL studies, which provided comparative cost figures and other data to the Air Force. Also, the following year, as we have seen, NASA backed up the Air Force during the noisy ETR-WTR controversy.

In addition to the above examples of cooperation, the two agencies coordinated their activities in several other areas. One involved the release and modification of certain NASA flight equipment for use in an Air Force pre-MOL flight test program. Another—which generated differing views before a compromise was reached—centered on the question of Air Force procurement of the Gemini B spacecraft.

NASA'S GEMINI AND THE GEMINI B CONTRACT

Several months before John Glenn became the first Mercury astronaut to orbit the earth in early 1962, NASA formally announced the initiation of the Gemini program. On 15 December 1961 it awarded a $25 million contract to McDonnell to begin design, development, and manufacture of 13 Gemini spacecraft. (The cost of these vehicles eventually ballooned to more than $790 million.[†]) NASA also assigned to SSD the job of procuring man-rated Titan II boosters to launch them. During 1962-1963, the development work proceeded satisfactorily and an important milestone was reached with the successful test firing on 21 January 1964 of the GT-1 (Gemini-Titan No. 1) launch vehicle.

Meanwhile, McNamara's announcement of 10 December 1963 that DoD would undertake the development of MOL made it apparent that certain under standings would have to be reached by NASA and DoD, since the system required a modified Gemini. On 23 January 1964, Drs. Seamans and Brown (then DDR&E) agreed the Air Force should negotiate a preliminary design study contract with McDonnell, with the arrangement to be subject to NASA review to assure McDonnell could do the work without interfering with the space agency program. The two officials also agreed the contract would not establish a pattern for any follow-on engineering or procurement contract relationship with McDonnell.[1] The Air Force contract subsequently was approved and, in June, McDonnell began a $1 million pre-Phase I Gemini B study which it completed by year's end.

In connection with this contract, the St. Louis firm was naturally eager to obtain additional Gemini business and retain the space engineering competency it had acquired during its work on Mercury and now Gemini. To support the latter, it had built up a Gemini team which included 441 personnel, 240 of them doing advanced engineering work. McDonnell advised the Air Force that it would need an early USAF commitment in order to keep the team intact. The firm's situation was discussed during the summer of 1964 by NASA and OSD officials, and they agreed that it was in the nation's interest to retain the newly-acquired industrial base. However, OSD was unable to make a commitment until it had decided whether or not to proceed with MOL development.[2]

Toward the close of 1964 several factors, including congressional pressures, conspired to push OSD toward such a decision.[‡] Thus, when Senator Anderson expressed concern to the President about duplication between NASA and DoD space programs and recommended cancellation of MOL, he was assured the two agencies were working closely together and would take advantage of each other's technologies and hardware. In January 1965 McNamara and Webb issued a joint statement touching on this point. "Duplicative programs," they declared, "will be avoided and manned space flight undertaken in the years immediately ahead by either DoD or NASA will utilize spacecraft, launch vehicles, and facilities already available or now under active development to the maximum degree possible."[3]

---

* NASA was allocated $5.2 billion in new obligational authority in fiscal year 1965, $5.1 billion in fiscal year 1966.

† In February 1963 NASA estimated the cost of the 13 Geminis, two mission simulators, five boilerplates, and other equipment at $456,650,062. By the end of the program, however, the cost of the spacecraft and ancillary equipment had risen to $790.4 million. [NASA Draft Chronology, Project Gemini: Technology and Operations, pp 108, 409]

‡ In March 1965 Congressman Teague of Texas expressed concern to the President that the valuable Gemini industrial team would be disbanded if a MOL decision was not made.

Figure 50. MOL Astronaut Practicing Movement
Source: CSNR Reference Collection

This policy statement encouraged the Air Force to seek the early release by NASA of its Gemini 2 spacecraft, successfully recovered from the Atlantic on 19 January 1965 following an unmanned suborbital test flight. On 9 February, during a meeting with Dr. Albert C. Hall (Special Assistant for Space, ODDR&E), General Evans mentioned the USAF requirement for the recovered spacecraft and he also suggested the Air Force be authorized to contract directly with McDonnell for development of the Gemini B. Dr. Hall approved both proposals and said OSD would contact NASA about them.[4]

On 3 March 1965, Dr. Brown wrote to Seamans about Gemini B. Refer ring to their agreement of the previous year, he advised that—in order "to preserve the option of proceeding at a later date with a configuration based upon Gemini B and Titan IIIC"—DoD planned to negotiate a second contract with McDonnell for design definition of Gemini B "to the point of engineering release." In response, Seamans reminded DDR&E that their 1964 agreement required the space agency's approval of any such follow-on contract. A second contract, he said, was "a matter of direct concern to us because of the possible effect it might have "on the fulfillment of NASA's Gemini Contract by McDonnell." To reach agreement on this matter, a meeting of top officials of both agencies was

scheduled. In the meantime, General Evans discussed MOL equipment requirements with Dr. George Mueller, Associate Administrator, NASA Office of Manned Space Flight and, in a follow-up letter, he forwarded a list of items of Gemini equipment—such as the Gemini spacecraft and Static Article No. 4—expressing hope they could be released to the Air Force as soon as possible.[5]

On 18 March, the DoD/NASA meeting to discuss the proposed Gemini B contract was convened at NASA Headquarters. Representing DoD were Drs. Hall and Flax, General Evans, and several others. Dr. Mueller, who headed the space agency contingent, began by reiterating NASA's concern about possible interference with the on-going Gemini activity at McDonnell. To avoid such disruption, he suggested that NASA be assigned responsibility for "the total spacecraft job for the Air Force for Gemini B/MOL." Dr. Hall agreed that interference with the Gemini program should, of course, be avoided or minimized, but he expressed doubt the space agency would be in a position to handle all the technical functions involved in the MOL program. To Hall's query whether NASA would make the GT-2 spacecraft and other Gemini hardware available to the Air Force, the NASA official made no response but returned to his original point: if the space agency was given responsibility for

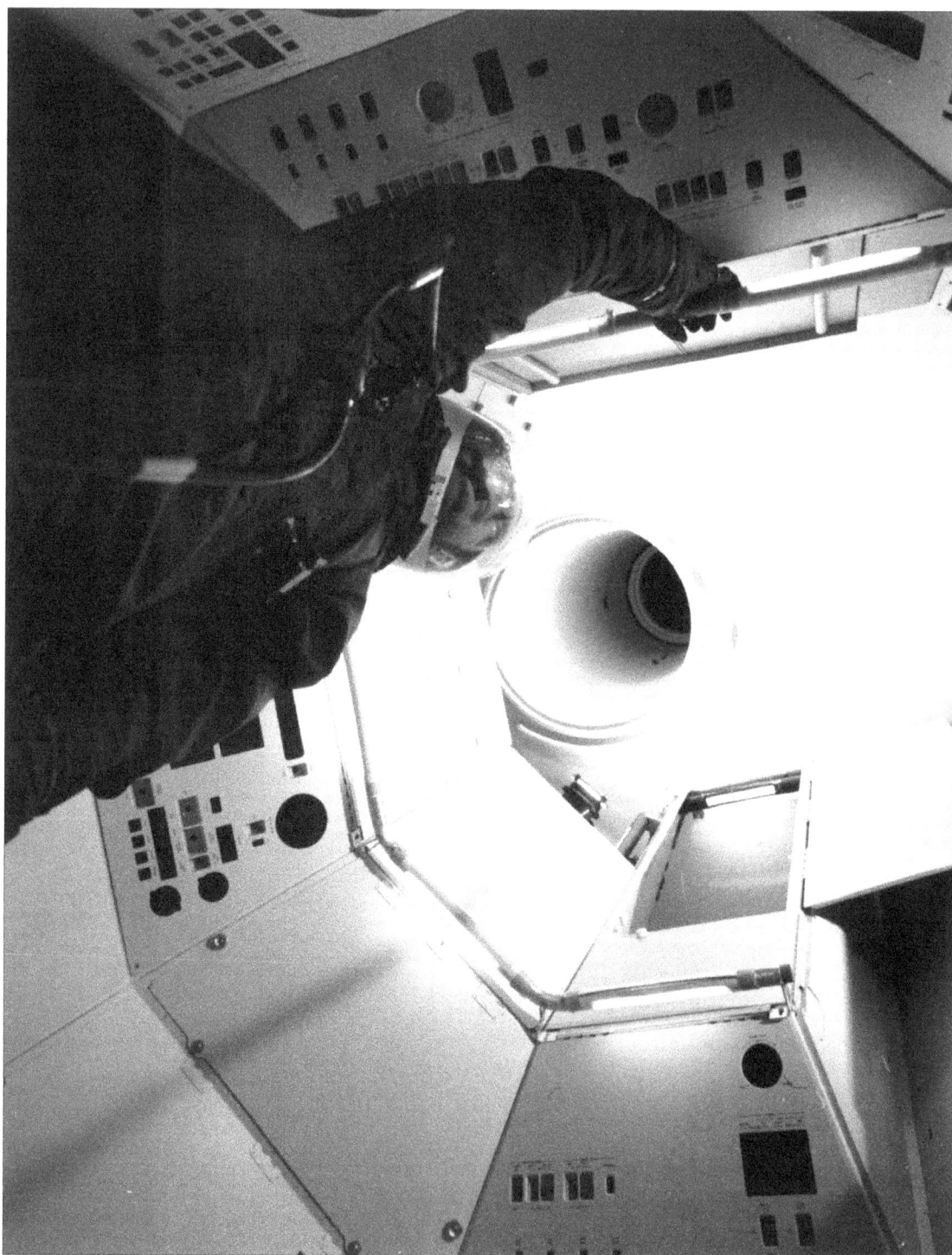

Figure 51. MOL Astronaut with Passageway
Source: CSNR Reference Collection

the technical direction and contracting for Gemini B, he could thus assure himself of minimum interference with the Gemini program.[6]

The conferees finally decided to dispatch a NASA-DoD task group to McDonnell on 22 March to determine the extent of such interference. After meeting with company officials, the task group returned to Washington to report to a reconvened conference—attended by Hall, Flax, Mueller, Evans, and others from both agencies—on 25 March. General Bleymaier, who had headed the DoD element of the task group, briefed the conference on the results of the survey, which he said indicated there would be little or no interference with the space agency's

program. This conclusion did not alter Mueller's view that executive management responsibility for design and acquisition of Gemini B should logically go to NASA.[7]

The issue remained unresolved during several follow-up meetings, including a separate conference on 30 March between Dr. Mueller and General Schriever. During one of the meetings in early April, NASA submitted for consideration a "Plan for NASA Support of the Air Force Gemini B/MOL Program." This plan assigned primary responsibility for Gemini B acquisition to the space agency, at least up to July 1966, at which time the Air Force would take over executive management. However, it remained unacceptable to OSD and Air Force officials.[8]

Figure 52. MOL Passageway
Source: CSNR Reference Collection

In early April General Evans prepared a counter-proposal for NASA's consideration. In forwarding the document to Mueller's deputy, Brig Gen David Jones, on 8 April, Evans said he believed it would minimize interference between the two Gemini projects, while assuring that the management of the highly integrated Gemini B/MOL system remained with the Air Force. According to this plan, NASA's Gemini Project Office—located at the Manned Spacecraft Center (MSC), Houston, Tex.—would be assigned full responsibility for modifying and refurbishing Gemini spacecraft required for MOL early test flights. On the other hand, the Air Force would retain responsibility for contracting, development, and acquisition of Gemini B.[9]

This compromise was accepted by the space agency. On 12 April, Drs. Seamans and Hall formally agreed the Air Force should proceed to negotiate the Gemini B contract with McDonnell. In light of this agreement, DDB&E wrote to Seamans to solicit suggestions "of technical or management methods" which could help DoD reduce Gemini B acquisition costs. Schriever followed up during a meeting with Mueller on 13 April. As a matter of policy, he said, the Air Force wanted and needed all the NASA technical help it could get, particularly in connection with refurbishment of the Gemini 2 spacecraft and other equipment for use in the MOL test flight program.[10]

During the next several days, Generals Evans and Jones worked out the details of the responsibilities of the two agencies. In their preliminary draft agreement, the Air Force assigned to NASA the responsibility for engineering, contract management, and procurement associated with refurbishment and modification of the GT-2 spacecraft and Static Article #4. The Air Force also agreed to provide several highly qualified personnel to participate in the above work. As for Gemini B, the Air Force alone would be responsible for its acquisition and would contract directly with McDonnell.[11]

A final, revised agreement—incorporating the major points agreed upon above—was signed on 21 April 1966 by Dr. Mueller for NASA and General Ritland for the Air Force. Simultaneously, the Manned Spacecraft Center designated Mr. Paul E. Purser as its main contact for NASA policy matters relating to MOL and Mr. Duncan R. Collins of the Gemini Spacecraft Office as the point of contact "for all technical assistance provided to the MOL Program.§"[12] Lt Col Richard C. Henry, USAF—assigned to NASA's Gemini Program Control Office at Houston—was named the Air Force's contact "for matters relating

to the transfer of materiel and equipment to the MOL program and for all matters pertaining to the HSQ (Heat Shield Qualification) program."

With this agreement the issue was resolved and the Air Force took steps to negotiate a Gemini B engineering definition contract with McDonnell. OSD provided seven million dollars for this work; the final contract, signed in May 1965, totaled $6,784,000.

TURNOVER OF NASA EQUIPMENT

After the President announced the Air Force would proceed with MOL, Dr. Seamans wrote to OSD on 23 September 1965 to offer again NASA's full support to the new program. His agency believed, he said, that "an experimental manned space flight program under the military" was justified and it was prepared to undertake joint planning "for the maximum practicable utilization by DoD of the NASA developed hardware and technology, our production, testing, processing facilities, and our management and operational experience."[13] Seamans' helpful offer was made against the background of three spectacularly successful NASA Gemini flights. On 23 March Gemini 3 was successfully launched, its astronauts achieving an important space "first" when they changed their orbit three times. Gemini 4, launched on 3 June, also made space history when Astronaut Edward H. White "walked in space" for 20 minutes and used a propulsion gun for the first time. Gemini 5, launched into an eight-day flight on 21 August 1965, shattered all existing space endurance records.

The newly formed MOL Program Office pursued the subject of turn- over of various items of Gemini equipment to the Air Force. After it had reviewed the subject, NASA on 4 October requested the Air Force to submit a complete list of all Gemini-associated equipment it needed for the MOL. This task was passed on to the MOL Systems Office, which by the end of October 1965 had compiled a list of items desired, including Gemini training boilerplates, flotation collars, mission simulators (one each located at Cape Kennedy and the Houston Center), crew station mock-ups and engineering mock-ups.

On 8 November General Evans forwarded the list to NASA and suggested the space agency designate a Houston official to work out details of the turnover with General Berg's representatives. On 29 November, after his staff had reviewed the items listed, Mueller directed Dr. Robert S. Gilruth, head of the Manned Spacecraft Center, to immediately transfer to the Air Force one Gemini training boilerplate, three Gemini flotation collars, Spacecraft Article #3, and Spacecraft #3A. He also

§   The space agency later assigned two employees at McDonnell to work full-time on the Gemini B.

requested the MSC to designate an individual to handle the transfer of such other equipment which it determined was "surplus to our manned space flight program." In the next several months a number of the above-mentioned items, plus 75 pieces of aerospace ground equipment (AGE) for use in support of the HSQ launch, were transferred to the Air Force.[15]

Subsequently, data from McDonnell's Gemini B engineering definition study became available and the MOL Systems Office in early 1966 was able to identify additional equipment needed. On 26 March General Berg forwarded to Dr. Gilruth a new detailed list which included: four boilerplate test spacecraft, three static test spacecraft, two recovered spacecraft, 25 ejection seat testing structures, six trainers, 199 mission recovery items, 34 long lead time AGE items, and 592 pieces of auxiliary equipment. Berg noted in his letter, that, while most of the equipment had been identified by name, some was general in nature due to his staff's inability "to inventory and establish firm, specific requirements in such areas as... components and vendor equipment at this time." He proposed that the best interests of both agencies would be served by adopting a general policy of transferring all NASA Gemini equipment "except that which is readily usable on the Apollo program."[16]

On 28 March 1966 Air Force and MSC representatives conferred at Houston to discuss the new list of requirements. The space agency's representatives—Mr. Purser and Colonel Henry—agreed "in principle" that Gemini equipment and materials not required by NASA would be released to the Air Force, at least on a shared basis, as they became available from the current test flight program. However, they indicated NASA had not made up its mind about certain "grey areas." These involved possible retention of equipment for use as artifacts at the Smithsonian Institution, the MSC lobby, worldwide travelling displays, etc. Concerning the Gemini simulator at the Kennedy Space Center, they startled the USAF representatives by stating the "NASA would release the simulator for a $5 million reimbursement." The conference adjourned with the understanding that the MSC would submit to the Air Force on a proposed procedure to govern transfer of certain items of equipment¶.[17] As subsequently received by the MOL Systems Office, this procedure proved acceptable.

On 23 April NASA headquarters authorized Houston to transfer 29 of 31 items of long lead time AGE. However, it delayed making a full response to General Berg's letter of 26 March. In early May General Evans wrote to

Dr. Mueller about the matter. After commenting on the various agreements they had reached and explaining that the Air Force expected soon to receive a Phase II proposal from McDonnell "conditioned on the use of Gemini equipment," Evans touched on a sensitive point:

> During the meeting at Houston on March 28, the MSC representatives proposed that the Air Force reimburse NASA $5 million in return for the mission simulator at Cape Kennedy. This came as quite a surprise in view of Dr. Seaman's statement to the Senate Aeronautical and Space Sciences Committee on February 24 that crew trainers and simulators would be made available to the MOL program "as soon as they can be scheduled for this purpose." We have not been advised of the terms of this latter qualification. In the same context, Dr. Seamans expressed a view of equipment availability on a nonreimbursable basis except where modification costs are incurred on a NASA contract as in the case of the HSQ spacecraft.[18]

After mentioning several other matters, including the possibility of obtaining assistance from NASA's resident/engineering/quality assurance personnel at McDonnell, Evans suggested to Mueller that they get together to discuss "the total subject of Gemini support to MOL." His complaint about the Kennedy mission simulator quickly produced results; NASA now determined it would be made available to the Air Force at no costs. Whereupon, on 3 June 1966, Evans advised Berg to arrange accountability and turnover of the simulator to McDonnell "for necessary refurbishment to Gemini B configuration."[19]

Early in July Dr. Mueller wrote to Evans to discuss the entire subject of NASA Gemini equipment transfer. Referring to his previous instructions to Gilruth to transfer 29 to 31 items of long lead time AGE to the Air Force, he now advised that the two items withheld also would be transferred. Further, he said, Houston had been authorized to work out a procedure with the MOL Systems Office to transfer a substantial portion of the equipment listed in Berg's letter of 26 March—to include the Kennedy simulator—plus Gemini peculiar components. Attached to Mueller's letter was a paper on "Procedures for Transfer of NASA Gemini Equipment to the United States Air Force for Utilizing on the Manned Orbiting Laboratory Program."[20] With this correspondence, the

¶ Air Force members at this meeting were Colonels Paul J. Heran and Russell M. Harrington, Lt Col Charles L. Gandy (of the MOL Systems Office), and Maj M. C. Spaulding of the MOL Program Office.

turnover of NASA Gemini equipment to the Air Force ceased to be an issue and, by the end of 1966, an estimated $50 million in space agency hardware had been transferred, or was scheduled for transfer, to the MOL program.[21]

DOD/NASA GEMINI EXPERIMENTS

When NASA's Gemini program was completed with the splashdown of Gemini 12 on 15 November 1966, USAF Program 631A—a series of military experiments performed during the flights—also ended. This program had origins in the McNamara-Webb agreement of January 1963 which established the Gemini Program Planning Board, whose mission was "to avoid duplication of effort in the field of manned space flight" between NASA and DoD.** In March 1963 the Board formed an ad hoc study group to review and recommend military experiments for inclusion in the Gemini flight program. It subsequently proposed a series of Air Force and Navy experiments and, on 25 August, an AFSC Field Office (Detachment 2, SSD) was established at Houston to manage their integration into the Gemini program. A technical development plan covering these Program 631A experiments was submitted to DDR&E and approved by him on 7 February 1964. Funding for the experiments totaled $16.1 million.[22]

Integration of the experiments was done to specifications developed by the Air Force and Navy experiment sponsors and SSD personnel as approved by the Gemini Program Office. Statements of work for the "Experiments Orders," written by SSD personnel, were submitted to the Gemini Program Office. Training, mission planning, test operations, and data collection were responsibilities of the Manned Spacecraft Center, assisted and supported by Detachment 2. However, in actual practice the detachment's project officer performed the dual role of experiment management for both the Manned Spacecraft Center and the Air Force.[23]

One of the most difficult management problems that emerged after the military experiments began flying— beginning with the GT-4 mission on 3 June 1965— concerned NASA's public information policy. According to the original ad hoc study group report which set up the program, military experiments were to be flown with the understanding that the results would be handled as classified information. However, prior to the flight of Gemini 5—which was scheduled to carry a number of photographic experiments—word was received in Washington that Houston officials had decided not to withhold information from the news media about

them. Whereupon, a team of USAF officers—including Major Robert Hermann of the MOL Program Office— visited the Manned Spacecraft Center to review NASA's information planning for the flight. They found space agency officials determined not to compromise NASA's information policy of full disclosure. Although MSC officials agreed to provide for special handling of photographs produced by the military experiments, this arrangement was never implemented.

Figure 53. Gemini 12 Recovery
Source: CSNR Reference Collection

The result was an upsurge of public criticism during the flight of Gemini 5 on 21-29 August 1965. When information was released on the DoD photographic experiments, "a hue and cry about NASA's peaceful image vs the military spy-in-the-sky implications" arose.[††] Part of the trouble according to Lt Col Wallace C. Fry, chief of the Space Experiments Office, Headquarters AFSC, was Detachment 2's failure to brief the astronauts on security aspects of the experiments. "As it developed," he said, "the crew of Gemini 5 apparently was not cautioned, and the astronaut who operated the D-6 lens made comments over the space-ground radio about the superb view and definition through the Questar telephoto lens. Also, post-flight handling of the film was open to suspicion and speculation [on the part of the news media] probably due to poor planning on our part." It did not take the Soviet Union very long to react and to accuse the United States of using Gemini 5 "to carry out reconnaissance from space." One result of the outcry was the withdrawal of the D-6 experiment from its second planned flight.[24] (For a complete list of the DoD experiments, see Chart on the next page.)

** See pp. 19-20.

†† See pp. 134-137.

Despite the above events, the cooperative DoD/NASA effort proved beneficial to both agencies and was considered a success. Experiments D-1, D-2, and D-6 clearly demonstrated the capability of man to acquire, track, and photograph objects in space and on the ground. Experiment D-3 showed it was feasible to determine the mass of an orbiting object (in this case, an Agena target vehicle) by thrusting on it with a known thrust and then measuring the resulting change in velocity. The mass as determined from the experiment procedure was compared with the target vehicle mass as computed from known launch weight and expendable usage to determine the accuracy of the method.[25]

Experiments D-4 and D-5 used two interferometer spectrometers and a multichannel spectroradiometer to successfully demonstrate the advantage of using manned systems to obtain basic celestial radiometry and space object radiometry data. Experiment D-8 successfully produced data on cosmic and Van Allen belt radiation within the Gemini spacecraft. Experiment D-9 demonstrated the feasibility of using a space sextant in an autonomous navigation system, the data comparing favorably with the accuracy of the spacecraft position computed from radar returns.

Experiment D-10 performed in an especially impressive manner on two flights. This ion sensing attitude control system experiment proved particularly useful during the flight of Gemini 12, after its fuel cells began acting up. With the cells turned off to conserve power, the crew relied upon the ion sensing control system and found it provided an excellent indication of attitude. D-12, the astronaut maneuvering unit (AMU), was not completed due to the inability of the astronauts to accomplish the tests on Geminis 9 and 12. Although carried on the Gemini 9, flight testing of the AMU was terminated when visor fogging obstructed the vision of the pilot during his extravehicular activity.

Experiment D-13—run in conjunction with NASA's visual acuity experiment (S008)—confirmed that the flight crew could discriminate small objects on the surface of the earth in daylight. Experiment D-14, involving UHF/VHF polarization measurements, was not completed, although the experiment equipment and technique was successfully demonstrated. Experiment D-15's image intensification equipment was used for the first time on Gemini 11 and demonstrated that, at night and under conditions with no moon, the crew could see bodies of water, coastlines, and rivers under starlight conditions. Experiment D-16 was not completed because the Gemini 8 mission was terminated early due to control problems, and pilot fatigue on Gemini 9 led to cancellation of its planned use.[26]

THE MANNED SPACE FLIGHT COMMITTEE

In his letter to DDR&E of 23 September 1965, in which he expressed NASA's desire to give full support to the MOL program, Dr. Seamans also raised a question about future top-level planning of the manned space programs of both agencies. He suggested that "joint planning and monitoring on the policy decision level" was needed and he proposed to Dr. Brown that they meet to discuss possible methods of conducting such reviews. Dr. Foster (Brown's successor as DDR&E) welcomed Seaman's suggestion and agreed to a meeting on the question of "coordination of our activities at a level which can determine policy."[27]

From this meeting, held in mid-October, and discussions of their staffs, emerged a tentative plan to establish an informal six-man DoD/NASA committee to review and solve manned space flight problems "not solvable by any other level." The committee, they agreed, would also serve to assure Congress that they were working closely together in the Gemini, Apollo Applications, and MOL areas. General Schriever, however, had been attempting to establish a close personal working relationship with Dr. Mueller and felt that the proposed committee would undercut his current effort. In a letter to Secretary Brown on 9 November 1965, he expressed the view that he and Mueller were "the appropriate level to resolve all problems except major policy problems" and there was no need for the committee. He asked the Air Force Secretary's support for his current negotiations with Mueller and "opposition to the committee arrangement."[28] Secretary Brown, however, felt that the committee would not interfere with Air Force management of MOL since it would deal with policy rather than program coordination questions.

The Seamans-Foster discussions, meanwhile, led finally to a plan to create a "Manned Space Flight Policy Committee (MSFPC)" to supersede the Gemini Program Planning Board established in 1963. The task of the new committee was to coordinate the manned space flight programs of DoD and NASA, resolve matters which could not be resolved at a lower level, make agreements involving top policy decisions, and facilitate exchange of view-points and information of importance. The formal agreement was signed in January 1966 by McNamara, Webb, Seamans, and Foster. The latter two were designated as Committee co-chairmen. Other members were Mr. Fink, OSD; Dr. Flax, Air Force; and Drs. Mueller and Homer E. Uewell, NASA.[29]

THE RECONSTITUTED NASA EXPERIMENTS BOARD

At the initial meeting of the new committee, held on 21 January, one of the several topics discussed was a proposed revision of the charter of NASA's Manned Space Flight Experiments Board (MSFEB) to include DoD membership.‡‡ The space agency subsequently distributed a draft Memorandum of Agreement on the proposed reconstituted Board and Dr. Flax solicited the views of Generals Schriever and Evans as to the membership. The MOL Program Director asked that he be a member, with Evans serving as his alternate. This suggestion was accepted.[30]

On 21 March 1966 Drs. Seamans and Foster signed the formal agreement establishing a reconstituted MSFEB "to coordinate experiment programs which will be conducted on DoD and NASA manned space flights." The Board was charged with the task of approving or disapproving experiments, recommending experiments for assignment to specific flight programs, setting priorities, reviewing the status of approved experiments, etc. The DoD membership included Mr. Fink and General Schriever; their alternates were Mr. John E. Kirk, Assistant Director for Space Technology, OSD, and General Evans. NASA's representatives were Drs. Mueller, Newell and Mac C. Adams; their alternates, James C. Elms, Dr. Edgard M. Cortright, and Dr. Alfred J. Eggers, Jr.[31]

Under terms of this agreement, before submitting proposed experiments to the Secretariat for consideration by the Board, the sponsoring agency was required to review them for scientific and technical merit and to establish its own list of priorities.[32]

DOD EXPERIMENTS FOR THE APOLLO WORKSHOP

As early as the spring of 1964 NASA had asked the Defense Department whether, as in the case of Gemini, it might also be interested in providing experiments for the upcoming Apollo spacecraft. In response to a request from DDR&E, the Air Force studied possible experiments and, at the close of 1964 and in early 1965, submitted three: Radiation Measurements, Autonomous Navigation, and a CO_2 Reduction System. All three were approved by DDB&E and accepted by NASA, which assigned them to Apollo-Saturn (AS) flights 207 and 209.[33]

While NASA was planning experiments for Apollo, it also was studying possible advanced manned missions to exploit the hardware being created by the lunar-landing program. In 1965, following these investigations, it outlined a plan for a series of post-Apollo flight missions "in earth orbit, in lunar orbit, and on the lunar surface." This follow-on Apollo Applications Program (AAP) it estimated would cost $1 to $3 billion a year.

On 17 January 1966, during a meeting of the MSFEB which General Evans attended as an observer,§§ one of the AAP "experiments " was outlined by a NASA official. He described a "S-IVB Spent Stage Experiment" whose purpose was to demonstrate the feasibility of providing a habitable, shirt-sleeve environment in orbit using already developed hardware. An airlock would be developed:

> ... using qualified Gemini flight hardware to allow docking of the Apollo Command Service Module (CSM) with the hydrogen tank of the spent stage of the S-IVB booster. Once docked, the airlock unit will provide ingress-egress capability, life support, electrical power, and the necessary environmental control required for pressurizing and maintaining the S-IVB stage hydrogen tank so that astronauts may work inside in a shirt-sleeve environment during a 14-day or greater mission.[34]

The "spent stage" experiment was to be scheduled for the SA-209 mission in the last quarter of 1967. Listening to this presentation, it occurred to General Evans that the early flight date might offer the Air Force a unique opportunity "to design experiments, directly supporting MOL development, to obtain information on crew activities in a large volume orbital vehicle."[35] On 21 March, at another meeting of the Board, he advised the NASA member that the Air Force was studying the possibility of conducting MOL-oriented experiments aboard the orbital workshop. Dr. Mueller welcomed the Air Force interest and said the Board would consider any experiments proposed.[36]

Following this meeting, the Vice Director, MOL, appointed an ad hoc group to study and recommend experiments for the Apollo Workshop. The group was chaired by Dr. Yarymovych and included representatives of the MOL Systems Office, Detachment 2 at Houston, and AFSC's Research and Technology Division (RTD) and Office of the Deputy Commander for Space. During April 1966 the

---

‡‡ Attendees were Seamans, Mueller, Newell, Foster, Flax and Fink.

§§ This was before the reconstitution of the Board.

ad hoc group met twice and identified nine experiments for consideration.[37] Subsequently, it selected six to be flown in the workshop: (1) Integrated Maintenance; (2) Suit Donning and Sleep Station Evaluation; (3) Alternate Restraints Evaluation; (4) Expandable Airlock Technology; (5) Expandable Structure for Recovery; and (6) Modular Assembly for Antennas. The MOL Systems Office sponsored the first three experiments, and RDT's Aerospace Propulsion Laboratories (APL) the last three. SSD's Detachment 2 was designated the Houston focal point and "management interface" for coordinating and integrating the experiments into the workshop.[38]

On 25 August, Evans submitted a report to Dr. Flax on the pro- posed experiments and recommended approval. He said they comprised a worthwhile effort and would enable the Air Force to test specific MOL equipment/ crew relations in time to incorporate the results into the MOL system, if appropriate. Evans estimated the cost of the three MOL experiments at $3.0 million, and the three APL experiments at $1.92 million. On 19 September Flax approved the first five experiments, deleting the modular assembly for antennas experiment. He advised Evans that the APL experiments were approved, "contingent upon RTD reprogramming internal funds to support the effort."[39]

That same day, 19 September, the five experiments were submitted to the Manned Space Flight Experiments Board and were accepted. At this meeting Dr. Mueller announced that the Orbital Workshop was firmly committed to AS-209, scheduled for flight in March 1968.[40] The next day Evans directed the MOL Systems Office to begin work on the experiments.¶¶ Several weeks later the MOL Program Office forwarded descriptions of the five Air Force experiments to NASA.[41]

In late October 1966, the MOL Program Office learned that NASA was reconsidering its basic plan for launch of the workshop. Instead of a single launch, NASA planners now proposed a rendezvous mode involving the addition of the SA-10 flight. On 26 October Evans wrote to Mueller to express concern about the effect the change might have on the schedule. He noted that the Air Force's basic objective in preparing the MOL experiments for the workshop was to test them "in sufficient time so that results can be incorporated into the MOL development." A delay in the launch date beyond December 1968 would not allow sufficient time to make inputs from these experiments into the MOL. He advised that, until he could evaluate the impact of a schedule change, he was directing a delay in the award of a contract for the experiments.[42]

¶¶ The RTD has some difficulty obtaining funds for the two APL experiments, but these were finally provided by AFSC reprogramming action.

In his response on 7 November, Mueller reassured Evans that the planned NASA launch would remain on or very near the original schedule. He said that while the space agency had not yet formally adopted a rendezvous mode for the workshop, it probably would do so since it would be a superior mode of operation. Even so, he stated that the mission would be flown "no later than July 1968" and he urged the Air Force to continue its experiment development "at the maximum pace possible."[43]

Several weeks later, at another meeting of the MSFEB at NASA head- quarters (attended by Kirk and Evans), the space agency described an even more ambitious Apollo Workshop plan, involving four launches. First, in June or July 1968, NASA would launch AS-209 into a 225 nautical mile orbit, the payload consisting of the Apollo Command Service Module and Mapping and Survey System. Five days later they would launch AS-210, unmanned, which would put the Orbital Workshop (S-IVB tank) into a one-year orbit. AS-209 would rendezvous and dock and the crew would enter and perform the workshop experiments. Reentry would take place 28 days after the launch of AS-209. Three-to-six Months later NASA would launch AS-211, manned, which would rendezvous and dock with the workshop. The crew would enter the tank, this time for a 56-day mission. One day later NASA would launch AS-212, unmanned, carrying an Apollo Telescope Mount, which would be tethered and docked for use as a manned orbiting telescope.

During this MSFEB meeting, which was held on 21 November, the Board approved seven scientific and technological experiments proposed by NASA for earth orbital flight, and one involving a lunar surface experiment. NASA officials also reported on, and Kirk and Evans approved, a priority listing of the DoD and NASA experiments for the workshop.[44] Eight days later, Evans directed the MOL Systems Office to proceed with its contract for the three experiments, aiming for a 30 June 1968 launch date.[45] At the close of 1966, the following eight experiments were scheduled for flight aboard NASA's Apollo spacecraft:

D-008 - Simple Navigation

D-009 - Radiation in Spacecraft

D-017 - Carbon Dioxide Reduction System

D-018 - Integrated Maintenance

D-019 - Suit Donning and Sleep Station Evaluation

D-020 - Alternate Restraints

D-021 - Expandable Airlock Technology

D-022 - Expandable Structure for Recovery

AAP USE OF TITAN III/MOL HARDWARE

Even as NASA was refining its "spent stage" experiment in early 1966, its post-Apollo proposals came under the scrutiny of the Bureau of the Budget. Searching for ways to cut federal expenditures, the Bureau requested NASA to consider several alternate approaches to a manned space flight program, such as possibly using the Titan III-MOL system plus the Saturn V-Apollo, or substituting the Titan booster for the Saturn IB at a certain point. The Bureau also asked the question whether the entire MOL system might not be used to perform the experiments NASA was planning for the Saturn IB-Apollo system.

The Bureau was not alone at this time in suggesting that MOL hardware might be useful in the NASA program. In a report published on 21 March 1966, the House Military Operations Subcommittee complained about "unwarranted duplication" between AAP and MOL and suggested that the greatest potential savings "would come from NASA participation in the MOL program." It noted that both NASA and the Air Force had talked about the possibility of accommodating NASA experiments on a non-interference basis on MOL "but to date little has been done to achieve this goal." Instead, the subcommittee said, NASA was proceeding with "a similar near-earth manned space project which will also explore the effects on man of long duration space flights and the capability of man to perform useful functions in space." The House unit urged the NASA and the Air Force to get together in a joint program which it said would save "billions of dollars."[46]

Meanwhile, DDR&E had learned of the Budget Bureau's request to NASA to study possible use of Titan III/MOL equipment. Anticipating the space agency would seek detailed information on the MOL system, Dr. Foster on 11 March directed the Air Force to undertake a study to provide answers to seven questions (which he provided) relating to NASA experimental use of MOL hardware, to include cost estimates and schedule impact on the MOL program.[47] Foster's request was the first of several submitted during the next several months, which led to extensive studies by the MOL Systems Office. On the basis of the Systems Office's initial investigation, whose results were presented to the MOL Policy Committee on 29 April 1966, Air Force officials concluded that MOL/Titan III hardware could handle the AAP experiments, although certain costly modifications and other changes would have to be made.[48]

However, it was not until mid-June that NASA approached DoD for information. At that time it submitted an initial set of questions relating to MOL equipment (other

questions followed) duly forwarded to the Systems Office or answers. By September the MOL Systems Office had compiled and the Air Force had provided the space agency "a wealth of technical and cost data" covering such matters as available payload weights and volume, electrical power, life support and environmental control based on the MOL 30-day baseline. In summing up this exercise, Dr. Yarymovych stated that the information indicated that "payloads in the order of 13,000 pounds in near-earth orbit at low inclination angles [launched] from ETR could conceivably, be supported by T-IIIM/MOL hardware subsequent to early 1970." Further, it appeared there would be no impact on the current MOL program by adding six more launches to the existing requirement, provided that the Air Force received "a near-future go-ahead for such a program."[49]

After reviewing over the voluminous Air Force data, NASA completed its final report on "Apollo Applications Program use of Titan III-MOL Systems" for the Bureau of the Budget. The report not surprisingly concluded that "the introduction of either the Titan IIIM launch vehicle or

Figure 54. MOL Test Launch Vehicle
Source: CSNR Reference Collection

Figure 55. MOL Test Launch
Source: CSNR Reference Collection

the Titan IIIM/MOL systems into the post-Apollo manned space flight program is neither technically desirable nor cost effective, and it could jeopardize the possible U.S. position in space by delaying by almost three years the low orbital application of proven U.S. space technology." NASA forwarded a copy of the report, dated 2 November 1966, to OSD with a draft of a joint NASA/DoD memorandum endorsing its conclusions. Before taking any action on this memorandum, Dr. Foster requested General Evans to critique the report.

On 6 December, after his staff had completed an intensive evaluation of the NASA document, Evans reported the MOL Program Office's findings to DDR&E. They were such "as to cast substantial doubt as to the objectivity, analytical thoroughness, and technical accuracy of the NASA report." Evans said that the report contained "undue bias against use of any hardware configuration other than Saturn-Apollo." Also, it took the position that the NASA study was "a sequel to and the converse of" the 1965 study on the possible use of Apollo systems for both NASA and DoD experiments, implying "a jointly planned and conducted study by DoD and NASA," which was false. The report further stated that the earlier study had indicated that Saturn/ Apollo systems could be used, beginning in 1968, to accomplish the DoD objectives assigned to MOL. Concerning this, Evans remarked that:

```
There is no doubt that technically,
if given sufficient resources and
time, Apollo systems could be used
in MOL. Similarly, under the same
assumption, MOL systems could also be
used in AAP. However, the assessment
of the desirability of use of one
specific system hardware in another
program must consider all cost
effectiveness factors, principally
those associated with performance,
schedule, and cost. The comment on
last year's study is incomplete and
is neither meaningful nor relevant.[50]
```

The Vice Director, MOL, consequently recommended that OSD "non-concur" in the 2 November 1966 NASA report.

Dr. Foster agreed. On 10 December he advised Dr. Seamans that he had reservations about the report in its "present form. Specifically, he said, the report did not represent "a joint" study, but rather was a NASA study on data provided by DoD on Titan III/MOL hardware. He also indicated there were other unsatisfactory aspects to the NASA report and he said he was prepared to

discuss the matter during the next meeting of the Manned Space Flight Policy Committee. Dr. Seamans accepted this suggestion.[51]

NASA's worries about the Bureau of the Budget inquiry—and DoD's position—were further exacerbated by the attitude taken by the President's Science Advisory Committee. In December 1966 PSAC circulated a draft report on "The Space Program in the Post-Apollo Period" which declared that:

```
Before substantial funds are
committed to the AAP plan to modify
Apollo hardware or to utilize the
orbital workshops for extended
periods, a careful study should be
made of the suitability, cost and
availability of Titan III/MOL systems
for biomedical studies of man for
periods up to 60 days. NASA should
also investigate whether delivery of
these components could be speeded
without interference with the MOL
program if additional funds were
contributed to MOL in the formative
years of the program

...Arrangements should be developed
between NASA and the USAF to use the
MOL Program as an importance source
of data on the capabilities of man
for space missions lasting 14 to 30
days, in addition to experience to be
gained in early Apollo Applications
missions.[52]
```

Secretary McNamara thought the PSAC report (published by the White House in February 1967), a very fine job "which perceptively addresses the important issues affecting NASA's future programs." It led in early 1967 to a Defense Department proposal that a joint DoD/ NASA study group to be set up to look into the entire matter, beginning with "an examination of the objectives of both Apollo Applications and follow-on MOL in low earth orbit."[53]

ENDNOTES

1. Memo for the Record (U), signed by Seamans 27 Jan, by Brown on 28 Jan subj: Gemini and Gemini B/ MOL Program.

2. Ltr (U), Seamans to Brown, 7 Aug 64; Memos (u), Brown to McMillan, 14 Aug 64, subj: Gemini Spacecraft Production; McMillan to Brown, 15 Sep 64, same subj.

3. Joint Statement by Secretary McNamara and Administrator Webb, issued 25 Jan 65.

4. Memo for the Record (S) , by Evans, 9 Feb 6.5, subj: Mtg of 9 Feb 65.

5. Ltr (U), Seamans to Brown, 9 Mar 65; Ltr (C), Evans to Mueller, 15 Mar 65.

6. Memo for the Record (U-AF Eyes Only), by Evans, 18 Mar 65, subj: Gemini B Procurement; Ltr (U), Seamans to Brown, 19 Mar 66; Memo for Record (C), by Evans, 29 Mar 65, subj: Gemini B Procurement.

7. Ltr (U), Evans to Schriever, 30 Mar 65, subj: Acquisition of AF Gemini B Spacecraft.

8. Memo for the Record (C), by Evans, 6 Apr•65, subj: Gemini B Management Proposal.

9. Ltr (U) Evans to Brig Gen David Jones, Hq NASA, 8 Apr 65, w/Atch, "Manage ment Plan for Dev and Acquisition of Gemini Spacecraft for MOL."

10. Msg (U), SAFSL to BSD (for Schriever), 13 Apr 65; Ltr (U), Brown to Seamans, 12 Apr 65; Memo for the Record (U), by Evans and Jones, 13 Apr 65, subj: NASA/ DoD Relations on Gemini B/MOL.

11. Msgs (U), SAFSL 63400, USAF to SSD (SSH), 14 Apr 65; SAFSL 63448, USAF to BSD (For Schriever), 14 Apr 65.

12. "Guidelines for the Dev of a Management Plan D fining NASA-DoD Relation ship for the Gemini/Gemini B MOL," signed by Mueller and Ritland, 20 Apr 65; Announcement, Manned Spacecraft Center, uniated, signed by Dr. Gilruth, MSC Director .

13. Ltr (U), Seamans to Brown, 23 Sep 65.

14. Ltrs (U), Schriever to Mueller, 24 Sep 65; Evans to Berg, 7 Oct 65, subj: MOL Support to be Obtained from NASA; Col Brady to Evans, 25 Oct 65, same subj .

15. Msg (U), Mueller to MSC (Gilruth, Matthews, Henry), 29 Nov 65, subj: Transfer of Gemini Equipment to MOL; Ltrs (U), Mueller to Gilruth, 20 Nov 65; Mueller to Evans, 20 Nov 65; Memo (U), Maj M. C. Spaulding to Evans, 30 Dec 65, subj: NASA Support of MOL.

16. Msg (U), SAFSL-6C-22879, Berg to NASA MSC (Lt Col Henry), 10 Mar 66, subj: Transfer of NASA Gemini Equipment to MOL; Ltr (U), Berg to Gilruth; 26 Mar 66, subj: NASA Equipment Needed to Suport MOL.

17. Memo for the Record (U), by Maj M. C. Spaulding, MOL Prog Office, 30 Mar 66, subj: NASA Gemini Equipment for MOL.

18. Ltr (C) , Evans to Mueller, 6 May 66.

19. Msg (C), SAFSL 96372, USAF to SSD (Evans to Beg), 3 Jun 66.

20. Ltr (U), Mueller to Evans, 7 Jul 66.

21. Ltr (U), Col Paul Heran, MOL Sys Office to Evans, 14 Dec 66, subj: Trans ferred NASA Equipment Costs.

22. Rprt (C) Preliminary Evaluation, Program 631A: Atch 9 Ltr (C), Col W. A. Ballentine, Cmdr, Det 2, SSD (AFSC), to Evans, 13 Sep 66, same subj.

23. Ibid.

24. Ltr (C), Lt Col W. C. Fry to Gen Hedrick, Hq AFSC, 1 Dec 66, subj: Experiments Management Relationship with NASA .

25. The summary of experiments is based on NASA report SP-138, Gemini Summary Conference, 1-2 Feb o7, Chpt 20, "DoD/NASA Gemini Experiments Summary," and Ltr (C), Lt Col Fry to Gen Hedrick, 1 Dec 66, subj: Experiments Management Relationship with NASA .

26. Ibid.

27. Ltr (U), Seamans to Brown, 23 Sep 65; Ltr (C), Foster to Seamans, 13 Oct 65.

28. Ltr (U), Schriever to SAF, 9 Nov 65, subj: NASA/ DoD Manned Space Flight Relationships.

29. Memo of Understanding Between DoD and NASA Concerning the Manned Space Flight Programs of the Trio Agencies, signed by Seamans, Webb, Foster, and McNamara between 9 and 14 Jan 66.

30. Min (C), 1st Mtg, Manned Space Flt Policy Cmte, 21 Jan 66, by D. J. Fink, OSD; Memos (U), Schriever to SAF, 25 Feb 66, subj: Manned Space Flt Activities; Flax to Schriever, 14 Apr 66, same subj.

31. Memorandum of Agreement (U), signed by Seamans and Foster, 21 Jan 65, subj: Establishment of a Manned Space Flt Experiments Board (MSFEB).

32. Ibid.

33. Chron (C), On-Board Experiments for NASA Apollo Program, atch to Evans (C), Schultz to Flax, l2 Jan 65, subj: Policy on Approval of Apollo On-Board Experiments; Memos (U), Flax to DDR&E, 3 Mar 65, same subj: Flax to DDR&E, 29 Mar 65, subj: AF Experiment on Apollo Flt; Fubini to Flax, 9 Apr 65, subj: AF Experiment for Apollo.

34. Min (U), MSFEB Mtg 66-1, 17 Jan 66.

35. Memo (S-DORIAN), Schriever to SAF, 4 Mar 66, subj: MOL Monthly Status Report.

36. Memo (S-DORIAN), Schriever to SAF, 8 Apr 66, s11bj : MOL Monthly Status Report.

37. Ltr (U), Evans to Berg, ltr, AFSC, 6 Apr 66, subj: DoD Experiments for NASA S-IVB Orbital Lab; Memo (TS-DORIAN), Schriever to SAF, 6 May 66, subj: MOL Monthly Status Reper.

38. Msg (U), SAFSL 84012 (Evans to Berg), 27 Jul 66; Ltr (S), Yarymovych to Col W. A.Ballentine (Det 2, SSD), MSC Houston, 29 Jul 66, subj: Transmittal of Program Plan re: AF Experiments for NASA S-IVB Workshop.

39. Memo (S-SAR), Evans to Flax, 25 Aug 66, subj: AF Experiments for NASA S-IVB Workshop; Memo (U), Flax to Evans, 10 Sep 66, subj: Approval of AF Experiments for NASA S-IVB Orbital Workshop.

40. Memo (S-DORIAN), Ferguson to SAF, 6 Oct 66, subj: MOL Monthly Status Report: Swmnary (C), MSFEB Mtg 66-5, 19 Sep 66, prpd by Brig Gen Walter R. Hedrick, 22 Sep 66.

41. Ltr (U), Col W. w.Sanders, MOL Prag Office to Mr. Haynes, Hq NASA, et. (C), 11 Oct 66, subj: Transmittal of NASA Forms 1138 for AF Experiment for S-IVB 0rbital Workshop.

42. Ltr (C), Evans to Mueller, 26 Oct 66; Msg (C), SAFSL 82424, Evans to Berg, 27 Oct 66.

43. Ltr (C), Mueller to Evan2, 7 Nov 66.

44. Memo (C) , Lt Col W. C. Fry , DoD Representative; MSFEB Working Group, to SCG, 25 Nov 66, subj: Summary of MSFEB Mtg 66-8, 21 Nov 66.

45. Msg (U), Evans to SSD (Berg), 29 Nov 66.

46. Twenty-third Report, Cmt on Government Operations, Missile and Space Ground Support Operations, 89th Cong, 2nd Sess (Report No. 1340), 21 Mar 66.

47. Memo (U), Foster to Flax, 11 Mar 66, subj: Titan IIIC/MOL Studies.

48. Memo (S-SAR), Evans to Flax, 22 Apr 66, subj: Titan IIIC/MOL Capabilities to Support NASA AAP: Mis (TS-DORIAN), AF MOL Policy Cmte Mtg 66-2, 29 Apr 66.

49. Ltr (U), W .E. Lilly, Hq NASA, to John E. Kirk, ODDR&E, 17 Jan 66; Memo (S), Yarymovych to Flax, 9 Sep 66, subj: Data Provided NASA on the Feasibility of T-III/MOL Support of NASA AAP .

50. Memo (S), Evans to Kirk 6 Dec 66, subj: Coordination of NASA Apollo Applications and DoD MOL Program.

51. Memo (S), Foster to Seamans, 10 Dec 66; Ltr (S), Willis H. Shapley, Associate Deputy Adminisrator, NASA, to Foster, 23 Dec 66.

52. Report of the President Science Advisory Committee, The Space Program in the Post-Anello Perici, February 1967, pp 24-25. The final draft of this report was circulated in December 1966.

53. Ltr (U), McNamara to the President, 7 April 1967.

MOL ASSEM[...]
INTEGRATION BUILD[...]

MATERIAL

PROPERTY OF THE
[U]NITED STATES GOVERNMENT

[I]F FOUND, DO NOT OPEN

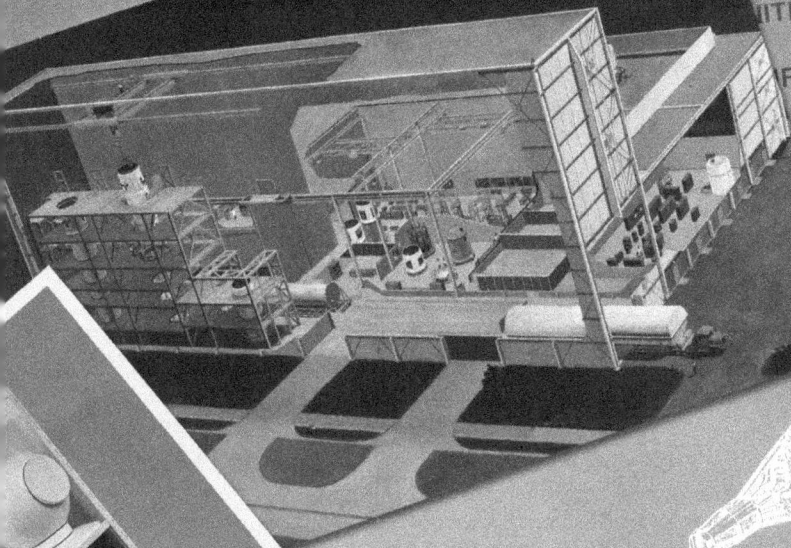

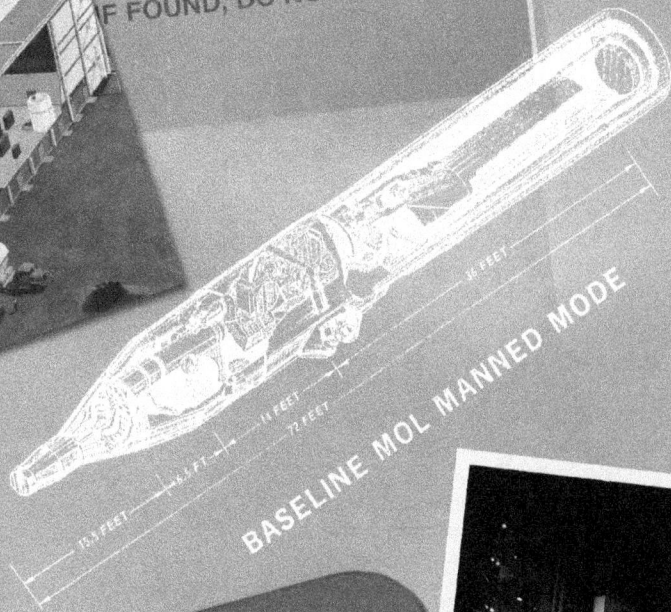

BASELINE MOL MANNED MODE

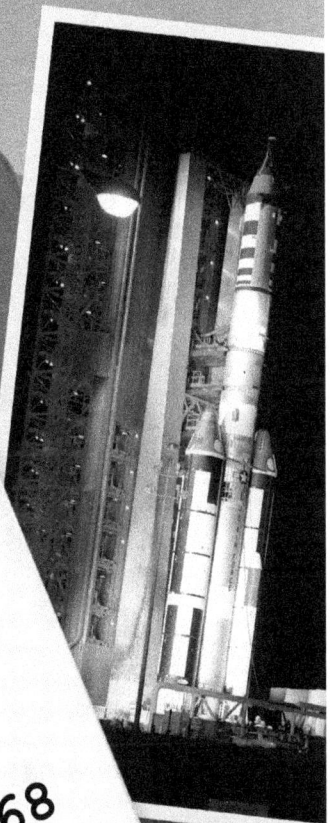

USAF MOL/KH-10[...]

THE DORIAN FILES REVEALED:
A COMPENDIUM OF THE NRO'S MANNED ORBITING LABORATORY DOCUMENTS

CHAPTER XIV:
NEW FINANCIAL AND
SCHEDULE PROBLEMS 1967–1968

BODY159

NEW FINANCIAL AND SCHEDULE PROBLEMS 1967-1968

As noted, from the earliest days of the program Air Force officials had felt a sense of urgency about getting the MOL into orbit at an early date. They were motivated in part by their experiences with "the B-70 and Dyna-Soar programs, which had been dragged out interminably—mainly due to lack of administration support—until finally cancelled. In both cases, the Air Force rationale had been disputed and neither was supported by the White House. MOL on the other hand, had been publicly endorsed by President Johnson and was assigned an important mission—very high resolution photographic reconnaissance. Still, Air Force officials felt it essential to get a laboratory vehicle into orbit as soon possible. They were consequently pleased when MOL entered Phase II engineering development in September 1966 and they had additional cause for satisfaction two months later when the Gemini B heat shield underwent a successful test flight launch from Cape Kennedy. The first unmanned MOL launch was set for April 1969 and the first manned flight for December 1969.*

Although engineering development work got under way officially in the late summer and fall of 1966, the MOL System Office at the time was still involved in hard contract bargaining with the Associate Contractors. The problem was a familiar one: the contractor's cost estimate and those of the Air Force were millions of dollars apart. McDonnell, for example, estimated its Gemini costs at $205.5 million for a fixed price, incentive contract whereas the Air Force offered $147.9 million. Douglas requested $815.8 million to develop and build the laboratory vehicles (fixed price incentive/cost plus incentive fee), the Air Force proposed $611.3 million. General Electric sought $198 million (cost plus incentive fee), the Air Force offered $147.3 million.[1]

With the government and contractors unable to reach an agreement, Air Force officials—with the approval of the MOL Policy Committee—in late November 1966 adopted a new "negotiating strategy." They directed the MOL Systems Office to reopen competition for those systems not already under contract and to halt the issuance of any further DORIAN clearances to contractor personnel. This tough stance broke the deadlock; by early December, the contractors had substantially reduced their cost estimates to bring them closer to the Air Force offers. On 4 January 1967—when Dr. Flax summarized the

results of these negotiations to Dr. Foster—he was thus able to report that total MOL development costs would run approximately $1.92 billion, a sum which included $295.0 million in "deferrals.†"[2]

THE FY 1968 BUDGET CRISIS

The December 1966 agreements with the contractors—completed to the handshake stage—were based on an understanding that OSD would release all deferred fiscal year 1967 funds and that at least $80 million would be provided the program in fiscal year 1968. The latter premise soon proved faulty. OSD—finding it needed huge sums to support U.S. operations in Southeast Asia—notified the Air Force on 7 January that it planned to request only $430 million from Congress in new obligating authority (NOA) for fiscal year 1968. This sum was $157 million below the amount the MOL Systems Office estimated was its minimum requirement and $381 million below the contractors' estimates.

McNamara's decision meant that the Air Force would have to renegotiate the prime contracts to reduce fiscal year 1968 fund requirements to $430 million NOA. On 15 February, in an effort to define a revised MOL baseline to fit the lower funding level, contractor and Air Force officials, including Dr. Flax, convened a meeting on the West Coast. It became evident during their discussions of ways to reduce the impact of the cut in funds that there would be additional slippage in the launch schedule. At the end of the conference, Dr. Flax directed the contractors and the MOL Systems Office to prepare new program schedules, using as their "Bogeys" the planning figures of $500 million and $600 million NOA's for fiscal years 1969.‡[3]

It was against this background that General Evans—soon to be succeeded as Vice Director, MOL, by Maj. Gen. James T. Stewart§—summed up "the current mess" and the "Pearl Harbor"-type crisis facing the program. In a memorandum to several aides, he reviewed various options they might examine in the future. One would accept a nine or 12-month program slip; another would

---

* This schedule constituted a year's slip from the launch dates announced by the President in August 1965. See Chapter XI.

† Items deferred included the data readout system, various spares, some test activities, and manpower requirements to support "out-of-plant" or field test operations for all of the contractors concerned.

‡ It was Dr. Flax's intention to reprogram other Air Force funds to meet the fiscal year 1968 deficit.

§ Stewart took over on 27 March 1968. General Evans retired to join an industrial concern.

accept "a fund-governed program" tailored to $228 million in fiscal year 1967, $430 million in fiscal year 1968, "and whatever the approved DoD figure is in FY 1969." Evans suggested MOL officials might consider a possible fourth option: "...decoupling the optics from the first manned flight with [the] objective being to provide a more completely man- rated system when the first- flight-qualified DORIAN package becomes available."[4]

He also recommended the MOL Program Office attempt to resolve several other issues, including determining the best approach to funding future contracts (commitment versus an expenditure basis), obtaining a DX priority for the program[¶], and reforming MOL management. This last subject was a sensitive one to Evans. For many months he had believed that existing arrangements—which had black MOL contract, development, management, and financial actions being accomplished by another Air Force activity—were ineffective. "The present management structure," he said, "is incapable of producing a well-integrated, well-managed large program such as MOL."[**5]

Some of the questions posed by Evans were discussed on 10 March 1967 during a management meeting attended by Dr. Flax, General Ferguson, and other MOL officials at Andrews AFB. Systems Office personnel briefed them on proposed funding and schedule revisions which they had been examining with the contractors. They advised that, even if a 12-month schedule slip was acceptable, MOL would require additional funds beyond the $500/600 million "Bogeys" proposed by Dr. Flax. If given only $480 million in fiscal year 1968, they foresaw a slip in the first manned flight of at least 15 months. This discouraging report was subsequently reviewed by Drs. Foster and Brown, who asked that new guidance be sent to the MOL Systems Office. Future funding, the Office was informed on 17 March, was to be on "a commitment basis" and, for planning purposes, totals were not to exceed $480 and $620 million in fiscal years 1968 and 1969. The Systems Office was directed to prepare a paper on the impact of those funding levels for presentation to Pentagon officials in Washington in April.[6]

The MOL schedule was reworked by MOL Systems Office personnel. They concluded that the 15-month slip in the first manned launch constituted "an optimum program from the standpoint of fund limitations" and

would ease development and hardware integration problems. However, even if funding was provided on a commitment rather than an expenditure basis, they saw the program requiring $518 million in fiscal year 1968. On 15 April 1967 their conclusions were presented to a MOL management meeting attended by Flax, Ferguson, Stewart, Bleymaier[††], Martin, Berg, and others. Dr. Flax's response was to challenge the cost estimates presented to him; it was his view that the contractor's costs were probably inflated, "at least for the scope of work as currently defined." He warned that the projected overall total MOL program cost rise of about half a billion dollars, if correct, might be "very detrimental" to its future and he directed the MOL Systems Office to take another look at its cost estimates.[7]

The cost figures were subsequently reworked on the West Coast and forwarded to Flax and Ferguson on 2 May by General Bleymaier and Colonel Heran. In their message, they stated flatly that the funding levels of $480/600 million in fiscal years 1968 and 1969 "positively eliminate any possibility of establishing a realistic 12 month slip program." Furthermore, those levels placed in jeopardy their ability to meet a 15-month slip, particularly if there were additional delays in contract negotiations over the new schedule and deciding on deferred items. They said that they had examined all possibilities and nothing could be further gained by efforts to hold the program to a 12-month slip. "The only hope in holding the slip to 15 months," they said, "lies in proceeding immediately with the negotiations of a 15 month schedule slide." The contractors "firm cost proposals," based on their detailed analysis of a 15-month slip, indicated a two-year funding need for $524 and $617 million. Accordingly, the $480 million fiscal year 1968 funding limitation would produce a $44 million deficit, Bleymaier and Heran further stated that:

> If you entertain the possibility of placing increased emphasis on those system elements that are critical technically and schedule wise, you will effectively reduce fund availability to continue the orderly development of the total system. To date, we have proceeded on a balanced funding approach with orderly development of interfaces and testing. To deviate would seriously impair the systems approach. Expediting part of the

---

¶   For many months Evans had sought to obtain approval for assignment of a DX industrial priority to the program. The request was never acted upon.

**   Gen Martin, previously opposed to Schriever's plan for integrated management, on 20 April 1967 agreed that "division of management responsibility and authority on the basis of security is totally unworkable." A major management reorganization followed, which placed all black contracting under the MOL Systems Office, effective 1 July 1967. [Ltr, Martin to Flax, Ferguson, 20 Apr 67, subj: Management Responsibilities for the MOL Program]

†† Bleymaier who officially took over as Deputy Director, MOL Program, on 1 July 1967, participated in various planning exercises during the spring of 1967. He was, at this time, serving as Commander, Air Force Western Test Range, Vandenberg AFB, Calif.

system could not in any case reduce
schedule slip since other elements
would be out of balance.[8]

After reading this pessimistic report, Flax wrote to Stewart: "There must be some schedule slip at which it is cheaper to stop some efforts, but we are informed that this is impossible because it would preclude 'orderly' development of everything."[9] After the two men discussed the situation, General Stewart sent a message to the MOL Systems Office advising that a meeting would be held in Washington on 11 May "to attempt to reach some understanding and agreement" on the program schedule and the "near-term contractual actions to be taken." The Systems Office would have to justify its unequivocal assertion that "all possibilities have been examined and nothing further can be gained" toward retaining the objective of a first all-up manned flight by the end of calendar year 1970. Stewart said that Dr. Flax desired the Systems Office's views on "the maximum reasonable curtailment of Martin and McDonnell work in FY 1968, short of termination, which would permit the continuance of minimum essential engineering interfaces and the maintenance of the minimum essential supervisory and technical teams for each."[10]

Further, he stated that while he and Dr. Flax appreciated the Systems Office's desire to have an "orderly development of the total system" and a balanced funding approach, the program "is now in a financial/schedule constraint not of our making and is in for more jeopardy than I seem to have impressed on you." If there was any feasible way in fiscal year 1968 to work toward a first manned launch by the end of 1970, he said, "it must be identified and pursued, recognizing its possible effect on 'orderly development of the total system.'" For their information, Stewart provided them the latest OSD planning NOA figure for fiscal year 1969—$661 million.[11]

This message apparently finally convinced both the contractors and the MOL Systems Office that $480 million was all that would be made available in fiscal year 1968. At the 11 May 1967 management meeting, they agreed they could live with it. However, $480 million was still $50 million shy of the amount the President requested from Congress. Dr. Flax indicated that at least a portion of the sum ($10 million) would be obtained from Air Force internal reprogramming action and the balance sought from the congressional appropriations committees. Concerning the two alternative schedule slips—12 or 15 months—Brown, Flax, and Stewart the next day—12 May— decided to proceed with a "compact 12" schedule, pointing toward the first all-up manned flight in December 1970 and working against fiscal year 1968 and 1969 "Bogeys" of $480 and $661 million. The

three officials agreed that the Phase II contracts and appropriate supplements should be signed as soon as possible to put the program on a sound basis. Instructions to this effect went out to the MOL Systems Office the same day.[12] On 22 May—after all the contracts had been signed—the Department of Defense announced to the public a total award of $855,072,744 to the McDonnell-Douglas Corporation for MOL engineering development work[‡‡]. The Douglas contract was for $674,703,744 and covered work on the manned laboratory vehicle; the $180,469,000 award to McDonnell covered preliminary design, development, and production of the Gemini B. On 29 May the contract award of $110,020,000 to General Electric for MOL experiment integration work also was announced.

Even as this important milestone was reached, OSD and Air Force officials became increasingly concerned over the effects of the war in Southeast Asia on the DoD budget. At a meeting of the MOL Policy Committee on 1 June 1967, Secretary Brown mentioned the possibility the Air Force would face severe cuts in its research and development funds. General Stewart interjected that, based on comments made to him during his recent MOL briefing to the chairman and staff of the House Appropriations Committee, he did not think Congress would reduce the MOL budget request. However, in reviewing for the Policy Committee the status of the program, he noted that overall program costs had risen to $2.35 billion.[13]

McNAMARA VISITS EASTMAN KODAK

During the summer of 1967 MOL officials began planning a proposed visit by McNamara to Eastman Kodak. An orientation and briefing on sensor development and the overall program was scheduled. To insure there would be no press coverage of his visit to Rochester, MOL, and other security personnel made extensive advanced preparations. One" cover" action was release of a short news item to the press stating that the Secretary of Defense would be inspecting various military installations over several weeks. The actual flight to Rochester was to 'be made under tight security wraps.

Also, to help make a case for continuation of the program, Colonel Battle and MOL Program Office personnel prepared a two volume report, titled "High Resolution Photography," to be provided the defense chief. The report analyzed the results of all G-series of unmanned reconnaissance satellite flights and the expected MOL product. The best photos from the KH-7

‡‡ During 1966 Douglas Aircraft ran into financial difficulties. McDonnell offered to merge with it and the corporate marriage was formally consummated late in the year.

UNCLASSIFIED
APPROVED FOR PUBLIC RELEASE
DECLASSIFIED BY DCI
25 OCTOBER 2000

Figure 56. KH-7 Imagery
Source: CSNR Reference Collection

missions, it noted, had resolutions of 21 inches with 20 percent having resolutions of more than 30 inches and 85 percent of all photos being four feet or more in resolution. In the case of the KH-8—still in the development phase—its photos were better than the mature KH-7 products. The report estimated that 93 percent of its photos "would be superior to 24 inches only about six percent of the time. On the other hand it declared that 99 percent of the MOL photos would fall in the better than 12 inch category and more than half would approach {an even better} class, while under ideal conditions {the best class of resolution} would be produced. The MOL report concluded, therefore, that the DORIAN system would be {many more} times as productive as photos in the 12 inch or higher class.[14]

McNamara's visit to Eastman Kodak took place on 14 September 1967. Flying to Rochester with him were Drs. Foster, Flax, Brown, and several others. During the flight he read the MOL Program Office report on "High Resolution Photography" and, as Stewart later remarked, "it apparently was the convincer that we needed the {best resolution} product." At Eastman Kodak, after touring the facilities, and being briefed by various officials, McNamara said that he had been concerned about the program, "particularly since he had noted an increase of 50 percent since the original estimate." He had wanted to assure himself on the status of the program. He commented that the presentations were excellent, he had received the information he required, and he thanked everyone for their efforts.[15]

The defense chief's visit seemed to have been a success. But the MOL program was still short $40 million in funds. At a staff conference on 28 September, General Stewart remarked that the program faced "a real crunch." He said other R&D programs—such as the C-5A and Minuteman III—were putting a squeeze on the Air Force budget and, with the MOL Program expending funds at the $480 million rate, unless an additional $40 million was made available "we might have to go unmanned." However, the Vice Director went on to say that he didn't think it would "come to this because the Air Force is emotionally committed to man in space."[16] Despite its commitment, however, it was unable to come up with even the $10 million Flax had earlier hoped to reprogram. The fiscal year shortage came to $50 million.

At this juncture, both Dr. Brown and General McConnell made informal requests to Chairman Rivers of the House Armed Services Committee soliciting a $50 million increase in the DoD budget. They were turned down, however. One reason—cited by Mr. Earl J. Morgan, counsel of a House subcommittee headed by Representative Melvin Price—was that Dr. Foster

had testified that $430 million would be adequate to support the program. Morgan also remarked that the Price subcommittee was "tired of getting the idiot's treatment" on MOL. A related reason, given by Mr. John R. Blanford., Rivers' chief counsel, was that the tight security surrounding the MOL created problems in the main committee.[17] The end result was that, when Congress completed its work in October on the defense appropriation bill for fiscal year 1968, the MOL was allotted only the original $430 million requested by the President in January.

The $50 million shortage required Air Force and contractor officials to revise the schedule again. On 19-20 October, Systems Office and contractor personnel met to discuss the problem. Their tentative conclusion was that "...an additional 12-week slip would have to be incorporated into the program, pushing the first manned launch date into 1971. However, they noted that even if another 12-week slip was accepted, the MOL contractors would be put into "an unacceptably exposed financial position." On 7 November Stewart directed Bleymaier to develop an appropriate adjusted schedule which would reduce or eliminate any contractor risk and which could be used as a departure point for fiscal year 1969 and 1970 funding projections.[18]

THE THIRD MAJOR SCHEDULE REVISION OF 1967

During late November and early December 1967 the third major program and schedule revision of the year got under way on the West Coast. On this occasion, the frustrated associate contractors asked Air Force officials to allow them to meet separately to devise a program which would meet both the goals and financial constraints placed upon MOL funding. The Air Force agreed, whereupon the contractors during the week of 30 November through 6 December met at McDonnell-Douglas's Huntington Beach facility. On 7-8 December, they submitted their revised MOL program to the Air Force management. Surprisingly, they recommended only minor changes and deletions in program content, and proposed a schedule keyed to a first manned MOL launch in August 1971. This schedule, which had the unreserved indorsement of all the major associate contractors, was based on the assumption that fiscal year 1969 funding would total $661 million.[19] On 8 December, a MOL Program Review Council, chaired by Dr. Flax, approved the proposed program revisions (with some minor exceptions). Both the government and industry were in agreement that they had a MOL plan which was "technically and financially sound" and promised to achieve all primary program objectives.[20]

However, scarcely had this agreement been reached than Secretary McNamara—the very next day, 9 December—dispatched a memorandum to Brown inquiring as to the feasibility of completing the program, at least in the first phase, "without a man" and limiting funding in fiscal year 1969 to $400 million. This astounding memorandum quickly doused Air Force hopes. On 15 December, Dr. Brown forwarded to McNamara a lengthy reply, in which he listed and discussed four alternative programs which might be adopted. They ranged from the current program to dropping man, to a severe stretchout of the MOL schedule. After examining each in some detail, Brown stated it was his opinion that they should proceed with the program "as presently constituted." He said:

> I believe the present MOL Program approach is worth the cost in terms of assurance of meeting the resolution goal and returning a worthwhile product at the earliest reasonable date, plus the verification and exploration of additional manned reconnaissance contributions such as target verification, target selection, weather avoidance, etc.
>
> I therefore recommend, as a first option, that we fund the present program in FY 69 at not less than $600 million. If that is not possible, then the program should be funded at not less than $520 million...in FY 69 and the resulting 5-6 month additional stretchout and increased total cost of the program be reluctantly accepted. We should do the latter only if we are willing to accept the $600 million cost in FY 70 and perhaps that much in FY 71. If we are not, we should terminate the MOL Program except for the Eastman Kodak and General Electric efforts and define a new unmanned system... In that situation approximately $400 million should be budgeted in the black for FY 69.[21]

Brown's first option—$600 million—was accepted by McNamara and incorporated into the President's budget, submitted to Congress in January 1968. However, it would not stand up during the new year, an election year, which found the United States beset by more grim events than it had experienced in several decades. First, on 24 January 1968, a bellicose North Korea seized

the U.S. intelligence ship, USS Pueblo, off the coast of Wonsan in the Sea of Japan, generating an international crisis leading the President to call up of the Air Force Reserve. More importantly, in Vietnam at month's end, the communists launched a powerful Tet offensive which carried North Vietnamese and Viet Cong troops into the heart of the country's major cities, including Saigon. Also, before a half year was out, President Johnson would announce his intention not to run for reelection, Sen. Robert Kennedy and the civil rights leader, Martin Luther King, would be assassinated, and riots would hit a dozen American cities.

All these events would eventually affect MOL directly or indirectly. It was, perhaps, just coincidence which led General Stewart; on 30 January 1968, to write to the Comptroller of the Air Force about MOL program cancellation costs. He advised that all available funds would, by the end of June, either have been spent or be needed legally "to cover noncancellable commitments made on the part of the MOL contractors." He reported special termination cost clauses had been included in all MOL contracts and that—if the program was cancelled late in the fiscal year—certain funds would have to be provided by the Air Force from sources outside the program He estimated such costs would total $46.7 million if the program was cancelled late in the fiscal year. He further advised that, while "no such action is contemplated," his memorandum's purpose was "merely to apprise you of the possible impact of MOL on administratively reserved funds should termination take place."[22]

The Director of the Budget, Maj. Gen. Duward L. Crow, subsequently replied that the Air Force had no specific administrative reserve of funds for special termination costs. If additional funds were needed, he said, "adequate unobligated funds would be available within the applicable appropriation, even though reprogramming from other approved programs might be necessary."[23]

CIA AND STATE DEPARTMENT OPPOSITION

In early 1968 a new factor entered the MOL picture—the growing opposition to the program of the CIA and State Department. On 14 February, during a meeting of the National Space Council—chaired by Vice President Humphrey—Ambassador Charles E. Bohlen of the State Department asked whether MOL was worth $2.4 billion. He noted that the government had spent "some $722 million on this project" and he suggested that a committee be set; up to "study the need before we go any further." He added that the system had never "been approved by USIB." Dr. Foster responded that it was not the role of the U.S. Intelligence Board to approve programs, but to establish national intelligence requirements. He said MOL's value and scope had been reviewed in detail and endorsed by McNamara, Schultze, Vance, Brown and Hornig before the President announced his 1965 decision. Further, he had personally reviewed the program and felt it would be valuable "to the future reconnaissance requirement." Foster suggested to Bohlen that the Defense Department brief him in detail on the program; the offer was accepted and a presentation made several days later.[24]

Figure 57. Richard M. Helms
Source: CSNR Reference Collection

Bohlen's criticism was similar to that expressed by Richard Helms, Director of the CIA. On 5 March 1968, he forwarded to Dr. Foster a statement summarizing his views, which he suggested be incorporated in a MOL Development Concept Paper ODDR&E was preparing with the assistance of Air Force officials[§§]. This CIA statement read:

> Mr. Helms, Director of Central Intelligence, has reservations as to the value of better [than one foot} resolution photography for national intelligence purposes. He recognizes that photography with resolutions better than that obtainable by the GAMBIT-3 system would be helpful but does not believe studies conducted

to date shown that the value of this increased resolution justified the expenditures associated with the MOL Program. He has initiated a review of these studies.[25]

The review by the CIA was completed by mid-May 1968, at which time Helms forwarded a summary of its conclusions to Ambassador Bohlen, Deputy Secretary of Defense Paul Nitze, and the Director of the Bureau of the Budget. It declared that there was no doubt very high resolution MOL photography "would make a valuable contribution to intelligence, particularly on detailed information relating to Soviet and Chinese weapons and programs." Satellite photography with {the best possible} to 12 inch resolution would help identify a larger number of small items or features beyond existing capabilities. It would increase U.S. confidence in identifying items "we can now [only] discern" and would reduce the error of measurement of such items. Higher resolutions also would improve U.S. understanding of some operating procedures and construction methods at Soviet military installations and technical processes and the capacities of certain industrial facilities.[26]

But, despite all the above cited advantages, the CIA paper concluded that no important agency estimates of Soviet or Chinese military posture, weapon performance, or size and composition of forces would be changed significantly by MOL photography. This conclusion, it said, was based partly on the judgment that some of the nation's outstanding intelligence problems were more likely to be solved by the acquisition of technical information from systems other than satellite photography. It noted, for example, that "electronic intelligence is needed for solving certain problems critical to our estimates of the capabilities of surface-to-air ABM systems..." Programs for the collection of such information were either under way or were scheduled for operation "by the time the MOL is operational." In summary, the CIA report stated that, while there was no question that satellite photography with ground resolutions of {the best anticipated} to 12 inches would provide useful intelligence, the "pivotal question" remained whether such additional intelligence was worth the costs.[27]

§§ The Development Concept Paper was a management device established by McNamara in September 1967. Its purpose was to "document the full military and economic consequences and the risks involved in each new major R&D program."

TROUBLE ON THE CONGRESSIONAL FRONT

The CIA's arguments reinforced Bohlen's doubts and opposition from the Bureau of the Budget, which had been unenthusiastic about MOL from the start. In addition, Air Force officials saw the program undermined somewhat, perhaps inadvertently, before key members of Congress. This occurred during an NRO briefing on 21 March 1968 for Chairman Mahon of the House Appropriations Committee and Congressmen Frank T. Bow and Glennard P. Lipscomb. Representing DoD were Drs. Foster and Flax,{ a NRO finance official}, and Col. David Carter of the Office of Space Systems, who made the formal presentation. Also in attendance were three CIA representatives. During his presentation, Carter referred at one point to the development of GAMBIT 3 and declared that its design goal was {better than one foot} ground resolution. The system's current performance, he said, was at the 13 to 15 inch level. At this point, Chairman Mahon remarked that the products looked so good that "we ought to be able to slip the MOL."[28]

After Carter concluded his briefing, Dr. Foster sought to allay any doubts raised about MOL by the presentation. He explained to the Congressmen that he had begun the November-December budget "scrubdown" with the intent of reducing MOL funds to the $400 million level in fiscal year 1969, which he had thought was feasible. However, he had finally become convinced that $600 million was the lowest level at which the DoD could have a viable program. Concerning the resolution question, he reiterated his belief that there was a need for MOL, that the {anticipated best resolution} product would greatly add to U.S. knowledge of many facets of Soviet and Chinese military capabilities. He also cited McNamara's view that the MOL system would be extremely useful in an arms agreement role.[29]

On 25 March, at a follow-up NRO program review with selected members of the Mahon committee, Mr. Robert L. Michaels, the chief staff assistant to the chairman, asked whether the intelligence community had "indicated a requirement" for the MOL program. Dr. Flax, the NRO Director and a strong MOL supporter, replied it had initially, but that there had been a change with the change of the top leadership in CIA and OSD. He said that when Mr. McNamara left office several weeks earlier, he had been convinced "that there was a vital need for the MOL system and the resolution that it will provide..." The MOL Program, Flax argued, was the only one "in which a high resolution reconnaissance system is being developed." Further, he believed that "the present resolution obtained from the GAMBIT 3 system is not sufficient to do the

complete analysis of the Tallinn Missile Systen." When Michaels pointed out that the MOL system would not be available for some time, Flax countered by saying that the United States would be facing "other Tallinn's in the future." At this point, Chairman Mahon specifically asked Foster and Flax if they were in favor of the MOL system. Both answered "Yes."[30]

Turning to the CIA representative, Mr. Carl Duckett, Mahon asked about the agency's views. Duckett replied that Director Helms had reservations "with respect to the value of the very high resolution in - view of the cost of the system," that he was not opposed to MOL, but felt the time had come to study the system again to confirm the value of its product versus the cost of acquisition. (The conclusions of this review, completed in May, have been noted above.) Toward the close of the meeting, when the discussion turned to MOL's fiscal year 1969 fund requirements, Dr. Foster remarked that the program was "not the softest spot in the DoD budget."[31]

Unfortunately, the financial pressures on the defense budget continued to increase and directly affect MOL. In the spring of 1968, before Congress agreed to pass the Revenue and Expenditure Control Act (the surtax) as requested by President Johnson, it required him to cut $6 billion from fiscal year 1969 government disbursements. Secretary of Defense Clark Clifford, McNamara's successor, subsequently established a special project aimed at getting the Department of Defense to absorb $3 billion of the reduction. DoD's cost-reduction effort seems to have stimulated the Congressional Quarterly, a private publication which covered the activities of Congress, to print an article on 28 June 1968 which declared that there was $10.8 billion in "fat" in the defense budget. Specifically, the Quarterly homed in on the MOL program as an area "ripe for cuts," and declared the time had come "in this period of revaluation of national priorities and objectives" to raise questions about that Air Force program.[32]

Greatly concerned about possible severe program cuts, General Stewart in early July 1968 met with a staff member of the Senate Committee on Appropriations, Mr. William W. Woodruff. The Vice Director explained he had no doubt that MOL would have to make its contribution to the general defense budgetary cuts. However, he feared the amount of the cut might seriously affect the schedule and total program costs. He reviewed the extensive reprogramming exercise which had been conducted in December 1967 and stated that, if the FY 1969 NOA was cut to $400 million, it would be necessary "to publicly terminate the program and continue payload development in the black." In his view, he said, $520 million would be the minimum figure at which the program "could remain viable." Stewart also expressed particular concern over the possible "double jeopardy" the program faced with regard to a Senate cut, followed by an OSD cut, to meet the $3 billion reduction goal. In response, Woodruff assured him the Senate cuts would be applied to the DoD quota and that Senator Russell could probably hold the committee "to a reasonable reduction in MOL."[33]

Subsequently, Congress authorized an FY 1969 MOL appropriation of $515 million, a figure acceptable to Foster, Brown, Flax, and Stewart. Months before the bill was passed, the MOL Systems Office was directed to restructure the program "based on a NOA of $515 million in FY 1969 and $600 million in FY 1970," which would require another slip in the launch schedule. On 15 July the MOL Systems Office convened a four-day conference of the associate contractors to once again readjust the program and schedule to fit the reduced funding. An observer, Lt Col Bertram Kemp, of the MOL Program Office, noted "a considerable amount of demurral" from the contractors over the schedule adjustments, which they apparently had not expected. In the end, they agreed to change the flight schedule to slip the first manned launch from August to December 1971. No changes were made in the program's technical content. The revised schedule was reaffirmed on 25 July at a meeting of contractor program managers at Valley Forge, Pa.[34]

By year's end, Air Force officials had successfully navigated MOL through the financial shoals of 1968 and still had a program which they considered viable. At this time, they were watching with interest the activities of the new President-elect, Richard M. Nixon, as he undertook to organize his administration. They were hopeful they would receive the DORIAN support of the new Chief Executive, whose campaign literature had pledged a strengthened military space program. At year's end General Stewart and his staff were making plans to brief the new defense chief and his aides and other officials who would take office in early 1969.

ENDNOTES

1. Mins (S-DORIAN), AF MOL Policy Cmte Mtg 66-5, 22 Nov 65.

2. Msgs (S-DORIAN), Charge 5790, Berg to Evans, 9 Dec 66; (S-DORIAN), Whig 5972, Evans to Berg, 9 Dec 66; Memo (S-DORIAN), Evans to Flax, 19 Dec 66, subj: Release of FY 67 funds; Memo (S-DORIAN), Flax to DDR&E, 4 Jan 67, subj: MOL Program Plan and Funding Rqmt.

3. Memo (S-DORIAN), Evans tc SAF, 6 Mar 67, subj: MOL Monthly Status Rpt.

4. Memo (S-DORIAN), Evans to Randall, Yarymovych, 5 Mar 67, subj: Mar 10 Mtg.

5. Ibid.

6. Msg (S), SAFSL 93344, Evans to SSD (SAFSL-1), 17 Mar 67; Memo (S-DORIAN), Stewart to SAF, 5 Apr 67, subj: MOL Monthly Status Rpt.

7. Memo for Record (S-DOEIAN), by Stewart, 28 Apr 67 subj: 15 Apr MOL Mgt Mtg.

8. Msg (C), SAFSL-1/38417, MOL Sys Off (Bleymaier & Heran) to OSAF (Flax and Ferguson), 2 May 67, subj: MOL Program Schedule.

9. Handwritten note, Flax to Stewart, 3 May 67, attacked to above msg.

10. Msg (S-DORIAN), Stewart to Bleymaier & Martin/Heran, 5 May 67.

11. Ibid.

12. Memo for Record (S-DORIAN), by Stewart, subj: Min, 11 May MOL Mgt Mtg; Msg (S-DORIAN), Whig 6437, Flax/Stewart to Martin/Reran, 12 May 66.

13. Min (TS-DORIAN), AF MOL Policy Cmte Mtg 67-1, 1 Jun 67.

14. MOL Program Office, High Resolution Photography, 7 Sep 67 (TS-DORIAN/ Talent Keyhole).

15. Memo for Record (TS-DORIAN), Lt. Col. James F. Sullivan, MOL Frog Office, 25 Sep 67, subj: Secy McNamara's Visit to EK Company, 14 Sep 67; also EK Note, undated, subj: Important Comments Made During Each Part of the Briefing, Doc F-022333-KH-001.

16. Notes, MOL Staff lilte; (TS-DORIAN), by Berger, 28 Sep 67.

17. Memo (S-DORIAN), Stewart to Flax, 23 Jan 68, subj: MOL and House Armed Services Cmte.

18. Memo (S-DORIAN), Stewart to SAF, 7 Nov 67, subj: MOL Monthly Status Rpt.

19. Memo (S-DORIAN-GAMBIT), Stewart to SAF, 6 Dec 67, subj: MOL Monthly Status Rpt.

20. Memo (S-DORIAN), Stewart to SAF, 4 Jan 68, subj: MOL Monthly Status Rpt.

21. Memo (TS-DORIAN), Brown to McNamara, 15 Dec 67, subj: MOL Program.

22. Memo (C), Stewart to Comptroller, Air Force, 30 Jan 68.

23. Memo (C), Crow to Stewart, 16 Feb 68.

24. Memo for Record (S-DORIAN), by Stewart, 19 Feb 68, subj: Feb 14 Space Council mtg; Memo (TS-DORIAN) Kirk to Foster, no date; Memo (TS-DORIAN), Foster to Flax, 16 Feb 66, subj: Space Council Mtg.

25. Memo (TS-DORIAN/GAMBIT), Helms to Foster, 5 Mar 68, subj: Intel Rqmts for the MOL Program.

26. Ltr (TS-DORIAN), Helms to Bohlen, 15 May 68, w/copies to Nitze and D/BOB also see Attached Memo, 15 May 68, subj: The Intelligence Value of the MOL Program.

27. Ibid.

28. Memo for Record (S-DORIAN/GAMBIT), by Col Ford, 22 Mar 68, subj: Congressional Contact with Chairman Mahon.

29. Ibid.

30. Memo for Record (S-DORIAN/GAMBIT), 27 Mar 68, subj: Congressional Contact with Chairman Mahon, Rep. Frank T. Bow, and Glennard P. Lipscomb, House Cmte on Appns.

31. Ibid.

32. The article was reproduced in the Congressional Record, 15 July 1968, p. S8633.

33. Memo for the Record (S-DORIAN) by Col Ford, 8 Jul 68, subj: MOL Briefing to Mr. Woodruff, Staff Member, Senate Cmte on Appns, 5 Jul 68.

34. Trip Rpt (S-DORIAN), Lt. Col. Bertram Kemp, 22 Jul 68, subj: AF/Contractor MOL Program Rescheduling Mtg; MOL Monthly Mgt Rpt, 25 Jun-25 Jul 68, submitted (Bleymaier to Stewart) 5 Aug 68; MOL Monthly Status Rpt, signed by Stewart, 6 Aug 68.

MOL ASSE[MBLY]
INTEGRATION BUILDI[NG]

[COU...] MATERIAL

PROPERTY OF THE
[U]NITED STATES GOVERNMENT

[I]F FOUND, DO NOT OPEN

BASELINE MOL MANNED MODE

USAF MOL/KH-10

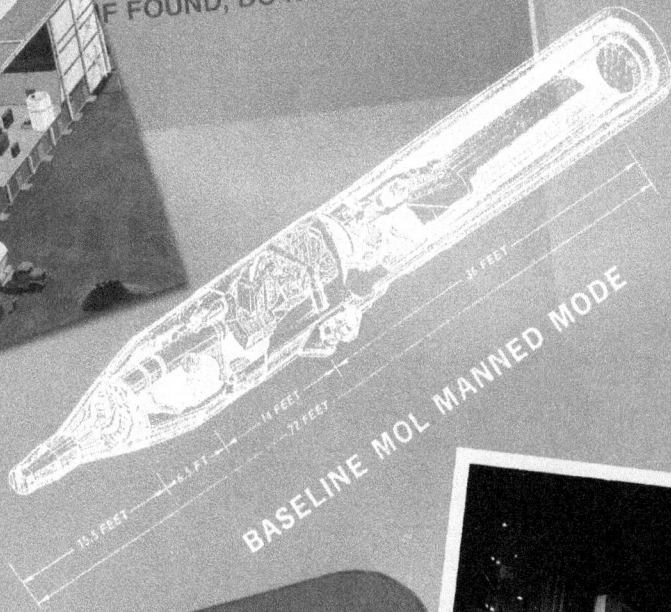

THE DORIAN FILES REVEALED:
A COMPENDIUM OF THE NRO'S MANNED ORBITING LABORATORY DOCUMENTS

CHAPTER XV:
THE PROJECT TERMINATED

THE PROJECT TERMINATED

In late November 1968—just prior to President elect Nixon's announcement that Congressman Melvin R. Laird (Rep., Wis.) would serve as his Secretary of Defense—a DDR&E ad hoc group completed work on the MOL Development Concept Paper (DCP).* It addressed three current "management" issues: should MOL be continued or terminated; if continued, should the unmanned capability be cancelled; if continued, what level of support should be provided in fiscal year 1970? The DCP reviewed the events leading to project approval in 1965, the recent criticism voiced by Bohlen and Helms and, in particular, the detailed CIA arguments contained in its May 1968 memorandum to the effect that MOL was not worth the cost.[1]

The DDR&E ad hoc group strongly disagreed with the CIA assessment. It declared that MOL's very high resolution (VHR) photography would "improve the accuracy and timeliness of performance estimates of enemy weapon systems over that provided by HR photography produced by a mature KH-8 system." MOL, it said, would produce photos containing sufficient detail to determine the performance characteristics, capabilities and limitations of important enemy weapons. It also could provide intelligence of {a highly important intelligence target} and contribute "to the monitoring of any arms limitation agreement." In periods of international crisis, it would prove especially helpful. As a case in point, the group cited the Cuban missile crisis of 1962, when very high resolution photos were required by the President "to provide the basis for verification of existence and removal of strategic system's from the island."[2]

After discussing MOL characteristics and technical aspects of the project, the group addressed three options that might be considered by the decision-makers. (1)Terminate the program; (2) continue it but terminate the unmanned version; or (3) continue the existing program (both the manned and unmanned versions), funding it in 1970 at levels of $442 million, $417 million; or $342 million (which assumed another $158 million would be added to each of the above totals for procurement support).

After he had reviewed the DCP, Dr. Foster concluded that the value of MOL photography to the Defense Department justified expenditure of the remaining costs to completion ($1.8 billion)and the estimated annual follow-on operating costs of $100-$200 million. On 4 December he recommended to Secretary of Defense Clifford that $575 million be "as low as we should go in FY 70 funding and the unmanned option should not be cancelled at this time." The OSD Comptroller, R. C. Moot, approved Foster's recommendation as did Deputy Secretary of Defense Nitze, on 6 December. Dr. Hornig, the President's Science Adviser, also concurred in the recommended $575 million MOL baseline program.[3]

The CIA, however, stuck to its guns. On 6 December 1968 Director Helms wrote to Nitze that his staff—after reviewing the DCP—believed it conveyed "an overly optimistic assessment of the potential value of the very high resolution photography anticipated from MOL." He said he continued to feel that while MOL-type photography would make a useful contribution to intelligence,'" it was not of sufficient importance "to justify the estimated cost."[4]

The Development Concept Paper and the study on "The Value of MOL and Very High Resolution Photography" were among several documents readied for submission to the new OSD team headed by Laird and his deputy, David Packard. The other documents included an updated "MOL Program Summary" prepared under General Stewart's direction, and two companion papers, "Man in MOL" and "Mission Value"—the last discussing the contributions very high resolution imagery could make to DoD decisions and operations.[5]

The outgoing Secretary of Defense—prior to the inauguration of the Nixon Administration on 20 January 1969—approved the assignment to MOL of a new, secondary mission {of high importance}. On 14 January, Stewart directed the MOL Systems Office "to proceed with the necessary action to incorporate {a new capability} into the MOL system..." Such a capability was to be provided on a non-interference basis with the primary MOL mission—earth reconnaissance—and no hardware modifications of the flight vehicles were to be made. The Systems Office was authorized to proceed with pre-contract award actions to procure {elements of the new capability} and to conduct a limited competition between General Electric, McDonnell Douglas, and TRW, Inc.[6]

---

* See also p 264. M0L Program Office personnel worked with the group and also participated in a study on "The Value of MOL and Very High Resolution Photography." The conclusions of this study were incorporated into the DCP. [Ltr, Foster to Brown, 12 Dec 68]

After key members of the Nixon Administration were sworn in—among them were the new, Secretary of the Air Force, Dr. Robert S. Seamans, Jr. and the new Director of the Bureau of the Budget, Robert P. Mayo—they were scheduled for a series of briefings on various defense programs. In the case of MOL, Air Force officials originally were allotted only 20 to 30 minutes to brief Deputy Secretary Packard. One of his staff members, Dr. Ivan Selin of OSD's Office of Systems Analysis (a holdover from the Johnson Administration) advised the new defense official that he did not believe MOL photography was very significant to the Defense Department or, if it was, there were far cheaper ways to get it. When they learned of this statement, Drs. Foster and Flax[†] recommended to Mr. Packard that he give them an opportunity to provide him a separate and more in-depth MOL review. He agreed and the briefing was scheduled for Saturday, 8 February 1969.[7]

The presentation on 8 February was made by General Stewart. Sitting in on the briefing were Drs. Brown, Foster, Flax, and Selin; Mr. Moot and General Carroll of the Defense Intelligence Agency. Both during Stewart's presentation and the ensuing discussion, Foster, Brown, Flax, and Carroll expressed favorable opinions on the value to DoD of the information "derivable from very high resolution photography" and strongly supported the existing MOL program for that purpose. They concurred that very high resolution photography "is of significant value to DoD in [making] multi-billion dollar R&D and force structure decisions." Stewart said that MOL was the best way to have a VHR photographic capability at an early date and that the Air Force had proceeded very deliberately to insure very high confidence in an operational system. He said that the program was "strictly dollar-paced" and that its status was such that "sizable dollars must be invested in FY 70 and 71 to avoid gross stretch-out, inefficiency and waste." Packard's reaction to this briefing was later reported to MOL officials as being "reasonably favorable."[8]

However, two other important agencies—the CIA and BOB—came forward with generally unfavorable or anti-MOL views. Based on information obtained from the CIA, Mr. Mayo on 13 February 1969 submitted a lengthy paper to Packard in which he questioned the value of MOL photography. He argued that other unmanned systems—such as GAMBIT-3 and SIGINT satellites—were or soon would be providing all the necessary information needed by DoD to make essential force decisions. He noted that the Director of Central Intelligence had "seriously questioned the benefits or value of the MOL's {anticipated resolution} photography "compared to that of the present

{better than 12 inch} photography of the G-3.[‡]" Mayo stated that improvements in the resolution and orbital life of the "proficient G- 3" made it "highly questionable that the MOL's marginal improvement beyond an already impressive capability is worth the huge cost."[9]

Assisted by the staff of the MOL Program Office, CSD officials were given an opportunity to critique the Mayo paper and counter its arguments. They remarked that, on a comparable basis, MOL would be far superior to the G-3 satellites in detecting aircraft, missiles, submarines, defense missiles, armed vehicles, and other enemy equipment. They also stated that MOL photography would make a significant contribution to policing an arms limitation agreement, that it had the potential to obtain VHR photos of targets and areas during periods of international crisis and tension, and that it would contribute to decisions on future force structures.[10]

THE FIGHT TO SAVE THE PROGRAM

Secretary Laird, meanwhile, had been scrutinizing the last defense budget (totaling $80.6 billion) submitted by President Johnson, with the idea of reducing it by at least $3 billion. In the case of MOL, he tentatively decided to drop the unmanned vehicles 6 and 7 and replace them with an additional manned vehicle. This was expected to produce an estimated saving of $20 million. On 19 February 1969, Stewart forwarded information on Laird's decision—subsequently made firm—to Bleymaier.[11]

The same day this message went out, OSD/BOB officials met to review Mayo's paper on MOL and GAMBIT-3. Afterwards, Packard asked the Air Force for additional information on MOL. Specifically, he wanted financial data on various alternative MOL programs, including one consisting of only three manned flights, and a "sustaining program which would minimize FY 1970 funding." A week later, on 26 February, Secretary Seamans forwarded information to Packard on four alternative MOL programs. In the case of the "minimum sustaining" one, he noted that:

---

† Dr. Foster was retained by Mr. Nixon as DDR&E. Dr. Flax departed to join the Institute of Defense Analysis.

‡ The tendency of officials to speak of a projected design capability in the present-tense was a familiar problem. During a briefing at Eastman Kodak in May 1969 given to a visiting group, when a project manager spoke of a GAMBIT-3 resolution of {better than 9"} without qualifications, Lt. Col. Daniel Lycan of the MOL Program Office objected. Based on his complaint to the briefer, an informal poll was taken of various members of the visiting group "to see what they understood would be the resolution..They answered {a resolution better than 12 inches} and then {and even better resolution.} The briefer then attempted to clarify the point that G-3 "was not now getting {the desired resolution}, but this figure was the design goal." Colonel Lycan, however, felt his statement was not "a sufficient qualifier." [Memo for the Record by Col Lycan, 26 May 69, subj: Visit of EXTRAND Gp to EKC.]

A program to minimize FY 70 funding might entail a 50 percent reduction in [the] work force[§], new material purchases, etc. Approximately $275-400 million would be required in FY 70 to maintain personnel competency and fast readiness. A delay of more than one year in development prior to the first manned flight would result and total program costs would increase more than $360 million.

In order to maintain a capability to pick up the present program pace at the beginning of FY 71, a smaller work force reduction would be appropriate, new material purchases limited, etc. Approximately $360 million would be required in FY 1970 for this alternative. The impact would be a one year stretchout in development prior to the first manned flight and a total program cost increase of approximately $260 million.[12]

As an alternative to the existing programs, Air Force officials recommended a four manned vehicle program (dropping the unmanned vehicles as already decided by Laird). Citing it as a "more clearly desirable" approach, Seamans said the four manned MOL program would protect until December 1970—with minimum commitment of funds—"a continuing very high resolution operational reconnaissance capability in the 1970s, provide time in which to carefully assess other options, and sustain a minimum cost development program leading to manned or unmanned operational systems." If this course of action were to be adopted, Seamans suggested that OSD fund the program "at no less than $556 million" in fiscal year 1970.[13]

OSD responded favorably, apparently influenced by former Congressman Laird, who had previously criticized one of his predecessors Secretary McNamara, for not putting enough resources into the MOL program. On 6 March 1969 Packard directed Foster to proceed with the four manned MOL program as recommended, with funding of $556 million in FY 1970. The first manned launch would take place in February 1972 and the fourth in September 1973. This program was expected to reduce total MOL development costs by approximately $200 million.[14]

However, the decision did not stand up, due in part to a growing revolt in Congress-fed by critics in the press and university communities—against the "military-industrial complex". Thus, a local publication, *The Washington Monthly*, published a lengthy article by a former DoD employee, Mr. Robert S. Benson[¶], titled "How the Pentagon Can Save $9 Billion." His article was placed into the Congressional Record on 26 February 1969 by a vociferous critic of military spending, Representative John Brademas (D-Ind). Benson's article led off with an attack on MOL, which he declared "receives a half a billion dollars a year and ought to rank dead last on any rational scale of national priorities." The program he declared, was "a carbon copy" of NASA's spacecraft operation and the only reason it was in the budget was "because the Air Force wants a piece of the extraterrestrial action, with its glamor and glory, and Congress has been only too happy to oblige." Benson recommended the program be terminated, thus saving the nation $576 million in fiscal year 1970.[15]

In addition to this public attack on MOL, the critics were out in full cry against President Nixon's decision to proceed with development of a modified version of the Army's Sentinel anti-ballistic missile (ABM) system. On 1 March he told a televised news conference that, instead of deploying ABM batteries around the nation's cities as originally proposed by the Johnson Administration, he had decided that the defensive missiles should be located to defend the Air Force's missile retaliatory force. This revised ABM program renamed Safeguard, became the focus of much anti-Pentagon sentiment in and out of Congress. Members of the President's own party vowed to fight passage of the authorization bill in the Senate.

Although the growing ABM debate dominated the news, General Stewart—who had been briefing key members of the House and Senate on MOL during February and March—soon was made aware that it was not being ignored by the critics. Thus, during a 5 March 1969 DORIAN briefing of Congressman Durward Hall (Rep., Mo.), the Representative urged him to release more information on the program. He expressed concern about the comments from Congressmen "who have not been completely informed on the MOL Program" and thought it might be "judged unfavorably by those who did not know its specific purpose."[16] This unrest also reached certain important staff members, such as John R. Blandford of the House Armed Services Committee, previously considered sympathetic. When Stewart during a briefing told him the cost of MOL would increase to $3 billion by fiscal year 1973, Blandford remarked that the

§ There were about 15,000 Associate Contractor personnel aboard at this time, about 85 percent of the expected peak work force.

¶ Benson served in the OSD Comptroller Office.

United States "could buy the Kremlin" for that amount. If the MOL was strictly for reconnaissance, he said, he could not support it.[17]

On the Senate side, members of the Armed Services Committee—during another DORIAN briefing—questioned of the value of {the desired resolution} photography as contrasted with the capability of the KH-8. Stewart responded with an effective exposition on MOL's very high resolution photographic capability, and climaxed his talk by showing them an excellent 12-inch photograph of a Soviet submarine with its ballistic missiles exposed. An Air Force observer later wrote that this photograph "did more than anything else to wet the appetites of the members present for the MOL system." That is, it became clear to them that additional resolution was required to define the capabilities and characteristics of the missiles. Unfortunately, another Senate staff member, Edward Braswell—who was not present when the photograph was shown—later told Stewart that: "In my opinion what you have to do is convince people that {the desired resolution} is essential, not just generally better but real gutsy examples of technical intelligence must be shown; an unbiased, objective, well informed champion of MOL must be found."[18]

There were many champions of the program in the Defense Department, but no one who could counter the Congressional criticism of military expenditures. The majority leader of the Senate, Mike Mansfield, announced that the Democrats were determined to cut at least $5 billion from the $77.6 billion in the fiscal year 1970 appropriations requested by the President in his revised budget." Declaring that the fight to curb military expenditures would not stop with a decision whether to deploy the Safeguard ARM system, Senator Mansfield listed 15 different defense programs as "economy targets," one of which was the Manned Orbiting Laboratory.[19]

MAYO TAKES THE MOL ISSUE TO THE PRESIDENT

Meanwhile, the Air Force learned that the Budget Bureau would not accept Secretary Laird's decision of 6 March to proceed with a manned only MOL Program as final. It was the BOB's view that "the proper decision is to terminate the whole program." During a meeting with budget officials, Stewart sought to convince them—without success—that MOL photographic resolutions would be {several times} better than GAMBIT-3. In reporting to Secretary Seamans on his meeting, Stewart said that Mayo planned to submit a "go no-go MOL Budget Issue" paper to the President.[20]

This news set off another round of OSD-Air Force paper studies on MOL. While Stewart coordinated with Mr. William Fisher of the Budget Bureau on the DoD portion of the BOB issue paper (required when a decision on defense projects was sought from the President), Packard asked Seamans to submit to him six possible FY 1970 MOL funding levels and descriptions of the programs associated with each. On 24 March, Seamans forwarded the information prepared by the MOL Program Office. He listed the various possible options—ranging from a "Zero FY 70 NOA"—which would require MOL's termination before mid-April "to permit all termination costs to be paid" within the current fiscal year—to funding levels from $150 million to $556 million. The Air Force Secretary pointed out that the MOL Program was spending at a high rate and that major reductions would be quite wasteful. "In my opinion," he said, "we should fund MOL at or near the presently approved $556 million level or terminate the program."[21]

Two days later, Stewart forwarded to Dr. John H. McLucas, Under Secretary of the Air Force, a draft of the proposed BOB/DoD issue paper to the President on MOL. In the Bureau argued that the improved GAMBIT-3 system (expected {better than one foot} resolution)

Figure 58. John H. McLucas
Source: CSNR Reference Collection

** * ** The Administration had cut $3 billion from the Johnson budget.

would provide in fiscal year 1972 most of the information on important weapon system characteristics discernible through photography. As for DoD's claim that MOL could help police a strategic arms limitation agreement, it stated that existing unmanned systems were capable of detecting changes in enemy weapon system deployments. Further, it argued that the more subtle qualitative improvements in enemy missiles—"such as accuracy/vulnerability (hardness, reliability and yield and type of warheads)"—were difficult to discern, "even with {higher resolution} photography.[22]

BOB also challenged—on the basis of GAMBIT-3 experience—the Air Force claim that the MOL would "achieve its goal for best resolution of {better than one foot.} If it failed, then "the improvement over GAMBIT-3 would be even more marginal." The savings provided through MOL's termination, it argued "would provide additional flexibility in future budgets to pursue other manned space projects or new types of intelligence capabilities such as warning..." The BOB rationale again cited the position taken by the CIA to back up its claim that, while MOL photography would be useful, it was not worth the very large cost involved.[23]

OSD's rebuttal, which would go forward with the BOB recommendation to the President, touched on many of the points previously made. Among other things, it argued that:

> MOL photography alone will enable the production of performance estimates of foreign weapon systems that are {several} times more accurate and 2-3 years sooner than from current all-source intelligence. Certain important performance parameters and characteristics of foreign weapons, systems, facilities and equipment can be derived with reasonable accuracy, timeliness and confidence from VHR imagery alone...
>
> MOL photography will be of considerable value in any strategic arms limitation agreement (along the lines of those now under discussion with the USSR) to provide very high confidence that the Soviets either are adhering to or violating the terms of the Treaty, and further to provide additional technical intelligence on subtle weapon improvements. The 1962 Cuban missile crisis is illustrative

of the need for convincing evidence when the President was reluctant to act on the basis of U-2 photography {...} but did act when low-level reconnaissance aircraft {photography} was secured...[24]

Before this issue paper was submitted to Mr. Nixon, Secretary Laird announced a further reduction of the program's 1970 budget. On 2 April 1969 he informed the House Committee on Armed Services that he was reducing MOL's budget from $576 million to $525 million. "A careful review of the work done to date," he said, "has convinced us that a total of six launches would probably be enough to accomplish all of the approved objectives. The elimination of one launch will save $20 million. The remaining reduction of $31 million will simply stretch out the program and delay the first launch by two to three months."[25]

Figure 59. HEXAGON on Factory Floor
Source: CSNR Reference Collection

This decision, however, was soon made academic. On 9 April—after reviewing the BOB/DoD Issue Paper with his budget chief—the President decided to reduce MOL FY 1970 funding to $360 million. Mayo passed this information along to Laird and also advised Helms that the President also had decided to terminate the HEXAGON unmanned photographic satellite development to save money[††].[26] The next day Dr. Foster requested Stewart to prepare suitable material that he and Dr. McLucas might use during a meeting with Laird or Packard on 14 April.

On 11 April Stewart forwarded to McLucas a lengthy paper prepared by his staff on the effect of a $360 million budget cut on the MOL schedule. It would delay the first manned flight as much as one year and increase total program costs at least $360 million and,

---

†† The HEXAGON system was initiated in 1964-1965 as a proposed replacement for the CORONA search system and as a possible partial substitute for the GAMBIT-3 spotting or surveillance system. The CIA was a strong proponent of this system.

in addition, would require approximately $550 million to $575 million in fiscal years 1971 and 1972. If this proved unacceptable, Stewart suggested continuing work on the camera system only aiming toward a possible unmanned application. In this case,.he would need up to $20 million in fiscal year 1969 funds to terminate work on all MOL contracts except for the camera. Another $200-$300 million would be needed in fiscal year 1970, depending on whether a decision would be made to proceed with an unmanned system. Still another option was to "terminate the entire program." This would require up to $30 million in additional funds to pay termination costs during the current fiscal year.[27]

In a separate attachment on MOL funding experiences and the schedule slips, Stewart remarked that:

> FY 1970 will mark the third straight year that MOL will have been funded at a level $84 million or below program needs for a reasonable development pace, and the third straight year that development will have been stretched out and finances manipulated on the premise that adequate funding would be available "next year."

> To minimize past development stretchouts and their related net increases in total program cost, the MOL Program has gradually moved toward an expenditure funding basis, and the maximum non-critical work (from a technological difficulty standpoint) has been deferred as far as possible into the future. As a result, there is no financial flexibility whatsoever in the program and the planned future workflow balance can be described as somewhat marginal.[28]

The above material was subsequently reviewed by OSD and Air Force officials, as well as other possible alternatives, including melding the DORIAN and HEXAGON equipment into a single system. They finally decided that a memorandum should be prepared for the President in a final effort to save the manned system. The draft of this memorandum, worked on by Drs. Seamans and McLucas and General Stewart, sought to make the point that astronauts in a manned system would increase the likelihood of obtaining very high resolution photographs sooner, that targets would be

covered in a more timely manner, and that the United States would have additional flexibility not practical in an unmanned system.[29]

As reworked toward the end of April, the proposed Laird memorandum to the President—sent to Packard by Seamans—began :

> Your expressed desire, as reported by Mr. Mayo, that we fund MOL at less than the $525 million now requested of the Congress for FY 1970 has resulted in our making a careful reappraisal of the program. I conclude that we either should fund MOL at a level commensurate with reasonable progress for the large amounts involved, or terminate the overt manned MOL program and continue only the covert very high resolution (VHR) camera system toward future use in an unmanned satellite...[30]

Meanwhile, General Stewart—who had become quite pessimistic about prospects for survival of the program—began drafting letters to be sent by Laird to chairmen of the House and Senate Armed Services and Appropriations Committees announcing MOL's termination. He also wrote draft letters to be forwarded for Senators and Representatives from states that would be most seriously affected by termination.[31]

While these activities were underway in the Pentagon, Budget Director Mayo and his staff were writing their own memorandum to the President urging that MOL be terminated. In this memorandum, which Mayo submitted to the President on 21 April 1969, he recalled that Mr. Nixon—before making his decision on the 9th (to terminate HEXAGON and slow down MOL)—had reviewed the option of "continuing the HEXAGON search system and the cancellation of MOL." On reflection, Mayo wrote, there might be additional reasons for reconsidering this option. "Politically,"‡‡ it might be desirable "to have the better performance of the HEXAGON search system to provide greater assurance, for example, to members of Congress who would be most concerned about our ability to police a strategic arms limitation agreement." In terms of added intelligence value, he said, "the MOL is the more questionable. Cancellation of MOL and continuation of HEXAGON would provide about the same savings below the presently proposed programs, both in FY 1970 and over the next five years, as your current decision."[32]

---
‡‡ Mayo's emphasis.

In an attachment to his memorandum, Mayo argued further that the urgency of achieving MOL objectives "has never been fully established." Therefore, he thought it would not be "a serious penalty to the nation" to defer the first manned launch by a year or more ($-165 million) or to reduce the MOL effort to that of optics and payload vehicle technology ($325 million). In this paper, Mayo listed comparative costs between GAMBIT, GAMBIT-3, and MOL for each launch, which indicated that each MOL mission would run about; $150 million a year as compared to GAMBIT-3's cost for each mission of $23 million. "The incremental value of the MOL {anticipated best} resolution," he concluded, "is not enough of an improvement over the present spotting system (GAMBIT-3) to justify the additional cost."[33]

THE WHITE HOUSE MEETING OF 17 MAY 1969

While the President reviewed these papers, there was a brief interregnum in early May. But news that the program 'was in possible danger reached at least one professional journal and members of Congress. On 5 May, in a story headlined "Budget Cuts Threaten MOL Project," Aviation Week reported that: "New financial digs into funding for Air Force's manned orbiting laboratory (MOL) reconnaissance satellite are raising questions as to the program's future as a whole. Dearth of funding as well as technological progress since the project's inception could spell an end or severe readjustment to the program..."[34] The following day, during an appearance before the Senate Committee on Aeronautical and Space Sciences, Dr. Seamans commented on MOL's financial problems. He said:

> It is my view that the MOL... has been underfunded the past few years. It is very difficult to run a program on a reduced budget and still have it meaningful, and it is even more difficult when the budgets are continually reduced to change the program to suit the budgetary needs. I believe that if the funding is reduced much below the present level, it would be very difficult to maintain progress and to keep up morale and achieve any meaningful results.
>
> So when this first came up after I joined the Department of Defense, the question was, should we not reduce the budget below President

Johnson's level of $576 million. I raised the question, should we not increase the budget.[35]

Not long after, OSD advised Seamans that the President had agreed to receive a personal briefing on the program at the White House before making his decision. On 9 May the Air Force Secretary met with members of the MOL Policy Committee to review the entire program. Among those in attendance were Dr. McLucas, {a NRO Representative}, and Generals Ferguson, Stewart, and Bleymaier, and several others. Ferguson reported he recently had directed a Board of Air Force officers—representing the best space management talent he had available—to review the program[§§]. Their conclusion, he said, was that MOL "was ready to go but it lacked the dollars necessary to proceed efficiently." Bleymaier, in a presentation of MOL's status, declared that the program was almost completely defined, test results to date had met or exceeded Air Force expectations, and that no technical or facility problems stood in the way of launching the first manned vehicle in mid-1972. He requested the Committee's support for "a firm commitment" to provide him $525 million in fiscal year 1970 and $625 million in 1971.[36]

Under the circumstances, Bleymaier's request was entirely unrealistic. {The NRO Representative} later remarked that he didn't think "anyone would give the program a firm commitment for $625 million for FY 71 in the present environment." He pointed out that they had sought budgets of over $600 million for a number of years "but the program had never made it." Toward the close of the meeting, Dr. Seamans requested he be provided copies of all the briefing charts for his meeting with the President. There would be no point in discussing budget details at the White House, he said, since his first job was "to save the program."[37]

On 17 May—a Saturday—Secretary Laird, Dr. Seamans, and General Stewart rode over to the White House to submit to. Mr. Nixon "the counter-case to the BOB proposal to terminate MOL." Among those attending this meeting were Mr. Mayo and an aide, and Dr. Henry Kissinger, the President's adviser on international affairs. Laird opened the session by stating he believed responsible DoD officials should have the opportunity to state their case to the President on difficult, complex Issues, which was why Dr. Seamans and General Stewart had been called in. After this introduction, the Air Force Secretary began his briefing. He reviewed the historical events leading to Dyna-Soar's cancellation in 1963, the initiation of the MOL program, and the two years of study

---

§§ Ferguson organized the Board in April 1969. The MOL Policy Committee Meeting of 9 May was convened at his request to hear the results of its review.

which followed and led to President Johnson's 1965 go-ahead decision. He reported about $1.3 billion had been spent on the MOL program to date, that another $1.9 billion would complete it, and that about 65,000 people were involved in the program (including the Associate Contractors and subcontractor personnel).[38]

Dr. Seamans described MOL's primary objectives and showed the President a missile picture montage to illustrate why very high resolution photography was needed to analyze weapon performance. He stated VHR photographs would help the United States determine weapon system performance and would be helpful during any future arms limitation arrangement made with the Soviet Union. He placed great emphasis on the activities of man in MOL, noting that the astronauts could identify and select "high-value" targets, fine-tune MOL equipment, read-out information to ground stations, and interpret film processed aboard the spacecraft. In his opinion, he said, MOL had "more value than anything under consideration by the President's Space Task Group.¶¶"

At the conclusion of his briefing, Dr. Seamans commented that the cancellation of MOL would be a "bitter pill" both for the Air Force and him personally to "swallow." If permitted, he said, he would find $250 million somewhere in the Air Force budget to continue the program. At one point, Secretary Laird recalled that, while in Congress, he had supported MOL and had once prepared a committee minority report criticizing McNamara for not putting more money into the program. After the formal briefing, the President asked General Stewart for his opinion.

The Vice Director, MOL, responded that if the United States should achieve an arms limitation agreement with the Soviet Union, he, the President, would be "pushing us to accelerate MOL" and would want even higher resolution photography to be sure the Russians were abiding by it. As the conferees walked out of the President's office, Seamans reminded Mr. Nixon that fiscal year 1971 actually would be the peak year for MOL. The President replied that he understood but the fiscal year 1970 was his immediate concern.[39]

The presentation seemed to have gone well and there was some hope for a favorable outcome. Exactly when the President made his decision is unknown. When he did, it was to accept Mayo's recommendation to terminate the MOL program and proceed with the HEXAGON project. In deciding so, Mr. Nixon apparently took into account Dr. Seamans' remark that cancellation would

be a great disappointment to the Air Force. To ease the pain, the President arranged to address the Air Force Academy during graduation ceremonies on 4 June 1969. His decision on MOL was still unknown to the Air Force when, at Colorado Springs, he lashed out at critics of the military, denouncing "the open season on the armed forces" and attempts to make them a "scapegoat." He also defended his current international policies and criticized as simplistic the slogans of the "isolationist school of thought" that "charity begins at home, let's first solve our problems at home and then we can deal with the problems of the world." Such a policy, he said, would be disastrous for the United States. "The danger to us has changed but it has not vanished. We must revitalize our alliances, not abandon them; we must rule out unilateral disarmament because in the real world it won't work." Further, he went on, "the aggressors of this world are not going to give the United States a period of grace in which to put our domestic house in order—just as the crisis within our society cannot be put on a back burner until we resolve the problem of Vietnam."[40]

Mr. Nixon's speech struck a responsive chord within the Air Force and the other services. However, buried within his address were also the following pertinent remarks on defense expenditures:

```
America's wealth is enormous, but
it is not limitless. Every dollar
available in the Federal Government
has been taken from the American
people in taxes. And a responsible
government has a duty to be prudent
when it spends the people's money.

There is no more justification for
wasting money on unnecessary military
hardware than there is for wasting
it on unwarranted social programs.
And there can be no question that we
should not spend unnecessarily for
defense...⁴¹
```

The day after this address, Dr. Foster phoned General Stewart to advise that the President had decided to terminate the project, except for the "automatic" camera system. This news set in motion a series of actions to publicly end the program. On 6 June, Stewart passed the news to Bleymaier and advised detailed guidance would be provided him. The next day he sent Bleymaier instructions to terminate all work on the Gemini spacecraft, the Titan IIIM, and the astronaut space suit and to cancel or reduce to a sustaining level work under other contracts. Military construction on the Vandenberg-

¶¶ Organized at the direction of the President in early 1969 under Dr. Lee A. DuBridge, his science adviser, the Space Task Group was directed to review the nation's space programs and to recommend future programs.

AFB launch facility was to be completed "to the minimum practical extent and mothballed" but other construction was to be halted as soon as possible. Since a public announcement was to be made on 10 June, after Congress was notified, Stewart directed Bleymaier to withhold information from the Systems Office staff until the close of the work day, Monday, 9 June.[42]

Meanwhile, Col Ralph J. Ford and Lt Col Robert Hermann of the MOL Program Office staff—working with OSD personnel completed a series of announcements connected with the President's decision. These included a press release on MOL's cancellation, sample questions and answers for the press, classified and unclassified letters to chairmen of key congressional committees, etc. A "termination scenario" was worked up as follows: (1) affected government officials would be notified informally; (2) former President Johnson and Secretaries McNamara, Zuckert, and Brown would be notified; (3) chairmen of congressional committees and individual congressmen whose states would be affected would be informed; (4) MOL contractors would be directed to terminate all efforts except covert camera activities applicable to an unmanned system. After these steps were taken, a press release might be distributed and a news conference held, if desired.[43]

On 9 June Packard formally directed Dr. Seamans to:

> ...terminate the MOL Program except for those camera system elements useful for incorporation into an unmanned satellite system optimized to use the Titan IIID. Directions to MOL contractors should be issued on Tuesday morning, June 10, at which time we will also notify the Congress and make a public statement that MOL is cancelled.
>
> Close-out costs for MOL, which I understand are approximately $75 million more than is now available to the MOL Program, should be included in the unclassified FY 70 Air Force budget. An additional $175 million should be included in classified NRP portions of the FY 70 Air Force budget. This will provide for development of the camera system at a reduced pace, for competition for a new spacecraft, and for possible initiation of system development late in FY 70.

> All future work on the camera and an unmanned system will be part of the NRP. As a security measure, appropriate elements of the MOL Project offices and the camera system contracts should be transferred to the Air Force NRP Special Projects Offices at an early date. Overt MOL activities should be phased out in conjunction with the closeout of MOL Program activities...[44]

The next day the classified and unclassified letters were delivered to key Senators and Representatives and, shortly after, Packard announced the termination to the press. On Capitol Hill, Secretary Laird told several Congressional committees that, "with the President's concurrence," he had decided to cancel MOL. He listed several reasons for the decision, including the need "to either drastically cut back or terminate numerous small but important efforts or one of the larger, more costly programs." Laird stated that "major advances have been made by both NASA and DoD in automated techniques for unmanned satellite systems...These have given us confidence that the most essential Department of Defense space missions can be accomplished with lower cost unmanned spacecraft." He also said:

> I wish to make two final points for the record. It should be clearly understood that termination is not in any sense an unfavorable reflection on MOL contractors. They have all worked very hard and have achieved excellent results. Likewise, MOL termination should not be construed as a reflection on the Air Force. The MOL goals were practical and achievable. Maximum advantage was being taken of hardware and experience from NASA and other Department of Defense projects, and the program was well managed and good progress was being made. Under other circumstances, the continuance would have been fully justified.[45]

Few regrets were voiced in Congress over the MOL decision. One Senator—Cannon of Nevada—was unhappy, however, and complained he had difficulty understanding "the logic of the Department of Defense." In one breath, he said, OSD officials claimed the United States was in the greatest mortal battle for survival—"a danger beyond any confrontation in our entire history as a Nation"—and at the same time they terminate "the most advanced surveillance system yet conceived." Noting that $1.3 billion had already been spent on MOL, he declared that to "scuttle this high investment for political expediency is unfair to the taxpayer and raises new questions concerning our national security."[46] A trade journal editorialized two days later :

> Someday the Department of Defense is going to find that it needs a manned military equipped space station positioned so that it can watch our adversaries 24 hours each day. We will spend billions for unmanned space-based detection and monitoring systems and Earth and space-based warning systems only to find that in the long run it will be more economical and reliable to place manned systems in fixed synchronous orbits over viewing our adversaries...[47]

These opinions, however, were in the minority. The critics of defense expenditures were pleased to see a major defense program ended. Also, the Administration apparently hoped that the decision would reduce some of the opposition to the President's Safeguard ABM program***. Thus, Secretary Packard—when he announced MOL's termination at a Pentagon press conference—suggested to the correspondents that its demise should satisfy the need for further major reductions in the DoD R&D budget. Senator Thomas J. McIntyre, chairman of the Senate Armed Services Subcommittee on research and development, disagreed. The cancellation of MOL, he said on 11 June, would not avert Congressional efforts to make additional cuts in the department's $8.4 billion R&D budget." At this point," he said, "I am not prepared to accept the idea that terminating MOL is enough economizing on research and development."[48]

*** Mr. Nixon won Senate approval of his ABM plan by a margin of one vote, on 6 August 1969.

ENDNOTES

1. Dev Concept Paper No. 59A, Manned Orbiting Lab (MOL), signed by Foster, 4 Dec 68.

2. Ibid.

3. Ibid.; Ltr (C), Hornig to Nitze, 5 Dec 68.

4. Ltr (TS-DORIAN), Helms to Nitze, 6 Dec 68.

5. MOL Program Summary (TS-DORIAN), 2 Jan 70.

6. Memo (S-DORIAN), Stewart to Bleymaier, 14 Jan 69, subj: no subject.

7. Memo (TS-DORIAN), Stewart to McConnell, 12 Feb 69, subj: MOL Briefing to the Dep Sec Def.

8. Memo for the Record (TS-DORIAN/RUFF/UMBRA), by Lt Col R. H. Campbell, 11 Feb 69, subj: Briefing to Mr. Packard on MOL and VHR Imagery Issues; Memo (TS-DORIAN), Stewart to McConnell, 12 Feb 69, subj: MOL Briefing to Dep Sec Def.

9. Ltr (TS-UMBRA), Mayo to Packard, 13 Feb 69, w/atch, "The Relative Value of MOL for U.S. Force Structure Decisions."

10. See Critique, atch to above ltr.

11. Msg (S-DORIAN), 1018, Stewart to Bleymaier, 1921072 Feb 69.

12. Memo (TS-DORIAN/GAMBIT), Seamans to Packard, 26 Feb 69, subj: MOL Prog Alternatives.

13. Ibid.

14. Memo (TS-DORIAN), I. Nevin Palley, Asst Dir (Space Technology) to DDR&E, 10 Mar 69, subj: MOL Prog Alternative Decision.

15. *Congressional Record*, House, 26 Feb 69, p. H1249. The article was also printed in the *Congressional Record*, Senate, 12 May 69, p S4886.

16. Memo for the Record (TS-DORIAN), by Col Ford, 5 Feb 69, subj: MOL Briefing, Congressmand Durward Hall, Rep., Mo.

17. Memo for the Record (TS-DORIAN), by Col Ford, 12 Feb 69, subj: Briefing to Mr. Russ Blanford and Mr. Earl Morgan, House Armed Services Cmte.

18. Memo for the Record (TS-DORIAN), by Col Donald Floyd, SAFLL, 5 Mar 69, subj: Briefing, Senate Armed Services Cmte Staff Personnel.

19. *N.Y. Times*, 21 Apr 69.

20. Memo (TS-DORIAN), Stewart to Seamans, 14 Mar 69, subj: Probable Presidential Budget Issue on MOL.

21. Memo (TS-DORIAN), Seamans to Packard, 24 Mar 69, subj: MOL FY 1970 Funding Options.

22. Memo (TS-DORIAN), Stewart to McLucas, 26 Mar 69, subj: Proposed DoD/BOB MOL Budget Issue Memo to the President, w/atch "BOB Recommendation and Rationale and Proposed OSD Rebuttal."

23. Ibid.

24. Ibid.

25. Quoted in Msg (U), SAFSL 001, Stewart to Bleymaier, 021456Z Apr 69.

26. Memo (TS-DORIAN/GAMBIT/HEXAGON), Mayo to the President, 21 Apr 69, subj: FY 70 Intelligence Program Savings.

27. Memo (TS-DORIAN), Stewart to McLucas, 11 Apr 69, subj: MOL FY 70 Funding.

28. Ibid.

29. Memo (TS-DORIAN), Stewart to Seamans, 17 Apr 69, subj: Draft Ltr to the President.

30. Memo (TS-DORIAN/GAMBIT/HEXAGON), Seamans to Packard, 30 Apr 69, subj: MOL FY 70 Prog Options, w/atch Draft Memo for the President.

31. Memo (TS-DORIAN/GAMBIT/HEXAGON), Stewart to McLucas and Tucker, 15 Apr 69, subj: MOL FY 70 Prog/Funding Options.

32. Memo (TS-DORIAN/GAMBIT/HEXAGON), Mayo to the President, 21 Apr 69, subj: FY 1970 Intelligence Program Savings.

33. Ibid.

34. *Aviation Week*, 6 May 69, p. 22.

35. Hearings, 6 May 69, before Senate Cmte on Aeronautical and Space Sciences, 91st Cong, 1st Sess, *NASA FY 70 Authorization*, Pt 2, pp. 825ff.

36. Memo for the Record (TS-DORIAN/HEXAGON), by Stewart, 13 May 69, subj: MOL Policy Cmte Mtg, 9 May 69.

37. Ibid.

38. Memo for the Record (TS-DORIAN/HEXAGON), by Stewart, 19 May 69, subj: Mtg with the President re MOL.

39. Ibid.

40. *N. Y. Times*, 5 June 69.

41. Ibid.

42. Msg (TS-DORIAN), 2390, Stewart to Bleymaier, 071528Z Jun 69.

43. Memo (TS/DORIAN/HEXAGON), Seamans and Foster to Laird, 8 Jun 69, subj: MOL Decision.

44. Memo (TS-DORIAN), Packard to SAF, Dir/NRO, 9 Jun 69.

45. Hearings before House Subcmte on Appns, 9lst Cong, 1st Sess, DoD Appns for 1970, Pt IV, pp 325-326.

46. Congressional Record, Senate, 12 Jun 69, p S6307.

47. *Space Daily*, 12 Jun 69, p 188.

48. *N. Y. Times*, 12 Jun 69.

MOL ASSEM...
INTEGRATION BUILDI...

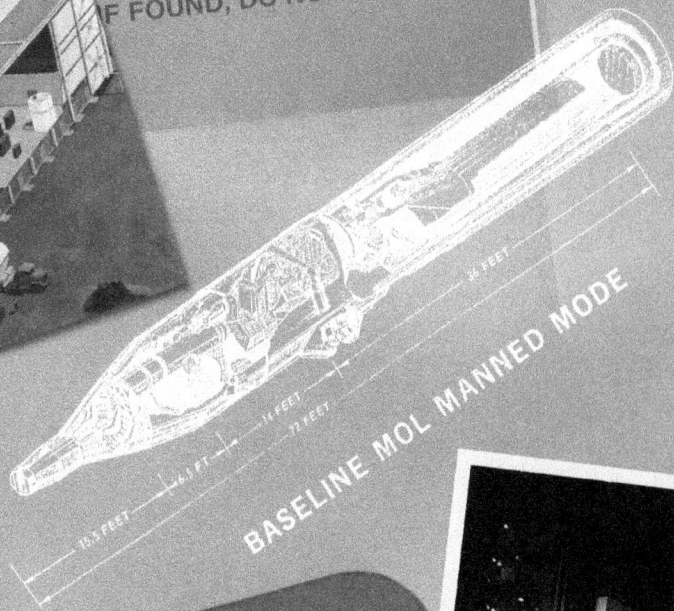

BASELINE MOL MANNED MODE

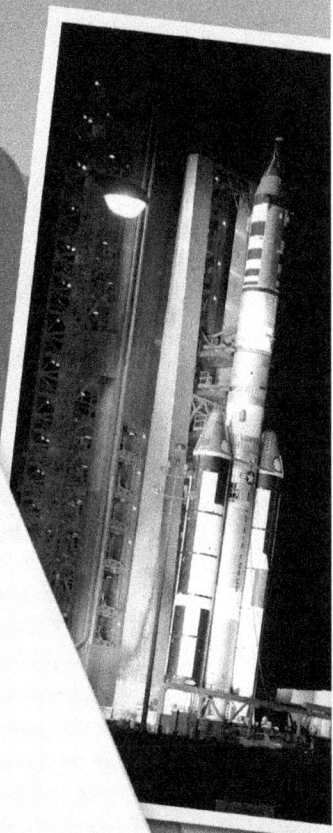

USAF MOL/KH-...

THE DORIAN FILES REVEALED:
A COMPENDIUM OF THE NRO'S MANNED ORBITING LABORATORY DOCUMENTS

CHAPTER XVI:
POST-MORTEM

POST-MORTEM

For military and civilian personnel closely associated with the program, including the MOL astronauts, the project's cancellation came as a distinct shock. Hopes, dreams, ambitions were suddenly disrupted. The Associate Contractors, immediately affected, were faced with the distasteful task of shutting down their MOL operations and laying off workers*. MOL's termination sent employees scurrying around to find new work as following letter, from a young high school girl in California, Susan Kasparian, written to the Secretary of Defense on 12 June, notes. Miss Kasparian wrote:

> The MOL program has been discontinued. I don't understand why and how the government can do something like that—cancel something which has taken years to start, that has taken so much money to continue and time from men who could have been more secure in another area of work. The past four years have been a waste to every man involved in the MOL program. How can the government say— all right, no more, find something else to do? I don't notice anyone cancelling the government.
>
> It's not a very pleasant experience to be out of a job. There's so much to worry about. My father is now looking for a job, we may move, because of the now extinct MOL program. He got up every morning at 6:30, sat behind a desk working for the government, came home at 5:30 and started the cycle again the next day. For what? Nothing, nothing at all. He has wasted his time, his effort and his intelligence on a whim of the government. Every single man and woman is like my father. What are they getting in return for this. The satisfaction of completing a job? The guarantee of another job in the same area. No, nothing—I don't understand what happens to all these people?

> ...I can't ask you to change your decision, so I'm just asking you— why?[1]

As Colonel Hermann of the MOL Program Office began drafting a reply explaining the government's need to reduce federal expenditures, portions of the project were already shut down. On 10 and 24 June, McDonnell Douglas halted all Gemini B and laboratory vehicle work at its St. Louis and Huntington Beach plants. On the 30th General Electric terminated work on the tracking mirror drive, camera controls, simulator, and other equipment. Eastman Kodak halted sensor R&D activity on the 30th[†]. The Martin-Aerojet General-United Technology Center group closed out all Titan IIIM work on 18 July.

By summer's end, contractor personnel assigned to MOL had been drastically reduced. From 6,263 personnel on 10 June, McDonnell Douglas cut its MOL staff to 369 by 30 September. General Electric went from 2,628 to 304 workers, Eastman Kodak from 1,684 to 84, and the Titan IIIM group from 2,391 to 140. Military and civilian personnel in the MOL Systems Office declined from 266 to 26. Eight of the MOL astronauts were subsequently reassigned to NASA, seven in crew duty; the others returned to their services. In Washington, the MOL Program Office at the end of December 1969 consisted of two officers one airman, and two secretaries, down from 25 personnel in June.[2] Colonel Ford, named assistant to Dr. Seamans for MOL to handle final termination and close-out activities, took charge of the Program Office after General Stewart was reassigned to Headquarters AFSC as Deputy Chief of Staff/Systems.

MOL PROGRAM COSTS

On 10 June 1969, the day MOL was publicly terminated, General Stewart appeared before a House subcommittee to discuss costs of the program since engineering development began in September 1966. He reported that $1.3 billion had been expended for MOL RDT&E and another $46 million was used to purchase Sudden Ranch and build various facilities at Vandenberg AFB. In addition, he informed the committee that the Air Force would require an estimated $125 million in fiscal

---

* The grim news produced a headline in the Wall Street Journal : "Mass Layoff Likely at McDonnell Douglas Over MOL Cancellation."

† On 1 September 1969 the EK contract and related activity were transferred to the Directorate of Special Projects.

Figure 60. MOL Vehicle Assembly
Source: CSNR Reference Collection

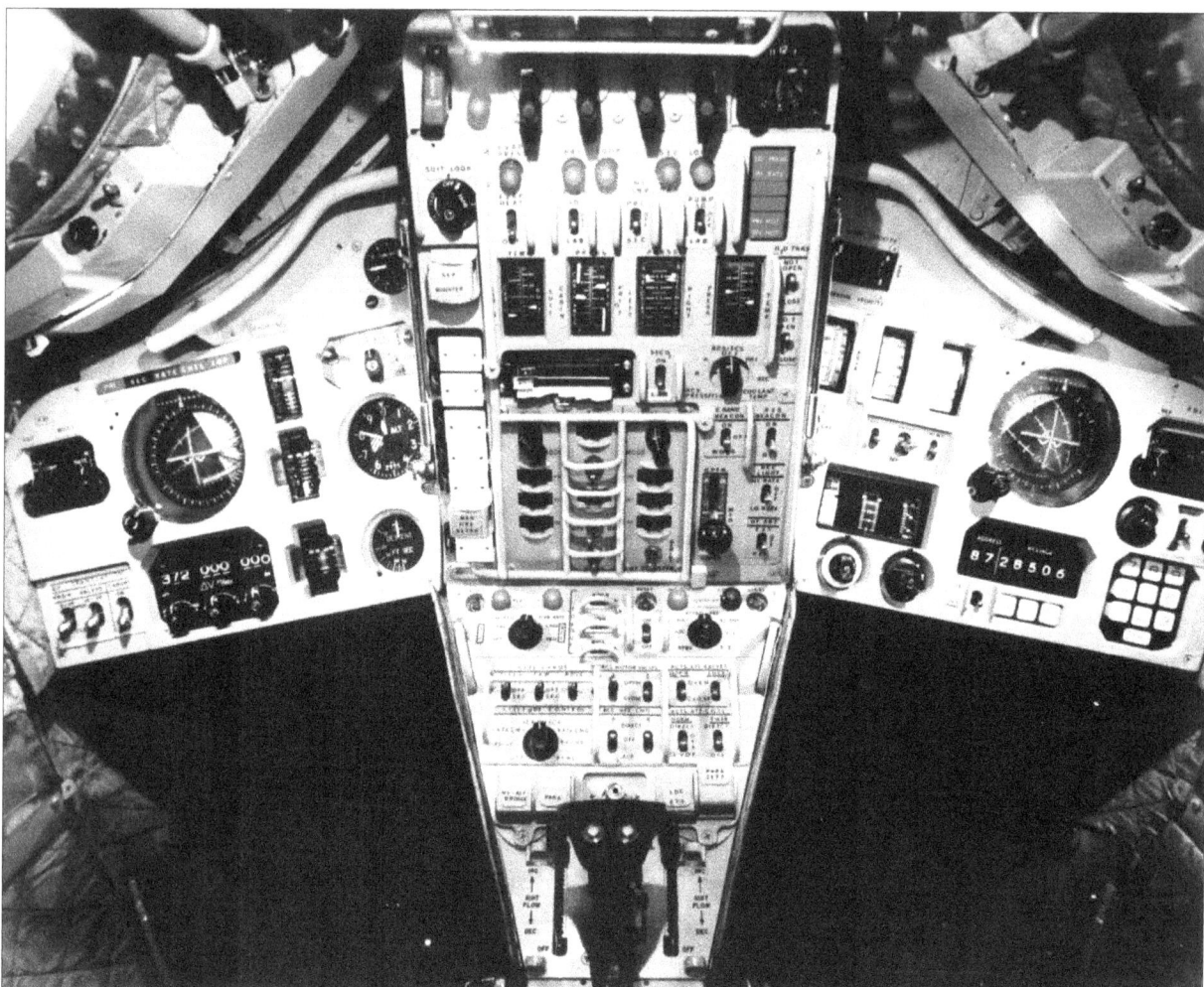

Figure 61. MOL Controls Model
Source: CSNR Reference Collection

year 1970 funds to pay contractor termination costs. Consequently the total overall cost of the MOL program would come t $1.54 billion.[3]

OSD subsequently submitted to Congress a change in its fiscal year 1970 MOL budget line item, reducing it from $525 million to $125.3 million to pay the termination costs. This sum was to be used for employee severance pay and relocation reimbursement, settlement expenses, and allowable post-termination activities such as contractor inventory, hardware, and equipment disposition, and plant maintenance. In mid-July, however, the MOL Systems Office advised Colonel Ford that the contractors' initial claims totaled $137 million. On his instructions, the Systems Office rejected their demands

for full fees and, by December 1969, the sum required had been reduced to $128 million. Toward year's end Dr. Seamans was informed that a further reduction in their closing costs was anticipated and that the $125 million fiscal year 1970 appropriation would be sufficient to satisfy all MOL program obligations.[4]

HARDWARE DISPOSITION

Along with an orderly phase out of the program, the Air Force initiated studies to identify MOL hardware or technology which might be useful to various USAF agencies or NASA. At the request of Dr. Seamans, an ad hoc group—chaired by Dr. Yarymovych of the Office of

Figure 62. MOL Controls Model
Source: CSNR Reference Collection

the Assistant Secretary of the Air Force for R&D‡—was formed at the end of June to do the work. The group held its initial meeting in the Pentagon on 1 July.§ At a second meeting on the West Coast on 10-11 July, Systems Office personnel and other officials briefed the group on the status of MOL hardware and equipment and their possible future use. During this second meeting, a plan for a final report was adopted and various individuals were designated to write certain sections.[5]

Associate contractors were invited to submit suggestions to the ad hoc group for disposition or utilization of MOL equipment and technology. After their presentations were made in Washington on 24 July, the group began work on the final report. It considered both the unclassified and classified MOL equipment. Among the latter were the tracking mirror control system, image velocity sensor, Bi-mat On-Board Film Processor, visual display projector, acquisition tracking scope, and the mission simulator. One of the group's preliminary findings, submitted to Secretary Seamans in a report dated 1 August 1969, was that decisions on disposition of most of the classified MOL equipment would have to await completion of additional studies, already begun.[6]

---

‡   Dr. Yarymovych had joined Flax's staff in 1968.

§   In attendance from the MOL. Program Office were Gen Stewart, Col Stanley C. White, Lt Col Donald L. Steelman, and Mr. Samuel H. Hubbard. Others present were: Brig Gen Raymond A. Gilbert,.AFSC; Brig Gen Louis L. Wilson, SAMSO; Col R. Z. Nelson, Dir/Space, Hq USAF; W. C. Schneider, Philip E. Culbertson, and M. W. Krueger, NASA; H. P. Barfield, ODDR&E; Lt Col Larry Skantze: MOL Systems Office; and Capt. Robert Geiger, Office of Space Systems.

Figure 63. MOL Solar Panels
Source: CSNR Reference Collection

In the case of the unclassified equipment, the group recommended—and Dr. Seamans authorized—the transfer to NASA of the MOL Laboratory Module Simulator (developed by McDonnell Douglas) and its specially modified IBM 360/65 computer. This equipment would be used in the space agency's AAP Workshop. Its original cost was approximately $30 million.

Dr. Seamans also authorized the turnover to the Air Force Office of Scientific Research (AFOSR) of one complete set of Computer Integrated Test Equipment (CITE) and its SDS 9300 computer. This equipment, developed by GE at a cost of approximately $7.5 million to check out the mission payload module, would be used to support the Tanden Van De Graff accelerator program at Florida State University's Nuclear Physics Laboratory.

Two other partially completed CITE sets were to be disposed of in a routine fashion with their associated automatic data processing equipment.¶

The Air Force Western Test Range was given one of two sets of Computerized Aerospace Ground Equipment (CAGE) and its two Sigma 8 computers. CAGE had been developed by Martin (at a cost of approximately $6 million) as an automated control and checkout system for the Titan IIIM. It would replace two existing leased Vandenberg Automatic Data Equipment systems in use at the AFWTR and save approximately

---

¶ These sets were subsequently transferred to the Army's Redstone Arsenal, after Army officials indicated they could make good use of the equipment and save $2 million in procurement monies. [Memo, Hansen to Ford, 11 Dec 69, subj: Disposition of MOL Program CITE Equipment]

$600,000 a year in leased costs. Sixteen remaining MOL computers were ordered reallocated to support Air Force/DoD requirements.

Dr. Seamans also approved the ad hoc group's recommendation transferring to NASA the MOL astronaut feeding system, pressure suit assemblies, waste management system hardware, and return the Gemini equipment previously provided to the Air Force. Other MOL hardware and technology—including fuel cells, attitude control engines, biotechnology and certain classified experiments hardware and equipment—were to be transferred to various Air Force laboratories and R&D activities.[7]

On 29 September, after the ad hoc group completed its study of possible uses of classified MOL equipment, Dr. Seamans approved the turnover to the Air Force Avionics Laboratory of the image velocity testing and sensor equipment, the Bi-mat On-Board Film Processor, visual display projector, acquisition tracking scope, and other items.[8]

Concerning the other classified MOL hardware and equipment, Air Force and NASA officials over a period of several months studied their possible use by the Space agency. This effort began as early as 6 June 1969, the day after word of MOL's termination was passed to the Air Force, when Secretary Seamans met with space agency officials and offered them any MOL equipment and technology that might be useful. Present at this meeting were Colonel Ford and Samuel H. Hubbard, Chief of MOL's Plans and Technology Division (a NASA assignee). Arrangements were made to brief space agency officials on the DORIAN equipment. Subsequently, NASA in September 1969 advised the Air Force it was strongly interested in General Electric's Acquisition and Tracking System (ATS) and Mission Development Simulator (MDS) for possible use in the space agency's earth sensing program. It also requested additional information about Eastman Kodak's facilities and capabilities to build large optics for its astronomy program. Mr. Hubbard arranged to accompany Dr. Henry J. Smith and W. S. Schneider of NASA to Rochester for a briefing on the firm's technological advances.[9]

During the fall of 1969, in accordance with NASA's expressed interest, a space agency contract was let to General Electric and Itek (through Air Force channels) for a study of the application of the ATS and mission simulator to the space agency's mission**. Col. Lew Allen of the Office of Space Systems was designated as NASA's focal point for the study. After NASA expressed interest in Air Force technical participation, Dr. McLucas

on 23 December 1969 designated Headquarters AFSC and Col. Stanley C. White, former MOL Assistant for Bioastronautics, as NASA's point of contact. Col. Benjamin J. Loret and Col. C. L. Gandy, Jr.: both formerly with the MOL Program, were named to assist Colonel White.†† [10]

Meanwhile, in August 1969 Dr. Smith and Mr. Schneider of NASA accompanied Hubbard to Eastman Kodak, where they were briefed on the significant advances the firm had made in sensor technology. A list of MOL equipment that possibly could be used in NASA's astronomy program was later provided them. Subsequently, at Hubbard's suggestion, NASA awarded a study contract to Eastman Kodak (20 January 1970) to undertake a rigorous analysis of what astronomical use could be made of MOL hardware. The equipment, meanwhile, was stored at the Eastman facility pending NASA's review of the study and its decision about a future approach.[11]

On 15 February, at the direction of Secretary Seamans, the MOL Program Office closed its doors. The Systems Office was scheduled to shut down on 30 June 1970. Thereafter, residual contractual matters were to be referred to AFSC and all other MOL matters to the Office of Space Systems for disposition.

COULD THE PROGRAM HAVE BEEN SAVED?

After MOL's demise, there was a post-mortem within and outside the Program Office on what steps might have been taken to save the project. One view—strongly held by some individuals—was that the Air Force managers had made a serious error trying to proceed with a full- equipped, "all-up" MOL system. That is, they argued the program might have survived if General Evans' suggestion of March 1966 had been pursued—decoupling the optics from the first manned flight in order to fly the "man-rated system" alone at an early date. If MOL had been flying, they believed, it might have had a better chance of surviving.

Another view—expressed as early as 1964—was even more pertinent. As noted in Chapter IV, when General Bleymaier briefed the Air Staff Board on 4 January 1964 on AFSC's proposed MOL implementation plan, General Momyer expressed concern about "putting all the Air Force man-in-space eggs in the reconnaissance basket." He recommended that other missions be examined.

---

†† Dr. McLucas informed Dr. Newell of NASA that the Air Force retained "a continuing high interest in the ATS and is enthusiastic about the possibility that it may be flown in AAP. The objectives that we hoped to achieve in MOL using this equipment remain valid and we would hope that some if not many of them could be accomplished in the workshop." [Memo, McLucas to Newell, 23 Dec 69]

** The contract was dated 17 November.

Unfortunately, the Air Force was unable to come up with another mission it could sell to OSD. As a result, during the late 1960s, MOL got caught between the extremely tight defense budget caused by the Vietnam war and the CIA/BOB arguments that unmanned reconnaissance vehicles could do the job cheaper.

The cancellation of MOL ended an Air Force dream of space flight that began in 1945, when General Hap Arnold spoke of the possibility of "true space ships, capable of operating outside the earth's atmosphere." After Sputnik, Air Force hopes and imagination soared, but its initial plan of early 1958 to get a man into space "soonest" was scuttled six months later with the creation of NASA. It then put its space flight hopes into Dyna-Soar, only to see that program terminated in December 1963 by Secretary McNamara.

Although the Defense Chief approved MOL as Dyna-Saar's successor, it took two years of paper studies before the Air Force was given the green light in 1965. Unfortunately, 1965 also was the year the United States sent military forces into South Vietnam to prevent that country's takeover by the Communist North. The cost of the Vietnamese war— which incredibly became the longest war in America's history—contributed directly to MOL's demise. MOL had the misfortune, as one observer put it, "of reaching a peak of financial need for full development and production at a time when the war in Vietnam was draining off all available assets."

During the summer of 1969—after hearing Dr. Seamans lament before his Senate Appropriations Committee that the decision to cancel MOL had been reached "over the objections of the Air Force, including the Secretary"— Senator Russell remarked: "I can understand the decision to postpone, but I did not know we had totally cancelled all military manned exploratory use of space. Because of what man is now doing in space, the control, knowledge, and utilization of space may well determine the course of future wars." Many airmen were convinced that this was so.

As the 1970s began, the Air Force had only the feeblest hope that a new joint effort with NASA—to develop a "reusable" space shuttle that could rendezvous with orbiting vehicles and return to land on earth a la Dyna-Soar—might provide it with the opportunity to get in the necessary "stick time" in space that it had sought for more than a decade.

ENDNOTES

1. Ltr, Susan Kasparian, Tustin, Calif., to the Secretary of Defense, 12 June 69.

2. Memo (U), Ford to Seamans, 3 Dec 69, subj: MOL Termination.

3. Heaings before House Subcmte on Appns, 9lst Cong, lst Sess, Part 4, DoD Appns for 1970, pp 753-762.

4. Memo (S-DORIAN), Ferguson to Seamans/McLucas, 23 Dec 69, subj: MOL Prcgram Close-Out Status.

5. Memo (U), Seamans to Grant L. Hansen, Asst SAF (R&D), 30 Jun 69, Mins (U), Ad Hoc Gp for MOL Residuals, Meeting of l Jul 69 and 10-11 Jul 69, prep by Lt. Col. Donald L. Steeban, MOL Program Office.

6. Memo (TS-DORIAN), Hansen to Seamans, 1 Aug 69; no subj, w/atch Report, Review of MOL Residuals, 1 Aug 69.

7. Memo (U), Seamans to Laird, 6 Oct 69.

8. Memo (U), Hansen to Seamans, 23 Sep 69; Seamans to Hansen, 29 Sep 69, no subj.

9. Memos (S-DORIAN), Ford to McLucas, 22 Sep 69; McLucas to Seamans, 24 Sep 69, Intvw, author with Samuel H. Hubbard, Chief, MOL Plans and Tech Div, 28 Jun 1970.

10. Ltrs (TS-DORIAN), Seamans to Newell, 29 Sep 69; Newell to McLucas, 7 Nov 69; McLucas to Newell, 23 Dec 69.

11. Intvw, author with Hubbard, 28 Jan 70.

MOL ASSEM...
INTEGRATION BUILDI...

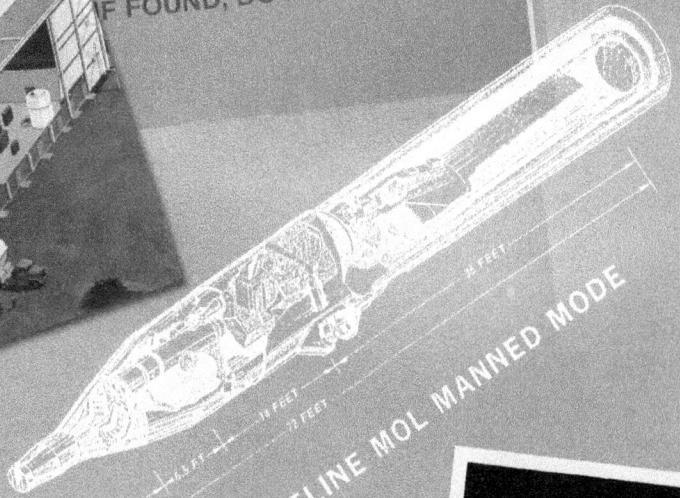

...COU...
MATERIAL

PROPERTY OF THE
...ITED STATES GOVERNMENT

...F FOUND, DO NOT OPEN

BASELINE MOL MANNED MODE

USAF MOL/KH-10...

THE DORIAN FILES REVEALED:
A COMPENDIUM OF THE NRO'S MANNED ORBITING LABORATORY DOCUMENTS

SUPPLEMENTAL
DOCUMENTS INDEX

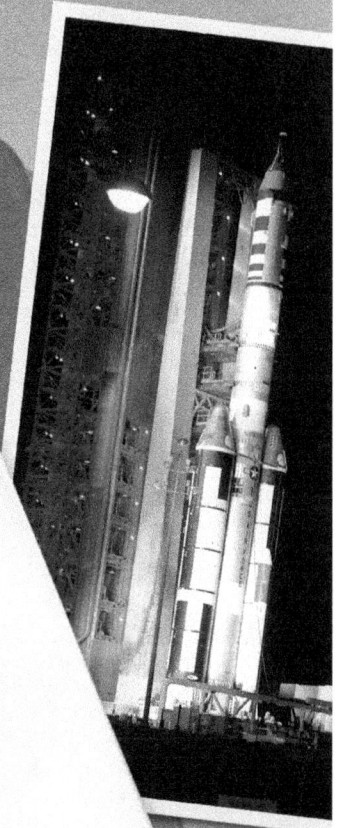

SUPPLEMENTAL DOCUMENTS INDEX

No.	DOCUMENT TITLE	DATE	No. OF PAGES
1	Memorandum for Director, MOL Program, Subject: Authorization to Proceed with MOL Program	8/25/1962	1
2	Memorandum for Director MOL by Mr. Zuckert, Subject: Authorization to Proceed with MOL	8/25/1962	1
3	Letter to Dr. Flax from Deputy Associate Admin for MSF, Subject: Second Phase of MOL Study	10/11/1963	3
4	The University of Michigan, Scientific Experiments for MOL	12/1/1963	275
5	New Program for the Development of a Near Earth MOL	12/10/1963	3
6	Press Release, Subject: Air Force to Develop MOL	12/10/1963	2
7	Aerospace Corporation, MOL Briefing, Management and Administrative Procedures During the Life of the Program	1/3/1964	117
8	Memorandum to Deputy Chief of Staff (R&D) from Under Secretary of the Air Force, Subject: Requirements and Objectives for MOL	1/15/1964	4
9	Aerospace Corporation, MOL Technical Presentation	1/17/1964	33
10	Memorandum for Director, Defense Research & Engineering Signed by Dr. Flax, Subject: MOL	1/18/1964	4
11	Letter to Gen. Schriever from Under Secretary of Air Force, Subject: Views on Management on MOL	2/6/1964	2
12	Notes on Draft Memorandum, Subject: MOL Experiments - Early Critical Comments on the Direction of the Program	2/10/1964	4
13	Letter to Gen. Schriever from Under Secretary of the Air Force, Subject: Confirmation About Details of MOL	3/3/1964	2
14	Security Briefing Statement Relating to Project DORIAN	3/5/1964	2
15	Experiments for MOL, and the Purpose of the MOL Program	3/5/1964	2
16	Memorandum for Under Secretary of the Air Force from Dr. Brown, Subject: Memorandum on Purpose of MOL Program	3/9/1964	10
17	Preliminary Technical Development Plan	3/10/1964	8
18	Security Restrictions on DORIAN	3/12/1964	3
19	Project DORIAN: Security Handling of Contract with Eastman Kodak (EK) Corp.	3/13/1964	1

No.	DOCUMENT TITLE	DATE	No. OF PAGES
20	MOL Technical Panel, First Preliminary Report	3/17/1964	305
21	Letter to John T. McNaughton from Jeffrey C. Kitchen, Subject: About Interest in the Development of Plans for MOL	3/25/1964	1
22	Candidate Experiments for MOL, Volume 1	4/1/1964	499
23	Public Affairs Plan for MOL	4/2/1964	2
24	Organization of MOL Program Management Office	4/6/1964	1
25	Letter to Gen. B. A. Schriever from Brockway McMillan about the Reconnaissance Aspects of MOL	4/6/1964	9
26	Schedule and Funding Information for MS-285	4/6/1964	3
27	Aerospace Support for MS-285	4/8/1964	2
28	Establishment of U.S. Navy Field Office for MOL	4/10/1964	1
29	Letter to Commander, USAF Systems Command from Chief, Bureau of Naval Weapons, Subject: MOL Navy Funding Requirements	4/22/1964	11
30	Memorandum for Secretary of the Air Force by Dr. Brown, Subject: Approval of USAF FY64 RDT&E MOL	4/29/1964	2
31	President's Foreign Intelligence Advisory Board Report: National Reconnaissance Program	5/2/1964	12
32	Cover Letter & Report for Mr. Bundy: National Reconnaissance Program	5/12/1964	5
33	Agreement Between NASA & DoD Concerning Gemini Program and MOL	5/20/1964	5
34	Eastman Kodak Corp. DORIAN Briefing to Dr. McMillan	6/2/1964	2
35	Preliminary Technical Development Plan for MOL	6/30/1964	223
36	Memorandum for Col. Schultz, Assistant for MOL (AFRMO) from Under Secretary of the Air Force, Subject: MOL Experiments on Electromagnetic Signal Detection	7/2/1964	3
37	MOL Experiments on Electromagnetic Signal Detection	7/8/1964	4
38	Aerospace Corporation, Evaluation of Geodetic Capability of MOL Optical Tracking System	7/24/1964	29
39	Letter to AFSC (MSF-1) from Gen. Kinney, Subject: Study of MOL Schedule Alternatives	8/7/1964	6
40	Aerospace Corporation Interoffice Correspondence, Subject: Crew Safety Briefing Charts	8/11/1964	49
41	Meeting 8/19/64 with IBM, Eastman Kodak, LMSC, and ITEK	8/19/1964	2
42	Cable to CSAF from AFSC Subject: Source Selection Board for MOL	8/27/1964	3

No.	DOCUMENT TITLE	DATE	No. OF PAGES
43	Navy Sea Surveillance Experiment for MOL	9/9/1964	1
44	Interface with MOL and MOL Contractors	9/11/1964	2
45	Memorandum for Director of Defense, Research & Engineering, Subject: MOL Program	9/18/1964	2
46	NASA Proposal to Study MOL ATS and DORIAN Technology	9/24/1964	2
47	Letter to Mr. Webb from Mr. McNamara, Subject: Air Force Effort on MOL During Last Several Months	9/25/1964	2
48	Letter to AFSC (SCK-3) from Mr. Best, Chief, Office of Procurement & Production, Subject: Request for Determinations and Findings	10/6/1964	14
49	Aerospace Corporation, MOL Rendezvous	10/15/1964	78
50	Performance and Design Requirements for the MOL System	10/19/1964	233
51	Aerospace Corporation Interoffice Correspondence, Subject: MOL Weight History	10/27/1964	6
52	Memorandum to AFCCS from W. K. Martin, Subject: Source Selection Board for MOL	10/30/1964	2
53	Aerospace Corporation, MOL Extravehicular Space Suit Data Book	11/1/1964	34
54	Problems Posed by the Image Velocity Sensor (IVS) Development and MS-285 Follow-On	11/19/1964	4
55	Preliminary Performance/Design Requirements for the MOL System	11/30/1964	232
56	Headquarters Space Systems Division, Air Force Systems Command Memo, Subject: Gemini/MOL Experiments	12/2/1964	12
57	Memorandum for Assistant Secretary of the Air Force (R&D) from Assistant Secretary of Navy (R&D), Subject: MOL Program Funding	12/24/1964	1
58	Procedural Considerations for MOL Program Management	1/1/1965	22
59	Memorandum for the Secretary of Defense, Subject: MOL Management	1/1/1965	45
60	Memorandum to AFSPD, Subject: Record of Performance of Selected Contractors	1/1/1965	26
61	Harold Brown to USAF on MOL	1/4/1965	4
62	Memorandum for the Record: Meeting with Dr. McMillan on MOL	1/6/1965	2
63	Study of Utilizing Apollo for the MOL Mission, Volume II: Subsystem Studies-Applied Mechanics Division	1/11/1965	304
64	Aerospace Corporation, MOL Mission Potential	1/16/1965	15
65	Qualified Contractors to Receive MOL RFP	1/18/1965	1

No.	DOCUMENT TITLE	DATE	No. OF PAGES
66	Memorandum from J.S. Bleymaier, Subject: MOL Source Selection Board	1/18/1965	7
67	Meeting Minutes of Briefing to DDR&E on 1/16/65, Subject: New MOL Objectives	1/18/1965	3
68	Memorandum for the Under Secretary of the Air Force, Subject: MOL Management	1/25/1965	6
69	DORIAN/BYEMAN and MOL Interface Under the New Security Ground Rules	1/30/1965	5
70	Pressure Suit and Extravehicular Performance Data for MOL	2/1/1965	55
71	Expandable Structures for Construction of Astronaut Transfer Tunnel	2/5/1965	52
72	Source Selection Board for MOL: Criteria the Board Used to Select Contractors for Early MOL Design Studies	2/8/1965	2
73	Memorandum for Chief of Staff, USAF from Brockway McMillan, Subject: Source Selection Board for MOL	2/8/1965	1
74	Aerospace Corporation, Evaluation of Apollo X for MOL Mission	2/10/1965	164
75	Memorandum for the Assistant Secretary of the Air Force, Subject: MOL	2/23/1965	2
76	Memorandum for SSGS from Jewell C. Maxwell, Subject: Code for MOL System Source Selection Board Report and Presentation	2/24/1965	1
77	MOL Policy Committee Meeting, Subject: Results of the MOL Laboratory Source Selection Board and Interactions with NASA	2/25/1965	2
78	Air Force MOL Policy Committee: Summary of Agenda Items from 2/25/65: Contains the Approved Security Classification Guide for MOL	2/25/1965	9
79	Proceedings of Air Force MOL Policy Committee Meeting, 2/25/65	2/25/1965	6
80	Information Responsibilities for MOL Program	2/26/1965	9
81	Memorandum for the Record, McMillan to Seamans, Subject: NASA MOL Security Policy	2/26/1965	1
82	Memorandum for Chief of Staff, USAF from Eugene M. Zuckert, Subject: Authorization to Award Contracts for MOL Preliminary Design Studies	2/26/1965	1
83	NASA/DOD Industry MOL Vehicle Requirements Briefing	2/27/1965	6
84	Security Policy and Procedures for NASA Participation in the DOD MOL Program	3/8/1965	7
85	Memorandum to AFCCS (Gen. Schriever) from Gen. McConnell, Subject: Selection of Aerospace Research Pilots for Assignment to MOL Program	3/15/1965	1
86	Primary Experiments Data for the MOL Program	3/31/1965	45
87	Objectives for the MOL Program	4/15/1965	2
88	Letter to Albert C. Hall from N.E. Golovin, Subject: Recent MOL Studies by Special Space Panel	4/21/1965	2

No.	DOCUMENT TITLE	DATE	No. OF PAGES
89	System Source Selection Board Oral Presentation for the MOL System	5/1/1965	65
90	Criteria to be Employed by the MOL Evaluation Group for Evaluation of MOL Preliminary Design Studies	5/8/1965	7
91	DOD/MOL Program Security	5/18/1965	10
92	Direction of MOL Program Resulting from Presentations and Discussions from May 17-19, 1965	5/20/1965	2
93	DORIAN Optical Studies	5/20/1965	2
94	Proceedings of Air Force MOL Policy Committee Meeting, 6/1/65	6/1/1965	2
95	Letter to Vice Adm. William F. Raborn from Cyrus R. Vance about MOL	6/12/1965	2
96	Memorandum for Gen. J.P. McConnell from Gen. B.A. Schriever, Subject: Objectives of the MOL Studies	6/12/1965	8
97	MOL Program Chronology	6/28/1965	2
98	Memorandum for the Secretary of Defense, Subject: Proposed MOL Program	6/28/1965	105
99	Information Needs for Satellite Photographic Reconnaissance	6/30/1965	23
100	Memorandum for SecDef by Mr. Hornig, Special Assistant Science & Technology, Subject: Special MOL Panel	6/30/1965	7
101	Memorandum for General Blanchard from Harry L. Evans, Subject: MOL Contractor Information	7/1/1965	9
102	MOL Management: 7/5/65	7/5/1965	10
103	Memorandum for Secretary of Air Force from Robert S. McNamara, Subject: Press Release on MOL	7/6/1965	1
104	Memorandum for the Executive Secretary, National Aeronautics and Space Council, Subject: MOL	7/7/1965	7
105	Memorandum for the Vice President, Subject: Organization and Public Position on MOL	7/8/1965	4
106	SAFSL/SAFSS Relationships, Functions of the NRO Staff with Respect to the MOL Program	7/9/1965	5
107	Memorandum for the Record, 7/13/65, Subject: MOL "Posture" Paper	7/13/1965	4
108	Memorandum for the Record- DOD MOL - A Consideration of International Political Factors	7/13/1965	2
109	Memorandum for the Record, Subject: MOL "Posture" Paper	7/15/1965	19
110	Memorandum for Robert McNamara, Subject: Space Council Meeting on 7/9/65	7/19/1965	1

No.	DOCUMENT TITLE	DATE	No. OF PAGES
111	Memorandum for the Record, 7/21/65, Subject: MOL "Posture" Paper	7/21/1965	5
112	Policy on Public Information Aspects	7/26/1965	14
113	Memorandum for the Record, from Paul E. Worthman, Subject: MOL Policy Paper	7/26/1965	15
114	Memorandum for the Record, Subject: Proposed MOL Press Release	7/27/1965	5
115	Memorandum to Dr. Hall from Paul Worthman, Subject: USIB Endorsement of MOL Photographic Capability	7/27/1965	2
116	Memorandum for the Record, Subject: MOL Policy Questions	7/27/1965	8
117	Letter to Mr. James E. Webb from Harold Brown, Subject: Policy on Public Information Aspects & International Reactions to MOL	7/28/1965	17
118	Memorandum for Dr. McMillan from John L. Martin, Jr, Subject: Selection of DORIAN Payload Contractor, Report To Survey Committee	7/30/1965	61
119	MOL Briefing Policy: 5 Alternative Security Solutions for a Problem Created by Conducting an Unclassified Program with a Classified Mission	8/1/1965	17
120	Government Plan for Program Management for the MOL Program	8/1/1965	189
121	To Mr. J. E. Webb from Harold Brown, Subject: Draft of Memo to President on MOL	8/9/1965	10
122	Memorandum for the Record from: Brockway McMillan, Subject: MOL Sensor Development	8/11/1965	6
123	Memorandum for the Record, from Paul E. Worthman, Subject: Memorandum to the President (MOL)	8/12/1965	10
124	Letter to Robert McNamara from Dean Rusk, Subject: 7/6/65 Space Council Discussion of the MOL Project	8/16/1965	2
125	Dean Rusk to McNamara, Subject: Information and Publicity Controls	8/16/1965	2
126	Note to Dr. Harold Brown from Donald F. Hornig, Subject: Draft Memorandum to the President on MOL	8/24/1965	2
127	Memorandum for Secretary of Defense by Mr. Zuckert, Subject: Memorandum of 28 June 65, Description of MOL	8/24/1965	1
128	Comments on Draft Memorandum for The President	8/24/1965	3
129	President Johnson's Statement on MOL-Press Conference	8/25/1965	1
130	Secretary of Air Force Order, Subject: Director of MOL Program	8/25/1965	2
131	Memorandum for Members, MOL Policy Committee by Mr. Zuckert, Subject: MOL Management	8/25/1965	4
132	Program Directive, Management of the MOL Program	8/25/1965	23

No.	DOCUMENT TITLE	DATE	No. OF PAGES
133	Program Directive, MOL Phase I Program Authorization	8/25/1965	24
134	Impact of Recent Publicity on the MOL Program	8/25/1965	45
135	Letter to Mr. James E. Webb (NASA Administrator) from Harold Brown (Dir Defense Research & Engineering), Subject: Draft of MOL Program	8/28/1965	28
136	Personnel Announcement for MOL Program Office	8/30/1965	2
137	Agreement Concerning MOL System Program Office and Program Management	9/1/1965	7
138	Titan III/MOL Compatibility Study, Performance Improvement Report Technical Summary	9/1/1965	327
139	Letter to Edward C. Welsh from Harold Brown, Subject: MOL Program	9/8/1965	2
140	Memorandum for Assistant Secretary of the Air Force from Harry L. Evans, Subject: Initiation of MOL Contract Definition Phase	9/10/1965	5
141	Outer Space Meetings—MOL or Gemini Issues	9/16/1965	9
142	NSAM on MOL	9/16/1965	7
143	State Department Guidance to U.S. Delegation, Outer Space Meetings-MOL or GEMINI Issues: Official Policy and Response to Soviet Characterizations of the MOL Program	9/17/1965	4
144	Letter to Dean Rusk from Cyrus Vance about the MOL Program, Subject: Public Statements about the MOL Program because of Potential Foreign Reaction	9/24/1965	1
145	Memorandum for Gen. Stewart, Your Relationship to the DORIAN Program	9/28/1965	4
146	DORIAN Security	9/29/1965	3
147	MOL Monthly Status Report: 9/65	10/7/1965	10
148	Memorandum for Secretary of Air Force, Subject: An Offer of Inspection of MOL	10/8/1965	12
149	A Report by the J-5 on an Offer of Inspection of MOL	10/12/1965	6
150	Security Inspection of MOL	10/12/1965	1
151	Airgram to Department of State from USUN Subject: Outer Space, MOL	10/12/1965	2
152	Proceedings of Air Force MOL Policy Committee Meeting, 10/14/65	10/14/1965	5
153	Highlight Summary of Air Force MOL Policy Committee	10/14/1965	28
154	Memorandum for Assistant SecDef for International Security Affairs from Harold Brown, Subject: Review - An Offer of Inspection of MOL	10/15/1965	2
155	Soviet Criticism of the US MOL Program	10/19/1965	9
156	For the Deputy Director, CIA, Subject: Offer of Inspection of MOL	10/19/1965	1

No.	DOCUMENT TITLE	DATE	No. OF PAGES
157	Public Affairs Guidance, Department of Defense MOL Program	10/21/1965	9
158	Public Affairs Guidance for the MOL Program, 10/22/1965	10/22/1965	4
159	Memorandum for Spurgeon Keeny from Alvin Friedman, Subject: NSAM on MOL	10/23/1965	3
160	Memorandum for Harold Brown from Alvin Friedman, Subject: MOL Inspection Proposal	10/23/1965	8
161	Memorandum for the Record, from John L. Martin, Jr. Subject: NRO and SAFSP Responsibilities Related to the MOL Program	10/25/1965	1
162	Martin Memorandum for the Record on 10/22 Discussion with Dr Flax	10/25/1965	1
163	Memorandum for Mr. Reber from Paul Worthman, Subject: MOL Inspection Proposal	10/25/1965	2
164	Memorandum for Commander, National Range Division, AF Systems Command, from Gen. Evans, Subject: Ship Support for MOL Orbital Insertion	10/26/1965	1
165	Offer of Inspection of MOL	10/26/1965	1
166	Public Affairs Guidance for the MOL Program, 10/29/1965	10/29/1965	13
167	Memorandum for Distribution from Arthur Sylvester, Subject: Public Affairs Guidance for the MOL Program	10/29/1965	1
168	Memorandum for Col. Worthman from Major Yost, Subject: P/A Guidance on MOL	10/29/1965	2
169	Annual Report to the Presidents Foreign Intelligence Advisory Board on the Activities of the NRP	10/31/1965	5
170	Fact Sheet, MOL Program: Chronology to Date and MOL Plan Moving Forward	11/1/1965	2
171	Semi-Annual Report to the PFIAB on the Activities of the NRP, 11/1/65 - 4/30/66	11/1/1965	5
172	Public Affairs Guidance for the MOL Program, 11/3/1965	11/3/1965	2
173	An Analysis of the Requirements for Very High Accuracy Image Motion Compensation	11/4/1965	91
174	MOL Monthly Status Report: 10/65	11/8/1965	11
175	Memorandum to AFSC Centers and Ranges from Gen. Schriever, Subject: Conduct of MOL	11/9/1965	2
176	Memorandum for the Record from: John L. Martin, Jr. Subject: 11/8/65 PSAC Reconnaissance Panel Roundtable Discussion on DORIAN	11/10/1965	6
177	President's Science Advisory Committee (PSAC) Meetings of 11/8/65	11/12/1965	8
178	Procedures for MOL Program Management	11/12/1965	7
179	To D/NRO (Dr. Flax), D/MOL (General Schriever) from John L. Martin, Jr., Subject: Unsatisfactory Performance by General Electric Missile and Space Division	11/15/1965	3

No.	DOCUMENT TITLE	DATE	No. OF PAGES
180	Memorandum to Mr. John Kirby, Subject: Soviet Air Attaché at Douglas Talk on MOL	11/18/1965	7
181	White House Memorandum, Subject: Reconnaissance Panel Views on the MOL Development Program	11/18/1965	2
182	Letter to Gen. Evans from George E. Mueller, Subject: Transfer of Certain Gemini Program Equipment to MOL	11/20/1965	4
183	Memorandum for Deputy Director, MOL from Gen. Schriever, Subject: Application of 375-Series Management Procedures to MOL	11/21/1965	1
184	Memorandum for Dr. Brown from Gen. B. A. Schriever, Subject: Disposition of General Electric MSD as a MOL Contractor	11/21/1965	3
185	Letter to Alexander Flax from Donald F. Hornig, Subject: Reconnaissance Panel Discussion on MOL	11/22/1965	1
186	MOL Management: Procedures for Washington Area	11/24/1965	2
187	Manned/Unmanned MOL Mission Assignments	11/29/1965	7
188	Proceedings of Air Force MOL Policy Committee Meeting, 11/30/65	11/30/1965	6
189	Highlight Summary of Air Force MOL Policy Committee Meeting, 11/30/65	11/30/1965	1
190	Memorandum for the Record, Subject: USUN Msg 533	12/2/1965	3
191	Preliminary Performance/Design Requirements for the MOL System, 12/3/65	12/3/1965	379
192	MOL Monthly Status Report: 11/65	12/9/1965	9
193	Memorandum for the Record by Col Worthman, Subject: MOL Relationship to Congress	12/17/1965	1
194	MOL-USIB Relationship	12/21/1965	2
195	Eastman Kodak, DORIAN Program Review	12/22/1965	83
196	Memorandum for Dr. Brown from Gen. B. A. Schriever, Subject: Disposition of General Electric (MSD) as an MOL Contractor	12/27/1965	1
197	Development Problems Inherent in an Unmanned DORIAN System	1/2/1966	20
198	MOL Monthly Status Report: 12/65	1/4/1966	11
199	Letter to Brigadier General Russell A. Berg from: Harry L. Evans, Subject: The Broadening of the MOL Program to Increase its Capabilities to Perform Useful Military Functions and Experiments in Space	1/4/1966	4
200	Soviet Orbital Rockets and the US MOL	1/6/1966	24
201	MOL Information Plan	1/10/1966	18

No.	DOCUMENT TITLE	DATE	No. OF PAGES
202	Memorandum for Distribution from E.B. LeBailly, Subject: MOL Information Plan	1/10/1966	18
203	Memorandum for Gen. Evans from Dr. Yarymovych, Subject: MOL Experiments Program	1/21/1966	3
204	Memorandum for the Record from Harry L. Evans, Subject: MOL SIGINT Study Program	1/24/1966	3
205	Summary Studies of Unmanned DORIAN System	1/28/1966	37
206	Preliminary Performance/Design Requirements for the MOL System, 1/30/66	1/30/1966	257
207	Proceedings of Air Force MOL Policy Committee Meeting, 2/8/66	2/8/1966	6
208	Program Directive, MOL SIGINT Study Program	2/9/1966	10
209	President's Science Advisory Committee Meetings of 2/9/66	2/11/1966	4
210	MOL Monthly Status Report: 1/66	2/12/1966	6
211	Press Guidance on Location of MOL Launches	2/14/1966	4
212	Memorandum for D/NRO from Arthur W. Barber, Subject: Project DORIAN Access	2/14/1966	1
213	Memorandum for Gen. Evans from Lewis S. Norman, Subject: Status of MOL Readout	3/3/1966	7
214	MOL Monthly Status Report: 2/66	3/4/1966	9
215	Memorandum for Deputy Assistant Secretary of Defense (Arms Control) from Alexander H. Flax, Subject: DORIAN Access	3/4/1966	2
216	Memorandum for Dr. Flax from Col. B. R. Daughtrey, Subject: MOL Program	3/9/1966	1
217	Memorandum to General Martin from Byron F. Knolle, Jr. Subject: Foreign Procurement	3/9/1966	1
218	Memorandum for Dr. Flax from Paul E. Worthman, Subject: U.S. Army Participation in MOL	3/10/1966	1
219	Memorandum for Chairman Revers from Harold Brown, Subject: Determination of the Launch Site for MOL	3/14/1966	2
220	Memorandum for Deputy Director, MOL Program, Subject: MOL Priorities	3/14/1966	9
221	Public Information Handling of the TITAN IIIC Space Launches from Cape Kennedy During 1966	3/19/1966	5
222	Approval of Special MOL Facilities at Eastman Kodak Corporation	3/21/1966	1
223	Request for Clearance of Proposed Material Containing MOL References: Decision to allow the Los Angeles Times to Include MOL as Part of a Larger Article on Current Space Projects	3/22/1966	3

No.	DOCUMENT TITLE	DATE	No. OF PAGES
224	Memorandum for Dr. Flax, Subject: MOL Priority	3/23/1966	4
225	Letter to Robert McNamara from Bureau of the Budget, Subject: Approval of MOL Programs	3/24/1966	1
226	Letter to Mr. Charles L. Schultze from BOB, Subject: Letter of 3/21/66, Review of Certain Factors in the MOL Program	3/25/1966	1
227	Memorandum for Director, NRO (Dr. Flax) from Gen. B. A. Schriever, Subject: Procedures for MOL Management	3/28/1966	4
228	Capabilities of the MOL System	4/1/1966	9
229	Memorandum for General Evans, SAFSL from Alexander H. Flax, Subject: MOL History	4/2/1966	5
230	MOL Program Review, 4/2/66	4/2/1966	102
231	MOL Program Management Review	4/2/1966	4
232	Memorandum for Director of Defense R&D from Dr. Flax, Subject: Eastman Kodak Corp. Facilities for MOL	4/4/1966	3
233	Approval to Proceed with the Procurement of DORIAN Long Lead Support Equipment	4/4/1966	1
234	Memorandum for the Director Defense Research and Engineering, Subject: MOL Program	4/5/1966	3
235	Program Review, MOL Systems Office	4/5/1966	4
236	Memorandum for the DDR&E, Subject: MOL Program	4/5/1966	3
237	Memorandum for the Record, from Richard S. Quiggins, Subject: Handling of MOL Security Review Matters	4/6/1966	1
238	Memorandum for Assistant Secretary of Army, R&D from Dr. Flax, Subject: MOL	4/6/1966	2
239	For Gen. Berg from Gen. Evans, Subject: Comparison of MOL with an Equivalent Wholly Unmanned System	4/6/1966	2
240	Memorandum for Director Defense Research and Engineering, Subject: Comparison of MOL to an Unmanned System of Resolution	4/6/1966	45
241	MOL Monthly Status Report: 3/66	4/8/1966	11
242	Costs and Schedules for the MOL Program	4/8/1966	1
243	Funding Estimates, from the NAVMOL Office, on the Navy's MOL Project	4/13/1966	7
244	Revision of Program Review, MOL Systems Office	4/13/1966	5
245	Establishment of a MOL Cost Review Team	4/15/1966	1

No.	DOCUMENT TITLE	DATE	No. OF PAGES
246	Final Report on the Role of Man in the MOL, 4/21/66	4/21/1966	1
247	Proceedings of Air Force MOL Policy Committee Meeting, Agenda Items for 4/29/66	4/29/1966	2
248	Proceedings of Air Force MOL Policy Committee Meeting, 4/29/66	4/29/1966	7
249	Memorandum for the Deputy Secretary of Defense from Alexander H. Flax, Subject: Foreign Procurement	5/2/1966	1
250	Memorandum for Director, NRO (Dr. Flax) from John L. Martin, Jr. Subject: Foreign Origin Procurement	5/2/1966	1
251	Proposal for a Domestic Source of Supply for High Quality Optical Glass	5/3/1966	5
252	MOL Monthly Status Report: 4/66	5/6/1966	6
253	Discussion Points for Eastman Kodak Company	5/10/1966	22
254	Paper from Air Force System Command, Operations Order for MOL, Subject: DoD Manager Relationship with MOL	5/11/1966	18
255	Memorandum to B.P. Leonard, W.F. Sampson, W.C. Williams, Subject: Reliability Growth and Cost Effectiveness Comparison of Manned and Unmanned MOL Systems	5/16/1966	32
256	Proceedings of Air Force MOL Policy Committee Meeting, 5/20/66, Position of Committee Members	5/20/1966	3
257	Proceedings of Air Force MOL Policy Committee Meeting, Agenda Items for 5/20/66	5/20/1966	1
258	Proceedings of Air Force MOL Policy Committee Meeting, 5/20/66	5/20/1966	5
259	MOL Leonard Briefing, 5/21/66	5/21/1966	29
260	MOL Policy Committee Meeting, 5/20/66	5/23/1966	1
261	Memorandum for Dr. Flax from Gen. Stewart, Subject: Unmanned DORIAN System Studies	5/23/1966	4
262	Final Report on the Role of Man in the MOL, 5/25/66	5/25/1966	2
263	MOL Monthly Status Report: 5/66	6/1/1966	7
264	Gen. Shriever Briefing Presented by Leonard, Manned System Performance Analysis	6/7/1966	107
265	Proceedings of Air Force MOL Policy Committee Meeting, Agenda Items for 6/9/66	6/9/1966	2
266	Aerospace Corporation MOL Simulation and Back-Up Analog Tracking Study	6/9/1966	19
267	Memorandum for Distribution from E.B. LeBailly, Subject: Annex 1, MOL Information Plan	6/11/1966	6
268	Army Participation in the MOL Program	6/13/1966	2

No.	DOCUMENT TITLE	DATE	No. OF PAGES
269	Memorandum to D/NRO from John L. Martin, Jr. Subject: Studies of a "Wholly Unmanned" DORIAN System	6/15/1966	3
270	Memorandum for Assistant Secretary Air Force (R&D) from Gen. Evans, Subject: MOL Launch Complex at Western Test Range	6/20/1966	11
271	MOL Program Suggested Changes to SS-MOL-1 Integrated 4/29/66: Review of System Specifications Contractor Comments and Baseline Changes	6/27/1966	91
272	Memorandum for Sec. Brown from Gen. Schriever, Subject: MOL Manpower Requirements	7/5/1966	2
273	MOL Monthly Status Report: 6/66	7/8/1966	6
274	Draft Memorandum to the President on MOL Revised to Reflect Dr. Hornig's Views	7/21/1966	8
275	Memorandum for Director, MOL Program from Harry L. Evans, Subject: Authorization for MOL Engineering Development	7/26/1966	5
276	Missile/Space Daily Article - MOL Information Policy not to be Affected by and U.S.-Soviet Accords	7/29/1966	1
277	MOL Monthly Status Report: 7/66	8/4/1966	6
278	Establishment of Working Group to Examine MOL Contributions to Space Astronomy	8/4/1966	1
279	Operation Order No. 66-3, Subject: Support of MOL	8/4/1966	43
280	Memorandum for Assistant Secretary of Navy (Research and Development) from Alexander H. Flax, Subject: MOL Ocean Surveillance Security	8/5/1966	1
281	DIAMOND II Study	8/5/1966	91
282	MOL Manned/Automatic and Automatic Systems Analysis, President's Science Advisory Committee Briefing	8/13/1966	36
283	Briefing to PSAC on MOL: A Briefing Presented to MOL Personnel Comparing MOL to the Unmanned DORIAN Reconnaissance System	8/17/1966	8
284	Memorandum for Dr. Flax from Harry L. Evans, Subject: Authorization for MOL Full-Scale Development	8/18/1966	3
285	Memorandum for Dr. Flax from Paul E. Worthman, Subject: PSAC Panel Comments on the MOL	8/19/1966	1
286	Memorandum for the Secretary of the Air Force, Subject: MOL Program Plan and Funding Requirements	8/20/1966	5
287	Manned/Unmanned Comparisons in the MOL	8/26/1966	47
288	Memorandum for Director, Defense Research & Engineering from Gen. Evans, Subject: Engineering Development Phase of MOL Program	8/29/1966	13
289	Secretary of Air Force Order 117.4, Director of MOL Program	9/1/1966	2

No.	DOCUMENT TITLE	DATE	No. OF PAGES
290	Government Plan for Program Management for MOL	9/1/1966	116
291	MOL Monthly Status Report: 8/66	9/7/1966	7
292	Memorandum for the Secretary of the Air Force, Subject: Engineering Development Phase of the MOL Program	9/7/1966	6
293	Memorandum to Assistant for R&D Programming from Gen. Evans, Subject: MOL Program Reclama on Draft Presidential Memorandum	9/9/1966	4
294	Memorandum to Dr. Flax from James T. Stewart, Subject: MOL Optical System Payload and Design	9/23/1966	30
295	DORIAN Lockheed Missile and Space Company, Offer to Sell Military Mission Simulation System	9/26/1966	3
296	MOL Mission Planning and Generalized Target Model	9/27/1966	9
297	Memorandum for Director Defense Research and Engineering, Subject: MOL Versus an Equivalent Wholly Unmanned System	9/27/1966	8
298	DORIAN Briefing, Possible Utilization of MOL Hardware for Long Duration Bioastronautics Test Missions	9/28/1966	41
299	Briefing to Presidents Science Advisory Committee on Possible Utilization of MOL Hardware for Long Duration Bioastronautics Test Missions	9/30/1966	22
300	Memorandum for Dr. Flax from Harry L. Evans, Subject: MOL Comparison Study Briefing for the Secretary of Defense	10/2/1966	1
301	Letter to Charles L. Schultze about the Studies Between the MOL Program and the Development of a Wholly Unmanned System of Equivalent Resolution	10/3/1966	2
302	Memorandum for Gen. Evans from Dr. Yarymovych, Subject: Comments on MOL SPO Letter on Subject of MOL Flight Test Objectives	10/4/1966	3
303	MOL Monthly Status Report: 9/66	10/6/1966	7
304	Memorandum to Deputy Director, MOL from Gen. Evans, Subject: DoD Participation in NASA Earth Sensing Programs	10/18/1966	8
305	Use of Easter Island for MOL Program	10/22/1966	3
306	MOL Program Instruction No. 1, MOL Management Meeting, Gen. Ferguson	10/27/1966	5
307	Memorandum for the Director of Defense, Research & Engineering, Subject: MOL/NRO Tasks for Ocean Surveillance	10/27/1966	3
308	Memorandum for Dr. Flax, Subject: Electromagnetic Pointing System for MOL	10/27/1966	21
309	MOL Monthly Status Report: 10/66	11/1/1966	5
310	Contingency Planning for MOL Flights 5, 6, & 7	11/9/1966	2

No.	DOCUMENT TITLE	DATE	No. OF PAGES
311	ITEK Corporation, Acquisition Telescope Preliminary Design Study	11/15/1966	15
312	Proceedings of Air Force MOL Policy Committee Meeting, 11/22/66	11/22/1966	5
313	Employment of the MOL Photographic Product	11/29/1966	3
314	Delegation of Authority and Designation as Head of a Procuring Activity, 12/1/66	12/1/1966	7
315	Memorandum to General Electric Company from Ward M. Millar, Subject: Information Policy on MOL Program	12/2/1966	1
316	MOL Monthly Status Report: 11/66	12/6/1966	7
317	Letter to Lt. Col. Ward M. Millar from R.C. Sharpe, Subject: Identification of the MOL Program	12/6/1966	2
318	Memorandum for Brig. Gen. Russell A. Berg from Harry L. Evans, Subject: MOL Follow-On Program Options	12/16/1966	4
319	Memorandum for Dr. Brown from Harry L. Evans, Subject: Status of MOL Program Contracts	12/16/1966	6
320	Memorandum for the Director MOL from Gen. Evans, Subject: MOL Program Plan and Funding Requirements	12/21/1966	6
321	Memorandum for Director MOL from Dr. Flax, Subject: Policy Relating to MOL Astronauts	12/28/1966	25
322	MOL Monthly Status Report: 12/66	1/6/1967	8
323	Memorandum to SecDef from Dr. Brown, Subject: MOL's Listing in DoD Master Urgency List	1/7/1967	2
324	Action Items from Last MOL Program Review: Documents Image Distortion Issues with the MOL Camera and the Acquisition and Tracking System	1/9/1967	5
325	Memorandum for Director, MOL Program from Alexander H. Flax, Subject: Authorization to Proceed with the Engineering Development Phase of the MOL Program	1/13/1967	1
326	NPIC Support for the DORIAN Program	1/13/1967	19
327	Minutes from 1/5/67, 3/10/67, and 4/14/67 MOL Management Meetings	1/16/1967	13
328	Memorandum for the Record, from Harry L. Evans, Subject: 1/5/67 MOL Management Meeting	1/16/1967	186
329	Briefing to General Evans, Approach to MOL Follow-On Planning	1/16/1967	29
330	Memorandum for Colonel Worthman from Richard S. Quiggins, Subject: Ambassador Goldberg being Briefed on MOL	1/16/1967	1
331	Memorandum for Dr. Flax from Harry L. Evans, Subject: Extended Lifetime Support Module for MOL	1/16/1967	6

No.	DOCUMENT TITLE	DATE	No. OF PAGES
332	Delegation of Authority and Designation as Head of a Procuring Activity, 1/20/67	1/20/1967	6
333	PSAC Panel Report, Subject: MOL Security	1/21/1967	2
334	Review of MOL/DORIAN Ground Test Planning	1/23/1967	81
335	Acoustic and Vibration Testing	1/26/1967	34
336	MOL Engineering Baseline Description to Mr. Gehrig and Others of the House Committee	1/30/1967	27
337	Lifetime Support Module for MOL	1/30/1967	1
338	Memorandum for Assistant Secretary of Air Force (R&D) from Dr. Foster, Subject: Extended Lifetime Support Module for MOL	1/31/1967	1
339	MOL Monthly Status Report: 1/67	2/7/1967	8
340	Memorandum for Dr. Flax from Gen. Evans, Subject: Astronomical Mission for MOL	2/7/1967	2
341	Application of MOL to Astronomical Observations	2/9/1967	8
342	Paper for Coordination Written by Capt. Goolsby, USN, Subject: Flight Objectives for MOL	2/9/1967	27
343	Photographic Readout System for Use in Reconnaissance Aircraft	2/9/1967	10
344	Memorandum for Dr. Flax from Gen. Ferguson, Subject: MOL Priorities	2/13/1967	11
345	Memorandum for Gen. Martin, Subject: Comments Related to the Benefits of the Presence of Man	2/13/1967	2
346	Review of MOL/DORIAN: Ground Test Planning	2/15/1967	27
347	MOL Management Meeting Minutes: Minutes from 2/15/67	2/15/1967	204
348	MOL/DORIAN Schedule/Cost Information	2/17/1967	2
349	MOL Program Directive: Security and Information	2/20/1967	18
350	MOL Monthly Management Meeting on 2/15/67	2/21/1967	3
351	Memorandum for Chief of Staff, USAF, Subject: MOL Requirement at Easter Island	2/24/1967	4
352	MOL Program Office Instruction No. 2, MOL Executive Council Management Meetings: Establishes Policy and Procedures for the MOL Executive Council Meetings	2/25/1967	6
353	MOL Directive No. 67-4, MOL Program Advanced Planning Approved by Gen. Ferguson	3/1/1967	6
354	Provisions of DRV and WBDL in MOL	3/1/1967	4
355	MOL Monthly Status Report: 2/67	3/6/1967	7

No.	DOCUMENT TITLE	DATE	No. OF PAGES
356	Photographic Intelligence Indoctrination Program for MOL Astronauts	3/13/1967	2
357	Memorandum for D/NRO from Harry L. Evans, Subject: Recommendations on Readout in the MOL Program	3/13/1967	7
358	Memorandum for the Record, from Richard C. Randall, Subject: 3/10/67 Meeting on MOL Revised Costs/Schedules	3/14/1967	3
359	Memorandum for Gen. Evans from Paul J. Heran, Subject: Decisions/Guidance for MOL	3/14/1967	2
360	MOL/DORIAN Funding Issues	3/22/1967	1
361	Letter to Gen. Evans from Gen. Hobson, Subject: Utilization of NASA Gemini Program Equipment on the MOL Program	3/24/1967	6
362	Handwritten note for Mr. Worthman and Gen. Stewart about MOL	3/28/1967	3
363	Aerospace Corporation, Application of MOL Hardware for Rendezvous/Resupply Operating Mode	3/31/1967	28
364	MOL Engineering Baseline Description Briefing	4/6/1967	51
365	MOL Proposed Plan to Reduce Early Fiscal Year and Overall Funding Requirements	4/10/1967	3
366	Memorandum for Deputy Director MOL Program from James T. Stewart, Subject: Advance Planning	4/11/1967	2
367	MOL Planning Briefing, 4/12/67	4/12/1967	237
368	DORIAN General Electric, Douglas Aircraft, and Eastman Kodak Briefing: Schedule Reprogramming and Technical Status	4/21/1967	53
369	Memorandum for Gen. Stewart from Walter W. Sanders, Subject: Convertibility in the MOL Program	4/27/1967	1
370	Convertibility in the MOL Program	4/27/1967	38
371	From MOL Systems Program Office to Douglas Aircraft Company, Subject: Request for Engineering Change Proposal	5/3/1967	9
372	MOL Monthly Status Report: 4/67	5/5/1967	5
373	MOL Monthly Status Report: 3/67	5/5/1967	11
374	Program Status: Fiscal Year 1967 Significant Milestones	5/10/1967	192
375	Memorandum for the Record from Richard H. Campbell, Subject: MOL Planetary Observations Program Directive	5/10/1967	2
376	Memorandum for the Record, Subject: Minutes, 5/11 MOL Management Meeting	5/11/1967	10
377	MOL Management Responsibilities	5/15/1967	1

No.	DOCUMENT TITLE	DATE	No. OF PAGES
378	Response to Questions Regarding MOL	5/15/1967	2
379	An Unmanned DORIAN System	5/16/1967	12
380	Cost of an Unmanned DORIAN Reconnaissance Satellite System	5/16/1967	12
381	MOL Management Meeting Minutes, 5/11/67	5/17/1967	5
382	Management Responsibility for Phase II MOL/DORIAN Activities at GE	5/18/1967	17
383	Memorandum for the Record, Presentation of Paper on Unmanned DORIAN System to Mr. Michaels, Staff Member of the House Appropriations Committee	5/19/1967	5
384	Memorandum for the Record, Subject: Manned Versus Unmanned MOL Cost Comparisons	5/19/1967	17
385	DORIAN Operations Concept for MOL Manned/Automatic Configuration	5/26/1967	53
386	Proceedings of Air Force MOL Policy Committee Meeting, 6/1/67	6/1/1967	7
387	MOL Monthly Status Report: 5/67	6/5/1967	6
388	Memorandum for Gen. Stewart, Subject: MOL Computer Program Management	6/12/1967	5
389	SSD Simulator Validation by Gemini Visual Acuity Experiment	6/12/1967	11
390	MOL Technical Director Memo for the Record, Subject: DORIAN Status Briefing to Dr. Hornig	6/12/1967	2
391	Questions on MOL from Office of SecDef	6/13/1967	4
392	An Operations Concept for the MOL/DORIAN: Manned/Automatic Configuration	6/15/1967	20
393	Memorandum for Dr. Flax from Dr. Michael I. Yarymovych, Subject: Competition on Beryllium Gimbal for DORIAN Payload	6/15/1967	1
394	MOL Program Plan, Volume 1 of 2	6/15/1967	119
395	MOL Program Plan, Volume 2 of 2	6/15/1967	137
396	Executive Session, MOL Program Review as of 6/14/67	6/17/1967	48
397	Delegation of Special Authority to the Head of a Procuring Activity	6/20/1967	5
398	Memorandum for D/MOL Program, Director of Special Projects from Alexander H. Flax, Subject: MOL Program Management	6/23/1967	20
399	Memorandum for Gen. Ferguson from Gen. Stewart, Subject: Recent MOL Events	6/27/1967	1
400	Memorandum for Dr. Flax from C.L. Battle, Subject: MOL Readout and Capsule Recovery	6/27/1967	2
401	Re-delegation of Procurement Authority	6/27/1967	4

No.	DOCUMENT TITLE	DATE	No. OF PAGES
402	Memorandum for the Record Written by Dr. Yarymovych, Subject: MOL Briefing to GAO	6/28/1967	1
403	DORIAN Minutes of MOL Electromagnetic Compatibility Control Board Meeting	6/30/1967	84
404	Personnel Transfers Memorandum of Understanding (MOU)	7/1/1967	3
405	Letter to Gen. Stewart from Executive Director NPIC, Subject: Recent Visit to Rochester & Philadelphia for a Fill-In on Status of DORIAN	7/3/1967	1
406	MOL Monthly Status Report: 6/67	7/6/1967	5
407	Lockheed DORIAN Resupply Study	7/6/1967	66
408	Review of Program Status & Problems for Vice Director, MOL	7/7/1967	226
409	Memorandum to Director, Procurement & Production MOL by Gen. Keeling, Subject: Re-delegation of Procurement Authority	7/8/1967	3
410	Briefings to General Stewart	7/10/1967	33
411	Memorandum for Gen. Bleymaier from James T. Stewart, Subject: 7/7/67 MOL Internal Management Meeting	7/20/1967	7
412	MOL Monthly Management Report: 5/25/67 - 6/25/67	7/21/1967	10
413	MOL/DORIAN Overview: Importance of the Program to the United States	7/24/1967	7
414	DORIAN General Electric & Eastman Kodak Briefing: Technical Status	7/27/1967	39
415	MOL Support Module Wideband Weight and Power Briefing for Program Review Council	7/27/1967	36
416	Memorandum for Dr. Flax from Maj. Gen. James T. Stewart, Subject: Resolution-Value Study for MOL	7/31/1967	2
417	MOL Monthly Status Report: 7/67	8/7/1967	4
418	MOL Planning Summary, 8/7/67	8/7/1967	22
419	MOL Management Directive, 8/8/67	8/8/1967	39
420	MOL VIP Presentation: High Level Update on Program	8/9/1967	66
421	Work Statement for DORIAN MP&E Development Phase	8/10/1967	2
422	MOL Generalized Target Model, Final Report, File Y Project Goodfellow	8/18/1967	9
423	Wood and Plastic Mock-Up of Acquisition Optics Subsystem Telescope	8/21/1967	4
424	MOL Management Directive, 8/23/67	8/23/1967	2
425	Memorandum for Gen. Ferguson, Subject: Monthly Management Report, 6/25/67 - 7/25/67	8/23/1967	8

No.	DOCUMENT TITLE	DATE	No. OF PAGES
426	Aerospace Corporation, MOL Program History	8/27/1967	8
427	Information Sheet for COMIREX, DORIAN Camera System	8/31/1967	3
428	Memorandum for Col. Howard from C.L. Battle, Subject: COMIREX Queries Concerning Gen. Stewart's DORIAN Briefing	8/31/1967	1
429	Presentation by General Stewart on DORIAN	9/1/1967	37
430	PSAC Review of MOL, 9/5/67	9/5/1967	8
431	MOL Monthly Status Report: 8/67	9/7/1967	3
432	President's Science Advisory Committee Briefings on Image Velocity Sensor (IVS) and Acquisition and Tracking System	9/7/1967	31
433	Basis for Confidence in Achieving the Objectives of MOL	9/8/1967	2
434	CCN for Phase I Support Module Study, The Automatic Mode Photographic Subsystem Of MOL/DORIAN Flight Vehicles 6 & 7	9/15/1967	6
435	Memorandum for Dr. Flax, Subject: Project Argo/DORIAN	9/19/1967	5
436	DORIAN Douglas Aircraft Corporation & Aerospace Corporation MOL Support Module Phase I Study Statement of Work	9/20/1967	18
437	Support Module Statement Of Work, Eastman Kodak Company	9/20/1967	10
438	ITEK Corporation, Preliminary Design Review Report, Acquisition Optics Subsystem, Volume I	9/25/1967	454
439	Observations on MOL Status: Cable Indicating the Funding Cutbacks Should be Considered as a Program Slippage, Not Misinterpreted as a De-emphasis of the Program	10/4/1967	1
440	Memorandum for the Record: Subject, 9/29/67, MOL Program Review Council Meeting	10/6/1967	7
441	MOL Monthly Status Report: 9/67	10/9/1967	4
442	Memorandum for Deputy Director MOL, Subject: MOL Advanced Planning by Gen. Stewart	10/10/1967	1
443	Optical Evaluation Procedures	10/10/1967	10
444	MOL Review of Graphics and Supporting Material	10/20/1967	3
445	High Resolution Photography, Volume 1	10/20/1967	56
446	Quick History of Aerospace Affects on MOL Payload	10/24/1967	3
447	Complementary Apollo Applications and MOL Program	10/24/1967	25
448	MOL Monthly Management Report, 8/25/67 - 9/25/67	10/25/1967	7

No.	DOCUMENT TITLE	DATE	No. OF PAGES
449	MOL Financial and Program Schedule Adjustment Review	10/27/1967	49
450	Aerospace Corporation Contributions of Man in the MOL/DORIAN System	10/31/1967	55
451	Talking Paper MOL/AAP Considerations	11/1/1967	42
452	MOL Monthly Status Report: 10/67	11/7/1967	4
453	Memorandum for General Stewart from: Ralph J. Ford, Subject: GAO Review of MOL	11/8/1967	1
454	Operation Order for Support of MOL (Program 632A) from AFSC	11/14/1967	43
455	MOL Monthly Management Report: 9/25/67 - 10/25/67	11/20/1967	7
456	MOL Program Review Committee Meeting Minutes for 11/17/67	11/27/1967	83
457	Notes on Selection of Targets for Viewing	11/28/1967	11
458	MOL Program Office Instruction No. 5, MOL Program Office Organization	12/1/1967	10
459	MOL Program Office Instruction No. 4, MOL Program Office Organization	12/1/1967	11
460	MOL Program Office Instruction No. 3, MOL Program Management Activities	12/1/1967	13
461	Auxiliary Memory Unit (AMU) Profile	12/4/1967	6
462	General Electric, Performance/Design and Product Configuration Requirements for Image Velocity Sensor	12/4/1967	61
463	MOL Program - Interface Documentation	12/5/1967	16
464	MOL Monthly Status Report: 11/67	12/6/1967	4
465	Memorandum for Dr. Leonard from James T. Stewart, Subject: ATS Resolution Capability	12/6/1967	4
466	ATS Resolution Capability	12/6/1967	5
467	MOL Monthly Management Report, 10/25/67 - 11/25/67	12/12/1967	9
468	Memorandum for Gen. Stewart from Dr. Yarymovych, Subject: Response to SecDef on MOL Program	12/13/1967	3
469	Memorandum for the Secretary of Defense from Harold Brown, Subject: MOL Program	12/15/1967	11
470	Memorandum for Dr. Flax, Subject: Deferral of Development of the Unmanned MOL System	1/1/1968	12
471	Position Description for the Assistant for MOL, NRO Staff	1/1/1968	4
472	DORIAN Optical Quality Factor	1/2/1968	4
473	MOL Monthly Status Report: 12/67	1/4/1968	6

No.	DOCUMENT TITLE	DATE	No. OF PAGES
474	Program Modifications, 1/8/68	1/8/1968	68
475	Memorandum of Understanding Between DoD Manager for Manned Space Flight Support Operations and the Director MOL Program	1/15/1968	4
476	MOL Monthly Management Report: 11/25/67 - 12/25/67	1/18/1968	6
477	Memorandum for Dr. Flax from James T. Stewart, Subject: MOL Program Structure Title Change	1/19/1968	12
478	Memorandum for Dr. Flax from Gen. Stewart, Subject: MOL Program Structure Title Change	1/19/1968	11
479	MOL Test Objectives Review Board Briefing	2/2/1968	59
480	Minutes of the MOL Test Objectives Review Board	2/2/1968	65
481	Memorandum for the Record, Subject: MOL/DORIAN Briefing to House Committee on Science and Astronautics Members	2/8/1968	3
482	Aerospace Corporation, Subject: Test Objectives Review Board	2/12/1968	9
483	Memorandum for the Record, by Col. Ford, Subject: MOL Program Briefing, DORIAN Level for Key Staff Members of the House Committee on Appropriations, Subcommittee on DoD (Mahon)	2/12/1968	4
484	Targeting Positioning for the MOL	2/13/1968	2
485	General Electric, Engineering Analysis Report Acquisition Subsystem for MOL	2/13/1968	612
486	Space Council Meeting: Meeting Minutes Discuss Some Confusion about Who Should be Reviewing the Value of MOL to the Reconnaissance Program	2/16/1968	5
487	Memorandum to DDR&E from AS/ST, Subject: To Inform Dr. Flax of the Discussions on MOL at the Space Council Meeting on February 14, Covering Brief	2/16/1968	3
488	Special Materials Support Requirements for MOL, 2/19/68	2/19/1968	2
489	Special Materials Support Requirements for MOL with Attachment, 2/19/68	2/19/1968	20
490	MOL/NSAM Briefings to Ambassador Bohlen and State Department Staff Members	2/19/1968	4
491	Memorandum for the Record, from: James T. Stewart, Subject: 2/14 Space Council Meeting, DORIAN	2/19/1968	3
492	Memorandum for the Record from James T. Stewart, Subject: MOL/NSAM 156 Briefings to Ambassador Bohlen and State Department Staff Members, DORIAN	2/19/1968	4
493	Memorandum for the Record, Subject: MOL/NSAM 156 Briefings to Ambassador Bohlen and State Department Staff Members	2/19/1968	3
494	MOL Program Review Council Meeting, 2/20/1968	2/20/1968	59
495	MOL and the National Intelligence Program	2/26/1968	3

No.	DOCUMENT TITLE	DATE	No. OF PAGES
496	Effects of Pointing and Target Tracking on DORIAN Photography	3/1/1968	16
497	Memorandum for Members of MOL Executive Council, Subject: Guidance for MOL Contractors in Dealings with Members of Congress	3/1/1968	4
498	Memorandum for the Record, by Gen. Stewart, Subject: MOL Briefings for House Armed Services Committee Members	3/1/1968	2
499	Briefings on MOL Error Budget and Target Location Uncertainties, 2/27/68	3/5/1968	4
500	Memorandum for Deputy DDR&E from Mr. Helms, Subject: Intelligence Requirements for the MOL Program	3/5/1968	1
501	Program Review Council Meeting Overview, 2/20/68	3/5/1968	5
502	Memorandum for Dr. Flax from Gen. Stewart, Subject: SAF Comments on MOL Monthly Status Report for February	3/11/1968	7
503	Mission Data Adapter Unit (MDAU) for MOL	3/12/1968	1
504	Special Materials Support Requirements for MOL, 3/12/68	3/12/1968	2
505	Development of Master Photo Chip and Increased ACIC Support for MOL	3/20/1968	1
506	MOL DORIAN Briefing to Chairman Chet Holifield	3/20/1968	2
507	Memorandum for Assistant Chief of Staff, Intelligence from Gen. Stewart, Subject: Special Materials Support Requirements for MOL	3/20/1968	1
508	MOL Monthly Management Report: 1/25/68 - 2/25/68	3/21/1968	9
509	Special Materials Support Requirements for MOL, 3/21/68	3/21/1968	1
510	MOL/DORIAN Requirements: As Stated by COMIREX	3/25/1968	5
511	Program Review Council Meeting, Los Angeles, California	4/2/1968	7
512	Minutes of Incremental Preliminary Design Review of Acquisition Subsystem	4/3/1968	375
513	ITEK Corporation, Rhomboid Alignment Procedure Scanner for the Slide Viewing System	4/4/1968	6
514	Letter to DIA from General Thomas, Subject: Special Materials Support Requirements for MOL	4/8/1968	12
515	Intelligence Information for Dr. Foster's Appearance Before the Senate Subcommittee on Military Preparedness	4/11/1968	5
516	Memorandum for Assistant Secretary of Defense (Public Affairs) from Dr. Flax, Subject: Public Presentation of Professional Papers Related to MOL	4/19/1968	4
517	Memorandum for Colonel Ford, Subject: Schedule Implications of Deferring Unmanned Vehicle Development Costs Until FY 1970 and Beyond	4/22/1968	15

No.	DOCUMENT TITLE	DATE	No. OF PAGES
518	Special Materials Support Requirements for MOL, 4/24/68	4/24/1968	1
519	MOL Monthly Management Report: 2/25/68 - 3/25/68	4/29/1968	9
520	MOL System Engineering for the MOL Program, Proposed DORIAN Part II	5/6/1968	125
521	MOL Flight Test and Operations Plan	5/8/1968	523
522	Contractor Allocations against Specific Numerical Requirements of SS-MOL-1B	5/9/1968	84
523	MOL Program Review Council Meeting 5/6/68	5/14/1968	8
524	MOL Program Office Directive 68-2, Program Requirement Document: Security Policy for MOL Contractors Desiring to use Fact of MOL Program Participation to Win New Business from the Federal Government	5/15/1968	8
525	MOL Planetary Observations	5/15/1968	8
526	MOL Program Phase II Management Activities	5/15/1968	64
527	The Intelligence Value of the MOL Program	5/15/1968	5
528	General Electric, Limited Search Mode Capabilities of the ATS	5/20/1968	21
529	MOL Monthly Management Report, 3/25/68 - 4/25/68	5/22/1968	9
530	Development of Master Photo Chip and ACIC Support of the MOL Program	5/24/1968	2
531	Subject: Action to Reduce MOL Program	5/24/1968	2
532	MOL Orbit Altitudes	5/24/1968	3
533	Memorandum for Dr. Flax from James T. Stewart, Subject: Deferral of Development of the Unmanned MOL System	6/1/1968	26
534	ITEK Corporation, Results of Qualitative Test Image Evaluation Resolution	6/3/1968	33
535	Electromagnetic Compatibility Control Plan Composite System	6/6/1968	26
536	Structural Criteria for Laboratory Vehicle for the MOL Program	6/6/1968	20
537	MOL Explosive Ordnance Systems Requirements	6/6/1968	29
538	MOL Orbiting Vehicle Power Utilization Control	6/6/1968	236
539	Memorandum for Dr. Flax from Mr. Goulding, Subject: Release of Technical Papers on MOL	6/10/1968	1
540	MOL Monthly Management Report: 4/25/68 - 5/25/68	6/17/1968	8
541	Talking Paper MOL/Apollo Applications Program (AAP) Considerations	6/18/1968	43
542	Memorandum for Dr. Flax from: James T. Stewart, Subject: NASA Interest in MOL	6/20/1968	2

No.	DOCUMENT TITLE	DATE	No. OF PAGES
543	Memorandum for Deputy Secretary of Defense from: John Foster, Jr. Subject: DOD/NASA Joint Program Review, Ways of Using MOL Hardware in Post-Apollo Program	6/27/1968	2
544	Memorandum to Dr. Flax from Gen Stewart, Subject: MOL Program Options	6/27/1968	5
545	DORIAN Goodyear Aerospace Corporation Technical Proposal, RFP 159	6/28/1968	60
546	Memorandum for Dr. Flax from Gen. Stewart, Subject: FY 69/70 MOL Program	7/1/1968	3
547	MOL Program Review Council Meeting, 6/24/68	7/1/1968	5
548	Memorandum for the Record, Subject: MOL Meeting with NASA AAP Representatives	7/1/1968	17
549	MOL Monthly Status Report: 6/68	7/3/1968	6
550	TRW Response to SAMSO Satellite	7/5/1968	2
551	Memorandum for Dr. Flax, Subject: NPIC Training for Astronauts	7/5/1968	58
552	Hycon, Volume II Technical Proposal, Beta System, Model HG-469B	7/5/1968	140
553	ITEK Corporation, Preliminary Design Review Report, A/O Scanner Assembly	7/5/1968	365
554	Mr. Steadman's Speech	7/12/1968	2
555	Memorandum for Dr. Flax from Gen Stewart, Subject: MOL Fuel Cell	7/12/1968	5
556	Aerospace Corporation, Subject: DORIAN Monthly Progress Report 6/1/68 - 6/30/68	7/15/1968	14
557	Aerospace Corporation, MOL Thermal Distortion Predictions for ULE and Cer-Vit Tracking Mirrors	7/15/1968	3
558	MOL Monthly Management Report: 5/25/68 - 6/25/68	7/16/1968	6
559	ITEK Corporation, MOL Design Report Slide Viewing System	7/16/1968	56
560	Aerospace Corporation Neutral Buoyancy Integrated Test Requirements, with Attachment Underwater Weightlessness Simulation	7/17/1968	10
561	Memorandum for the Record, Subject: NASA/MOL Meeting, 7/10/68	7/17/1968	14
562	ITEK Corporation, Optical Alignment Procedure K Rotator Assembly, Drawing 122430	7/21/1968	8
563	Trip Report - Air Force/Contractor MOL Program Rescheduling Meeting	7/22/1968	4
564	Memorandum to Col. Battle from Richard A. DeLong, Subject: MOL Readout Capability	7/23/1968	1
565	DORIAN Aerospace Corporation Bench Test Contrast Data	7/26/1968	2

No.	DOCUMENT TITLE	DATE	No. OF PAGES
566	Aerospace Corporation & General Electric ITEK Corporation Field Curvature Bench Test Results	7/26/1968	3
567	Aerospace Corporation, Use of Anamorphs in Simulator	7/26/1968	3
568	Aerospace Corporation, Optical Surface Quality Specification	7/26/1968	11
569	Memorandum for the Record, Subject: MOL Electrical Power System Status	7/26/1968	2
570	ITEK Corporation, Technical Proposal Image Velocity Sensor with Image Intensified Vidicon	7/29/1968	54
571	DORIAN Aerospace Corporation Briefing Charts: Stick Transfer Function Briefing	7/31/1968	20
572	Aerospace Corporation, MOL Monthly Progress Reports, 7/1/68 - 7/31/68	7/31/1968	11
573	Aerospace Corporation, MOL Soft Switching in the PSS	8/1/1968	5
574	Aerospace Corporation, Briefing MOL Support by SCF	8/2/1968	11
575	MOL Laboratory Vehicle Low Level Vibration Acceptance Tests	8/3/1968	3
576	MOL Monthly Management Report: 6/25/68 - 7/25/68	8/5/1968	6
577	MOL Monthly Status Report: 7/68	8/6/1968	6
578	Memorandum for Dr. Flax from James T. Stewart, Subject: MOL Data Requested by NASA, List of Documents and Studies Provided NASA	8/6/1968	3
579	Memorandum for Gen. Stewart from Gen. Bleymaier, Subject: Schedule and Funding Problems Associated with the Acquisition of COMSEC Hardware, SCF Hardware and SCF Engineering Support for MOL	8/6/1968	5
580	ITEK Corporation, Eyepiece Assembly Test Procedure, Drawing No. 14911	8/6/1968	4
581	Minutes of Technical Direction Meeting for the Beta System	8/9/1968	6
582	ITEK Corporation, Optical Alignment Procedure Lower Rhomboid Arm, Drawing No. 122433	8/12/1968	6
583	Issuance of RFPs on MOL Fuel Cells	8/13/1968	4
584	DORIAN ITEK Corporation Index to Alignment Procedures and Test Procedures	8/14/1968	2
585	ITEK Corporation, Primary Reticle Assembly Procedure, Drawing No. 149107	8/15/1968	6
586	ITEK Corporation, Interim Report Point Spread Function (PSF) Measurements Made on Slide Viewing System	8/16/1968	37
587	Aerospace Corporation, Trip Report—FAMS Light Location Review at Eastman Kodak	8/16/1968	4
588	A Summary of The Paul-Beta Sensor Operation	8/23/1968	14

No.	DOCUMENT TITLE	DATE	No. OF PAGES
589	Memorandum for Mr. Palley, Subject: MOL Development Costs	8/27/1968	17
590	DORIAN Image Velocity Sensor Primer	8/28/1968	8
591	Memorandum for Deputy Director, MOL from Gen Stewart, Subject: MOL Image Velocity Sensor (IVS) Ad Hoc Review Group	8/28/1968	5
592	Acquisition Subsystem On-Orbit Performance Prediction	9/1/1968	53
593	Memorandum for Dr. Flax from Gen. Stewart, Subject: MOL Development Costs	9/6/1968	6
594	Memorandum to Maj. Wolfsberger from Col. Merritt, Subject: Space Electric Power Requirements for Advanced MOL Missions	9/6/1968	2
595	MOL Monthly Status Report: 8/68	9/9/1968	6
596	DORIAN Aerospace Corporation Guidance and Control Laboratory Justification Direct Hardware Support and Hardware Studies for Air Force Satellite Programs and Activities in Support of MOL	9/10/1968	3
597	Introduction to the DORIAN IVS System	9/10/1968	36
598	Letter to Colonel Yost from James T. Stewart, Subject: Arms Limitation Agreement	9/11/1968	9
599	Capabilities and Computational Methods of the TWONDER Study Program for MOL	9/12/1968	363
600	MOL Baseline Document for Statistical Program of On-Board Decisions, Computer Program to Simulate Target Visibility Activities	9/18/1968	22
601	Special Materials Support Requirements for MOL, 9/19/68	9/19/1968	2
602	MOL DORIAN Image Velocity Sensor Sub-System, AD/HOC Review Group Final Report 9/19/68	9/19/1968	10
603	Goodyear Aerospace Corporation Subcontract #029B25006, Performance Prediction Techniques Final Technical Report	9/20/1968	32
604	NRO Proposals for Meeting World-Wide Positioning Requirements KH-4B System to Incorporate a Transit Beacon	9/20/1968	2
605	Image Velocity Sensor Subsystem Review Presentation Charts, Ad Hoc Review Group Final Report	9/20/1968	212
606	Image Velocity Sensor Subsystem, Ad Hoc Review Group Final Report	9/20/1968	11
607	MOL Monthly Management Report: 7/25/68 - 8/25/68	9/24/1968	9
608	Memo for COMIREX, Subject: Program for Planning the Exploitation of Reconnaissance Imagery	9/25/1968	13
609	Memorandum for Dr. Flax and Gen. Ferguson, Subject: MOL Image Velocity Sensor (IVS) Ad Hoc Review Group	9/26/1968	3
610	Paper to Gen. Ferguson from Gen. Stewart, Subject: Update on MOL Happenings	9/27/1968	2

No.	DOCUMENT TITLE	DATE	No. OF PAGES
611	MOL Monthly Progress Report, 9/1/68 - 9/30/68	9/30/1968	8
612	Special Materials Support Requirements for MOL, 10/2/68	10/2/1968	1
613	General Electric, Mary Dynamic Null Study to Explain How the Hycon Image Velocity Sensor Works	10/2/1968	17
614	MOL Monthly Status Report: 9/68	10/7/1968	4
615	DORIAN Aerospace Corporation Performance Calculation	10/7/1968	2
616	Aerospace Corporation, Acceptance Technical Readiness and FACI Briefing	10/8/1968	12
617	MOL Monthly Management Report: 25 August - 25 September 68	10/9/1968	5
618	DORIAN Aerospace Corporation Image Velocity Sensor (IVS) Comments MOL	10/9/1968	12
619	Statement of Work - MOL System	10/16/1968	7
620	Memorandum for Dr. Flax & Gen. Ferguson from James T. Stewart, Subject: Eastman Kodak Study of Possible Future MOL Camera System Improvements	10/16/1968	3
621	Mission Development Simulator Performance Design Requirements for MOL	10/18/1968	185
622	General Electric, Contract End Item Detail Specification (Prime Equipment) Performance/Design and Product Configuration Requirements	10/18/1968	62
623	Special Materials Support Requirements for MOL, 10/23/68	10/23/1968	2
624	DORIAN Aerospace Corporation ATS Alignment Accuracy Requirements For Tracking Assembly	10/23/1968	2
625	DORIAN Aerospace Corporation MOL PSPP Update, Advance Plans Section	10/24/1968	9
626	Aerospace Corporation, Subject: A Proposal for a Vehicle Based Alignment System	10/24/1968	35
627	Aerospace Corporation, MOL Briefing Charts, Pointing and Tracking Simulator	10/24/1968	24
628	Program Review Council Agenda, Part 1, 10/24/1968	10/24/1968	50
629	Program Review Council Agenda, Part 2, 10/24/1968	10/24/1968	61
630	DORIAN Aerospace Corporation Work Statement Review, Recommended Changes To Contractor's Statement of Work for MOL	10/28/1968	17
631	Aerospace Corporation, Relationship Between Mirror Gimbal Angles and Stereo and Obliquity Angles	10/28/1968	19
632	Memorandum for Gen. Bleymaier from Gen. Stewart, Subject: Schedule and Funding Problems for MOL Secure Communications	10/31/1968	2
633	DORIAN, Aerospace Corporation MOL Monthly Progress Report for 10/1/68 through 10/31/68	10/31/1968	4

No.	DOCUMENT TITLE	DATE	No. OF PAGES
634	PAUL-BETA Subsystem, Engineering Prototype Evaluation Model, Part 0, Final Report	10/31/1968	114
635	Program Review Council Meeting, 10/24/68	11/2/1968	5
636	MOL Monthly Status Report: 10/68	11/8/1968	10
637	MOL Monthly Management Report: 9/26/68 - 10/25/68	11/14/1968	8
638	Aerospace Corporation, Simulators Status and Plans	11/19/1968	32
639	Memorandum by John Kirk, Past Experience on the MOL Program	11/22/1968	4
640	ITEK Corporation, Quarterly Technical Progress Report, Dual Strip Camera Back Study Phase IV	11/22/1968	58
641	Aerospace Corporation, Manpower Support to MOL	11/25/1968	183
642	Eastman Kodak Company, Engineering Analysis Report, Photographic Payload and Related Support and Test Equipment for MOL/DORIAN System, Volume 1	11/27/1968	382
643	Eastman Kodak Company, Engineering Analysis Report, Photographic Payload and Related Support and Test Equipment for MOL/DORIAN System, Volume 2	11/27/1968	313
644	General Electric, Mission Development Simulator, Phase 0, System Test Requirements, Revision A	12/2/1968	64
645	General Electric, Mission Development Simulator, Phase 0, System Test Plan	12/5/1968	87
646	MOL Monthly Status Report: 11/68	12/6/1968	5
647	Letter to Paul H. Nitze from Richard Helms, Subject: Comments on the Development Concept Paper Relating to the MOL Program	12/6/1968	2
648	DORIAN Successes with SO-121 Color Film	12/9/1968	2
649	Aerospace Corporation, Viewer Requirements: Associated Crew Tasks and Capabilities	12/17/1968	7
650	Memorandum for Gen. Stewart from Col. Ford, Subject: Funding for MOL Support Requirements	12/18/1968	2
651	DORIAN Aerospace Corporation MOL Progress Report for 11/1/68 through 11/30/68	12/19/1968	10
652	MOL Monthly Management Report: 10/26/68 - 11/25/68	12/20/1968	8
653	MOL Briefing by Vice Director, MOL Program to Dr. Townes Space Task Group	12/24/1968	8
654	White House Correspondence, Nixon Staff Task Force of Space	12/27/1968	1
655	DORIAN Aerospace Corp Block II Study, Improved Resolution	12/31/1968	3
656	General Electric, Initial Capability Readout System for Early Implementation into the MOL/DORIAN System	12/31/1968	187

No.	DOCUMENT TITLE	DATE	No. OF PAGES
657	NASA Letter to Air F orce Regarding use of Acquisition Tracking System	1/1/1969	2
658	Man in MOL, 1st Edition	1/2/1969	33
659	User's Manual for TSPOOND	1/7/1969	78
660	Aerospace Corporation MOL Technical Status Summary Briefing to General Ferguson, 1/9/69	1/9/1969	29
661	MOL Monthly Status Report: 12/68	1/13/1969	4
662	Nitze Comments on MOL Development Paper and the DDR&E/DIA Study of Very High Resolution (VHR) Imagery	1/13/1969	6
663	Aerospace Corporation, MOL Monthly Progress Report, 12/1/68 - 12/31/68	1/20/1969	10
664	Aerospace Corporation, Simulator Requirements Identified in MOL	1/23/1969	12
665	Selin Comments on MOL Development Paper and the DDR&E Study of Very High Resolution (VHR) Imagery	1/24/1969	14
666	General Electric, Mass Properties Status Report	1/25/1969	33
667	MOL Monthly Management Report: 11/26/68 - 12/25/68	1/27/1969	7
668	MOL DCP and the DDR&E/DIA Study of the Need for High Resolution	1/29/1969	6
669	MOL DCP	1/31/1969	1
670	MOL Program and Value of Very High Resolution Imagery	1/31/1969	5
671	TSPOOND Mathematical and Subroutine Description	2/3/1969	177
672	Intelligence Targets for MOL Crew Training, 2/4/69	2/4/1969	2
673	Intelligence Targets for MOL Crew Training, 2/5/69	2/5/1969	10
674	Packard Briefing, The MOL Program and Very High Resolution (VHR) Issues	2/6/1969	2
675	Aerospace Corporation Briefing, Wide Band Data Read Out	2/6/1969	20
676	MOL Monthly Status Report: 1/69	2/7/1969	6
677	MOL Monthly Management Report: 12/26/68 - 1/25/69	2/11/1969	7
678	Briefing on MOL and VHR Issues Given to Deputy SecDef on 2/8/69	2/11/1969	51
679	Letter to David Packard from Robert P. Mayo, Subject: Relative Value of the MOL for U.S. Force Structure Decisions	2/13/1969	11
680	Memorandum for Deputy Director, MOL from James T. Stewart, Subject: Designation of MOL as the KH-10 Photographic Reconnaissance Satellite System	2/14/1969	3
681	Memorandum for Director, Procurement & Production, MOL, Subject: Re-delegation of Procurement Authority	2/19/1969	1

No.	DOCUMENT TITLE	DATE	No. OF PAGES
682	Memorandum for Dr. McLucas - NASA Proposal to Study MOL ATS and DORIAN Technology	2/20/1969	6
683	ITEK Corporation, Final Report, Image Velocity Sensor Program	2/21/1969	132
684	Aerospace Corporation, Monthly Progress Report for MOL Program, 1/1/69 - 1/31/69	2/24/1969	16
685	Aerospace Corporation & Eastman Kodak Company Review Comments of The Eastman Kodak Engineering Analysis Report	2/25/1969	38
686	Memorandum for Dr. Seamans, Dr. Flax from James T. Stewart, Subject: MOL Program Alternatives	2/25/1969	38
687	MOL Monthly Status Report: 2/69	2/28/1969	8
688	Letter for DNRO from Vice Director/MOL Program, Subject: MOL Book for Upcoming Congressional Hearings	3/5/1969	9
689	MOL Monthly Management Report: 1/26/69 - 2/25/69	3/7/1969	7
690	Memorandum for DDR&E from I. Nevin Palley, Subject: MOL Program Alternative Decision	3/10/1969	3
691	Aerospace Corporation, MOL Monthly Progress Report: 2/1/69 - 2/28/69	3/19/1969	13
692	General Electric, Mission Development Simulator, Phase 0, System Test Procedure	3/20/1969	179
693	General Electric, Viewgraph Presentation Alpha Subsystem Component Status	3/21/1969	109
694	Memorandum for Dr. Seamans from James T. Stewart, Subject: MOL Funding	3/26/1969	7
695	Aerospace Corporation, Last Quarter FY 1969 and FY 1970 Manpower Review Engineering Directorate	3/28/1969	54
696	Man in MOL, 2nd Edition	3/28/1969	37
697	MOL Monthly Status Report: 3/69	4/1/1969	5
698	Letter for Gen. Bleymaier from Gen. Stewart, Subject: The Development Effort Toward an Unmanned MOL Reconnaissance System in the Present Program has been Deferred	4/1/1969	3
699	Authority and Direction to Notify MOL Contractors	4/1/1969	2
700	Memorandum for Dr. McLucas from James T. Stewart, Subject: SAF-Level MOL Management	4/2/1969	22
701	Memorandum for Dr. McLucas, Subject: SAF-Level MOL Management	4/2/1969	7
702	MOL Backup Material for Congressional Hearings	4/3/1969	26
703	General Electric, Image Velocity Sensor	4/4/1969	27

No.	DOCUMENT TITLE	DATE	No. OF PAGES
704	MOL Monthly Management Report: 2/26/69 - 3/25/69	4/10/1969	8
705	MOL: Outlines the Case Against the MOL Program	4/14/1969	3
706	MOL Program Funding	4/14/1969	3
707	Briefing to Dr. McLucas, General Program Orientation	4/15/1969	108
708	MOL Program Review Council Meeting	4/15/1969	126
709	Draft Memorandum for the President, Subject: MOL	4/17/1969	7
710	Memorandum for Dr. Seamans from James T. Stewart, Subject: MOL FY 70 Program Options	4/22/1969	11
711	Computer Program Contract End Item Detail Specification Performance/Design Requirements, Mission Planning Software for MOL/DORIAN Program 632A	4/22/1969	80
712	PSAC Review of MOL	4/25/1969	2
713	Aerospace Corporation Memo, Subject: Current Ephemeris Error Estimates and Related Discussion	4/25/1969	13
714	Memorandum for The President from Melvin R. Laird, Subject: MOL	4/28/1969	24
715	Memorandum to Gen. Ferguson from James T. Stewart, Subject: Federal Budget for MOL	4/28/1969	3
716	Memorandum for Deputy SecDef from Robert C. Seamans, Jr, Subject: MOL FY70 Program Options	4/30/1969	25
717	General Electric, Stick Experiment for the Primary Optics System	5/1/1969	62
718	MOL Management; SAFRD and SAFUS	5/5/1969	2
719	ICRS Briefing on DORIAN Targeting Data Requirements	5/7/1969	16
720	DORIAN Aerospace Corporation MOL Four Charts Rendezvous Orbiting Vehicle Evolution	5/7/1969	4
721	Aerospace Corporation, Subject: Trip Report to Establish Working Relationship with Error Validation and Error Control People at General Electric	5/8/1969	7
722	Memorandum for Dr. Foster from Herbert D. Benington, Subject: Immediate Alternatives for MOL	5/8/1969	2
723	MOL Status Report: 5/9/69	5/9/1969	68
724	General Electric, Final Report, Survey Study of the Data Acquisition Potential of the "Enhanced" MOL/DORIAN Baseline System During the Block II (7/74 - 1/76) Time Period	5/9/1969	108
725	MOL Briefing to the House Armed Services Committee	5/13/1969	2

No.	DOCUMENT TITLE	DATE	No. OF PAGES
726	Point Paper on Man in MOL	5/14/1969	11
727	Aerospace Corporation, Subject: Trip Report to Inspect Alpha System Bench Test	5/14/1969	3
728	MOL Monthly Management Report, 3/26/69 - 4/25/69	5/16/1969	6
729	Memorandum for the Record from James T. Stewart, Subject: Meeting with The President	5/19/1969	5
730	Internal Management Audit of MOL	5/21/1969	10
731	Memorandum for Dr. McLucas from James T. Stewart, Subject: Internal Management Audit of MOL	5/21/1969	10
732	Aerospace Corporation, MOL Monthly Progress Report For 4/15/69 - 5/15/69	5/22/1969	14
733	The Rivers Committee and MOL, Congressional Hearing FY 70 RDT&E Authorization Request	5/28/1969	11
734	Memorandum for the Record, Subject: The Rivers Committee and MOL	5/28/1969	1
735	The Roles of Man in MOL, Volume II - Illustrations	6/1/1969	43
736	MOL Program Perspective	6/6/1969	6
737	Terminate MOL Except for the "Automatic" Camera System	6/7/1969	5
738	MOL Program Background	6/8/1969	9
739	Memorandum for SecDef from Robert C. Seamans, Jr., Subject: MOL Decision	6/8/1969	22
740	Memorandum for Gen. Bleymaier from James T. Stewart, Subject: MOL Termination Guidance	6/8/1969	7
741	Memorandum to Gen. Bleymaier from James T. Stewart, Subject: Future Plans for DORIAN Camera System	6/8/1969	3
742	The Department of Defense has Terminated the Air Force MOL Program	6/9/1969	6
743	Letter to George H. Mahon from David Packard, Subject: Primary Purpose of MOL	6/9/1969	8
744	Memorandum for Secretary of Air Force, D/NRO from David Packard, Subject: Termination of MOL Program	6/9/1969	1
745	Cable: Termination of MOL Program	6/9/1969	4
746	Program Schedule Status, Program Control Directorate, 6/10/69	6/10/1969	41
747	Memorandum for Dr. Seamans from James T. Stewart, Subject: MOL Termination	6/10/1969	3
748	MOL Program Termination	6/10/1969	3
749	DORIAN Aerospace Corporation, Flash Cleaning of Optical Surfaces in Vacuum	6/16/1969	3
750	DORIAN Camera System Continuation	6/18/1969	4

No.	DOCUMENT TITLE	DATE	No. OF PAGES
751	Cable, from Dr. McLucas to General's Martin & Bleymaier, DORIAN Camera Use in Unmanned Satellite	6/18/1969	7
752	Termination of MOL Contracts	6/23/1969	2
753	Memorandum for Alan M. Eldridge from Daniel L. Lycan, Subject: DoD's Termination of the Air Force's MOL Program	6/24/1969	4
754	Termination of MOL Program	6/25/1969	2
755	Aerospace Corporation, Technical Status of the DORIAN Payload at the Time of Termination	6/27/1969	16
756	MOL Security Policy on Contractor's Solicitation of New Business	6/30/1969	1
757	Aerospace Corporation, Technical Evaluation of General Electric, Performance of MOL	6/30/1969	57
758	Memorandum to Dr. Seamans from James T. Stewart, Subject: Ad Hoc Group to Review Residual MOL Hardware	6/30/1969	4
759	Cancellation of MOL Program	7/2/1969	3
760	Memorandum for Dr. Seamans from James T. Stewart, Subject: MOL Termination Status Report	7/3/1969	6
761	Development of Target Model for the DORIAN System	7/24/1969	118
762	DORIAN Memorandum for Dr. Seamans from Grant L. Hansen, Subject: Establishment of Ad Hoc Group to Review MOL Hardware	8/1/1969	1
763	Review of MOL Residuals	8/1/1969	180
764	Letter To Robert C. Seamans from Homer E. Newell Subject: DORIAN Elements	8/28/1969	25
765	Memorandum for Dr. McLucas from Ralph J. Ford Colonel, USAF, Subject: NASA Proposals to Study MOL, ATS and DORIAN Technology	9/1/1969	2
766	Memorandum for Security Advisor, NRO Staff from E.J. Kane, Subject: BYEMAN Security Policy Guidance Close Out of Project DORIAN	9/3/1969	3
767	Close-out of DORIAN Documentation and Materials	9/5/1969	1
768	Review of MOL Residuals-MOL 4 Pi Extended Performance Flight Computer	9/8/1969	2
769	Applications of the MOL Acquisition/Tracking System to NASA Space Missions	9/10/1969	37
770	NASA Proposal to Study MOL	9/12/1969	2
771	NASA Astronomy Program Considerations of DORIAN Technology	9/15/1969	13
772	The Rationalization of Very High Resolution	9/15/1969	5

No.	DOCUMENT TITLE	DATE	No. OF PAGES
773	NASA Proposals to Study MOL Acquisition Tracking System and DORIAN Technology	9/22/1969	8
774	Memorandum for Dr. Seamans, Subject: NASA Proposals to Study MOL Acquisition and Tracking System (ATS)	9/24/1969	6
775	Relation of MOL to Eight Card	9/25/1969	2
776	Letter to Homer E. Newell, Subject: NASA Interest in Exploring Utilization of Certain MOL Developed Hardware and Related Technology	9/29/1969	2
777	MOL Systems Office Post Termination Report	9/30/1969	148
778	Disposition of MOL Residuals	10/1/1969	9
779	MOL Howen Facility Photos	10/1/1969	2
780	MOL Inventory	10/2/1969	1
781	Cable, for Gen. Bleymaier, Disposition of MOL Residuals	10/3/1969	2
782	Memorandum to Mr. I. Nevin Palley from Floyd J. Sweet, Subject: MOL Acquisition and Tracking System and Mission Development Simulator	11/6/1969	7
783	Letter to John L. McLucas from Homer E. Newell, Subject: Study of Potential Application to NASA Mission of the MOL Acquisition & Tracking System	11/7/1969	8
784	ITEK 12" Brief to NASA	11/12/1969	3
785	NASA/GE Covert Study: Potential NASA Use of DORIAN ATS and Mission Development Simulator	11/18/1969	2
786	Memorandum for the Record, from Bertram Kemp, Subject: Disposition of Covert MOL Equipment	11/20/1969	2
787	NASA use of MOL Classified Residuals	12/1/1969	11
788	Disposition of MOL Program CITE Equipment	12/11/1969	7
789	Memorandum to Deputy Director, Procurement and Production, MOL from Gen. Higgins, Subject: Re-delegation of Procurement Authority	12/12/1969	1
790	Letter to John L. McLucas from Homer E. Newell, Subject: Examination of the Applicability of the MOL Developed Optical Technology & Facilities to Our Space Astronomy Program	12/17/1969	1
791	Reporting of MOL Excess ADPE	12/22/1969	2
792	Memorandum For Distribution Subject: Air Force Participation in NASA Study of MOL Acquisition and Tracking System	12/23/1969	3
793	Memorandum for Dr. Seamans, Dr. McLucas from James Ferguson, Subject: MOL Program Close-Out Status	12/23/1969	11

No.	DOCUMENT TITLE	DATE	No. OF PAGES
794	General Electric Company Briefing Charts, Advanced MOL Planning; Missions and Systems	12/31/1969	60
795	MOL Program: Cancelation of Pentagon Activities	1/8/1970	1
796	Memorandum from Secretary of Defense to D/NRO, Subject: Termination of MOL	1/8/1970	1
797	Response to Report of MOL Excess ADPE	1/9/1970	2
798	MOL 4PY EP Computers	1/10/1970	3
799	Storage And Maintenance of Government Furnished Equipment for NASA	1/21/1970	2
800	History of MOL	2/1/1970	356
801	NASA Technical Feasibility Study at Eastman-Kodak	2/4/1970	2
802	MOL/General Electric Study	3/6/1970	2
803	MOL 4PI EP Computers	3/31/1970	2
804	MOL Systems Office Termination	3/31/1970	8
805	Transfer of MOL Residuals to NASA	4/8/1970	20
806	Memorandum for Mr. Krueger and Mr. Sweet, Subject: Guidelines for NASA Earth-Sensing Activity	4/17/1970	7
807	Turnover of the MOL Mission Development Simulator to NASA	6/16/1970	2
808	Letter to Dr. Newell from John L. McLucas, Subject: MOL Acquisition and Tracking Scope, Mission Development Simulator & Drive A System	6/16/1970	1
809	Turnover of MOL Residuals	6/18/1970	1
810	MOL Program: Close of MOL Systems Office in LA	7/29/1970	2
811	MOL Excess Computers	7/30/1970	17
812	Memorandum for the Record, from Frederick L. Hofmann, Subject: MOL Program Office Close Out	9/18/1970	1
813	MOL Residual Computers	9/22/1970	15
814	Classified MOL Residual Hardware	10/8/1970	2
815	Total MOL Program Funding Requirement Forecast	1/7/1971	21
816	Project Colt Proposal by Dr. Meinel, University of Arizona	1/19/1971	2
817	General Electric MOL Computer Return	2/25/1971	4
818	MOL Mission Development Simulator Computers	5/25/1971	1

No.	DOCUMENT TITLE	DATE	No. OF PAGES
819	Memorandum for Dr. Naka from Harold S. Coyle, Jr., Subject: Manned Space Flights over the Soviet Union	8/24/1971	3
820	Memorandum for the Record from Frederick L. Hofmann, Subject: MOL Mission Development Simulator & Associated Computers Which Were Loaned to NASA	2/16/1972	3
821	Memorandum For The Record from Frederick L. Hofmann, Subject: MOL Residuals at ITEK	2/1/1973	1
822	Memorandum 6/5/73 - MOL Status	6/5/1973	1
823	Note For Dr. Yarymovych from Frederick L. Hofmann, Subject: MOL Residuals Transferred to NASA	10/15/1973	7
824	Information about SAMOS, Lunar Orbiter, and MOL Program Involvement	2/9/1976	2
825	MOL/DORIAN Overview Slide	11/13/1978	1

www.ingramcontent.com/pod-product-compliance
Lightning Source LLC
Chambersburg PA
CBHW050410110426

42812CB00006BA/1852